Lecture Notes in Computer Scien

Commenced Publication in 1973
Founding and Former Series Editors:
Gerhard Goos, Juris Hartmanis, and Jan van Leeuwen

Shanchieh Jay Yang Ariel M. Greenberg
Mica Endsley (Eds.)

Social Computing, Behavioral-Cultural Modeling and Prediction

5th International Conference, SBP 2012
College Park, MD, USA, April 3-5, 2012
Proceedings

 Springer

Volume Editors

Shanchieh Jay Yang
Rochester Institute of Technology
Department of Computer Engineering
Rochester, New York 14623-5603, USA
E-mail: jay.yang@rit.edu

Ariel M. Greenberg
Johns Hopkins University
Applied Physics Laboratory
Research and Exploratory Development Department
Laurel, MD 20723-6099, USA
E-mail: nau@cs.umd.edu

Mica Endsley
SA Technologies
3750 Palladian Village Drive, Building 600, Marietta, GA 30066, USA
E-mail: mica@satechnologies.com

ISSN 0302-9743   e-ISSN 1611-3349
ISBN 978-3-642-29046-6   e-ISBN 978-3-642-29047-3
DOI 10.1007/978-3-642-29047-3
Springer Heidelberg Dordrecht London New York

Library of Congress Control Number: Applied for

CR Subject Classification (1998): H.3, H.2, H.4, K.4, J.3, H.5

LNCS Sublibrary: SL 3 – Information Systems and Application, incl. Internet/Web
and HCI

Typesetting: Camera-ready by author, data conversion by Scientific Publishing Services, Chennai, India

Printed on acid-free paper

Springer is part of Springer Science+Business Media (www.springer.com)

Preface

Welcome to the proceeding of the 2012 International Conference on Social Computing, Behavioral–Cultural Modeling and Prediction (SBP 2012). Continued from 2011, the SBP 2012 conference was the result of merging two successful international conferences on closely related subjects: the International Conference on Social Computing, Behavioral Modeling, and Prediction (SBP) the and International Conference on Computational Cultural Dynamics (ICCCD). The overall goal of the conference is to bring together a diverse set of researchers/disciplines to promote interaction and assimilation for the better understanding of social computing and behavior modeling. Social computing harnesses the power of computational methods to study social behavior within a social context. Cultural behavioral modeling refers to representing behavior and culture in the abstract-sense, and is a convenient and powerful way to conduct virtual experiments and scenario planning. Both social computing and cultural behavioral modeling are techniques designed to achieve a better understanding of complex behaviors, patterns, and associated outcomes of interest. In 2008, the first year of SBP, we held a workshop and had 34 papers submitted; in 2011 we had grown to a conference and had 85 submissions from 18 countries. This year, we had 76 submissions representing works in health sciences, computing and information sciences, military and security, economics, methodology and others. We truly hope that our collaborative, exploratory research can advance the emerging field of social computing and behavioral–cultural modeling.

The overall SBP 2012 conference program encompassed pre-conference tutorials, keynote presentations, high-quality technical paper presentations and posters, a cross-disciplinary round-table session, a modeling challenge session and a panel featuring program staff from federal agencies discussing potential research opportunities. This year, SBP 2012 included a new dimension with the Modeling Challenge session aiming to demonstrate the real-world and interdisciplinary impact of social computing. The challenge was geared to engage the social computing research community in solving relevant, interesting, and challenging research problems contributing toward theoretical, methodological, and applicational advancement of the area.

The accepted peer-reviewed papers cover a wide range of interesting problem domains, e.g., economics, public health, and terrorist activities, and they utilize a broad variety of methodologies, e.g., machine learning, cultural modeling and cognitive modeling. Over the past several years of SBP conferences, we have started to see the increasing participation of the human and social sciences in social computing, as well as the active collaboration between such fields and science and engineering fields. Disciplines represented at this conference include computer science, electrical engineering, psychology, economics, sociology, and health sciences. A number of interdisciplinary and applied research institutions were represented.

SBP 2012 could not be run successfully by a only few. We would like to first express our gratitude to all the authors for contributing an extensive range of research topics showcasing many interesting research activities and pressing issues. The regret is ours that due to the space limit, we could not include as many papers as we wished. We thank the Program Committee members for helping review and providing constructive comments and suggestions. Their objective reviews significantly improved the overall quality and content of the papers. We would like to thank our tutorial and keynote speakers for presenting their unique research and visions. We deeply thank the members of the Organizing Committee for helping to run the conference smoothly: from the call-for-papers, the website development and update, to proceedings production and registration.

Last but not least, we sincerely appreciate the support from the University of Maryland and the following federal agencies: Air Force Office of Scientific Research (AFOSR), Air Force Research Laboratory (AFRL), Army Research Office (ARO), National Institute of General Medical Sciences (NIGMS) at the National Institutes of Health (NIH), National Science Foundation (NSF), and the Office of Naval Research (ONR). We also would like to thank Alfred Hofmann from Springer and Lisa Press from the University of Maryland. We thank all for their kind help, dedication, and support in making SBP 2012 possible.

April 2012   Shanchieh Jay Yang
Ariel M. Greenberg
Mica Endsley

Organization

Conference Co-chairs

V.S. Subrahmanian	University of Maryland, USA
Nathan Bos	Johns Hopkins University, Applied Physics Lab, USA

Program Co-chairs

Shanchieh Jay Yang	Rochester Institute of Technology, USA
Ariel Greenberg	Johns Hopkins University, Applied Physics Lab, USA
Mica Endsley	SA Technologies, USA

Steering Committee

Huan Liu	Arizona State University, USA
John Salerno	Air Force Reasearch Laboratory, USA
Sun Ki Chai	University of Hawaii, USA
Dana Nau	University of Maryland, USA
V.S. Subrahmanian	University of Maryland, USA
Patty Mabry	National Institutes of Health, USA

Advisory Committee

Rebecca Goolsby	Office of Naval Research, USA
Joseph Lyons	Air Force Reasearch Laboratory, USA
Jeffrey C. Johnson	Army Research Lab/Army Research Office, USA
Fahmida N. Chowdhury	National Science Foundation, USA
Patty Mabry	National Institutes of Health, USA

Poster Session Chair

Lei Yu	Binghampton University, USA

Tutorial Chair

Anna Nagurney	University of Massachusetts, USA

Challenge Problem Chair

Nitin Agarwal	University of Arkansas, USA

Workshop Co-chairs

Fahmida N. Chowdhury   National Science Foundation, USA
Bethany Deeds   National Institutes of Health, USA

Sponsorship Committee Chairs

Huan Liu   Arizona State University, USA

Student Arrangements Chair

Patrick Roos   University of Maryland, USA

Publicity Co-chairs

Donald Adjeroh   West Virginia University, USA
Gerardo Simari   Oxford University, USA

Webmaster

Damon Earp   UMIACS, University of Maryland, USA

Technical Program Committee

Myriam Abramson   US Naval Research Laboratory, USA
Nitin Agarwal   University of Arkansas at Little Rock, USA
Geoffrey Barbier   Arizona State University, USA
Joshua Behr   Old Dominion University, USA
Lashon Booker   The MITRE Corporation, USA
Nathan Bos   John Hopkins University (APL), USA
Elizabeth Bowman   Army Research Laboratory, USA
Jiangzhou Chen   Virginia Tech, USA
Xueqi Cheng   Chinese Academy of Sciences, China
Alvin Chin   Nokia Research Center
David Chin   University of Hawaii, USA
Hasan Davulcu   Arizona State University, USA
Bethany Deeds   NIDA, USA
Brian Dennis   Lockheed Martin, USA
Mica Endsley   SA Technologies, USA
Richard Fedors   AFRL (RI), USA
Laurie Fenstermacher   AFRL (RH), USA
William Ferng   Boeing
Clay Fink   John Hopkins University (APL), USA
Anthony Ford   AFRL (RI)
Wai-Tat Fu   University of Illinois at Urbana-Champaign, USA

Geovanni Giuffrida	University of Catania, Italy
Ariel Greenberg	John Hopkins University (APL), USA
Michael Hinman	AFRL (RI), USA
Terresa Jackson	US Navy, USA
Ruben Juarez	University of Hawaii, USA
Byeong Ho Kang	University of Tasmania, Australia
Douglas Kelly	AFRL (RH), USA
Jang Hyun Kim	University of Hawaii, USA
Masahiro Kimura	Ryukoku University, Japan
Jonathan Kopecky	John Hopkins University (APL), USA
Ee-Peng Lim	Singapore Management University, Singapore
Mitja Lustrek	Jozef Stefan Institute, Slovenia
Joseph Lyons	AFOSR, UK
Patricia L. Mabry	National Institutes of Health, USA
Achla Marathe	Virginia Tech, USA
Edgar Merkle	University of Missouri, USA
Hiroshi Motoda	AFOSR/AOARD, UK
Sai Motoru	MIT, USA
Anna Nagurney	University of Massachusetts at Amherst, USA
Keisuke Nakao	University of Hawaii at Hilo, USA
Kouzou Ohara	Aoyama Gakuin University, Japan
Bonnie Riehl	AFRL (RH), USA
Patrick Roos	University of Maryland, USA
Christopher Ruebeck	Lafayette College, USA
Kazumi Saito	University of Shizuoka, Japan
John Salerno	AFRL (RI), USA
Antonio Sanfilippo	Pacific Northwest National Laboratory, USA
Arunabha Sen	Arizona State University, USA
Vincient Silenzio	University of Rochester, USA
Michael Spittel	NIH, USA
George Tadda	AFRL (RI), USA
Lei Tang	Yahoo Labs
Shusaku Tsumoto	Shimane University, Japan
Changzhou Wang	Boeing, USA
Haiqin Wang	Boeing, USA
Zhijian Wang	Zhejiang University, China
Rik Warren	AFRL (RH), USA
Paul Whitney	Pacific Northwest National Laboratory, USA
Xintao Wu	University of North Carolina at Charlotte, USA
Laurence T. Yang	STFX, Canada
Shanchieh Jay Yang	RIT, USA
Bei Yu	Syracuse University, USA
Lei Yu	Binghamton University, USA
Philip Yu	UI Chicago, USA
Daniel Zeng	University of Arizona, USA
Inon Zuckerman	Ariel University Center of Samaria, Israel

Table of Contents

Consensus under Constraints:
Modeling the Great English Vowel Shift

Kiran Lakkaraju[1], Samarth Swarup[2], and Les Gasser[3]

[1] Sandia National Laboratories*
klakkar@sandia.gov
[2] Virginia Bioinformatics Institute,
Virginia Tech
[3] Graduate School of Library and Information Science,
University of Illinois at Urbana-Champaign

Abstract. Human culture is fundamentally tied with language. We argue that the study of language change and diffusion in a society sheds light on its cultural patterns and social conventions. In addition, language can be viewed as a "model problem" through which to study complex norm emergence scenarios.

In this paper we study a particular linguistically oriented complex norm emergence scenario, the Great English Vowel Shift (GEVS). We develop a model that integrates both social aspects (interaction between agents), and internal aspects (constraints on how much an agent can change). This model differs from much of the existing norm emergence models in its modeling of large, complex normative spaces.

1   Introduction

A society can be viewed as a system of mutually-constraining norms. They range from simple norms like manners of greeting, to complex ones like marriage customs and rules for inheritance of property. These norms structure the way people interact by making behavior more predictable. Once established, they are generally self-reinforcing in that people prefer to conform, and violations are met with varying degrees of sanctions [1].

Loosely speaking, norms are collective behavioral conventions that have an effect of constraining, structuring, and making predictable the behaviors of individual agents—that is, they are conventions that can exert a *normative force* that shapes individual agent behaviors—removing some behavioral possibilities from consideration and encouraging others—without a centralized "enforcement agency." We are particularly concerned about the emergence of norms and language. We view language as a type of norm. Existing models of norm emergence

* Sandia National Laboratories is a multi-program laboratory managed and operated by Sandia Corporation, a wholly owned subsidiary of Lockheed Martin Corporation, for the U.S. Department of Energys National Nuclear Security Administration under contract DE-AC04-94AL85000.

S.J. Yang, A.M. Greenberg, and M. Endsley (Eds.): SBP 2012, LNCS 7227, pp. 1–8, 2012.

(e.g. [2,3]) can be applied to linguistic cases. However, the language case introduces three intricacies that enrich the general study of norms:

Large Normative Option Space. The multiplicative interaction between lexicon and grammar allows for an enormous number of possibilities. A population must converge on a shared lexical and grammatical convention within this very large space.

Complexly Structured Option States. A language has many interactions between component elements, such as lexical and grammatical constraints. Thus language as a normative case is quite different from other convention phenomena studied earlier (e.g. [2]) that have a simple binary normative option space.

Ambiguity and Imperfect Knowledge. Agents have no direct access to others' preferred linguistic choices, but rather infer from limited samples. Again, this differs significantly from binary option spaces in which knowledge of other agents' states can be inferred perfectly: $\neg 1 = 0$ and $\neg 0 = 1$.

The general aim of this research is to understand the impact of these three issues on the emergence of norms by using language as a "model problem". In this paper we focus on studying the well known phenomena of sound change in language (such as the Great English Vowel Shift (GEVS) in England or the Northern Cities Vowel shift in the U.S. [4]). Modeling sound change requires addressing the intricacies listed above.

Section 2 describes the sound change phenomena and describes our Agent Based Model (ABM). Section 3 describes our simulation results; and Section 4 concludes with a discussion of future work and conclusions.

2   The Great Vowel Shift

While language changes continuously, at some points there have been large, significant changes in language, one of these is known as the *Great English Vowel Shift* (GEVS).

The GEVS took place roughly from the middle of the fifteenth century to the end of the seventeenth century. It was a change in the pronunciation of certain vowels; "the systematic raising and fronting of the long, stressed monopthongs of Middle English" [5]. For example, the pronunciation of the word "child" went from [čild] (*"cheeld"*) in Middle English to [čəild] (*"choild"*) to [čaɪld] (*"cha-ild"*) in Present Day English.

The GEVS is often seen as an example of a *chain shift* – a situation where one vowel changes pronunciation, thus "creating space" for another vowel to "move up". This causes a chained shift for a set of vowels – starting from a change by one vowel [4].

Figure 1 is a graphical depiction of the GEVS. The trapezoid is an abstract representation of the vowel space – the space of possible pronunciations of a vowel. The symbols represent vowels (in the standard *International Phonetic Alphabet* notation). Note that the topology of the vowel shift is linear, and thus we can model it in a linear array.

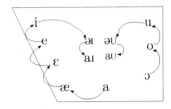

Fig. 1. The space of vowels, and the shift that occurred during the GEVS

It is still not clear *how* the GEVS happened. Some theories suggest purely internal, linguistic factors; while others suggest that the interaction between different social groups with regional dialectal variations caused a shift [5,6,7].

Our objective in this work is to evaluate the latter theory using an ABM, and through it study the emergence of a norm in a complex system. We will show that a chain shift can occur as a consequence of the interaction of social and internal forces. Agents try to increase communicability with each other when they interact by trying to align their pronunciations (social), but are constrained by a *phonetic differentiability* requirement, i.e. they have to maintain sufficient *spacing* between their phonological[1] categories to be able to distinguish the vowels from each other (internal). The interplay of these two processes leads to a chain shift.

2.1   A Model of Population Wide Language Change

The ABM will have *agents* that represent individuals; each agent has a language that represents vowels and their pronunciations. Agents will be embedded in a social network that determines who the agents can interact with. In the next section we describe the model.

A language will consist of 5 vowels, represented as 5 variables $\{v_0, v_1, v_2, v_3, v_4\}$. The pronunciation of a vowel is characterized by the principal frequency components of the sound. The lowest frequency component is called the "first formant", the second lowest is the "second formant", etc. Due to the low resolution of human perception the first two formants chiefly disambiguate a vowel [8]; for simplicity we focus on a single formant model, following [9]. Each vowel can take on an integer value between $0 \ldots (q-1)$ which represents the possible values of the first formant.

A language must satisfy the following constraints:

Phonetic Differentiation Constraint. The vowels (in a vowel system) should have pronunciations that vary enough to allow reasonable differentiation between them. Intuitively, if two vowels are very similar they will be mistaken for each other. Two vowels x, y are said to be *differentiable* if $|x - y| - 1 \geq d$ for $0 < d < q$.

---

[1] *Phonetics* refers to the study of the (continuous) sound signal; *phonology* refers to the study of the (discrete) categories into which this signal is perceptually mapped.

Ordering Constraint. There is a total ordering on the vowels. This means that two vowels cannot swap their locations, $v_i < v_j \forall i, j \ s.t. \ i < j$.

At each time step in the simulation we simulate an interaction between individuals via a *language game* [10] – an interaction in which agents exchange knowledge of their language with each other. Agents change their language as a result of these interactions.

A language game consists of two agents, a speaker (a_s) and a hearer (a_h). One of the five vowels is chosen as the *topic* of the game. The speaker then communicates to the hearer the value of this vowel. The hearer changes the value of its own vowel to match that of the speaker while still satisfying the constraints described above. If matching the speakers vowel would violate the constraints, the hearer does not change at all.

Note that this seemingly unrealistic interaction is an abstraction of a more natural interaction, where an agent utters a word that contains the vowel which we are calling the topic above. Further, it is generally possible to guess the word from context, even if the vowel pronunciations of the two agents disagree. Thus, we assume that the hearer knows both which vowel the speaker intended to utter, and which one (according to the hearer's vowel system) he actually uttered. The hearer then changes his own vowel system based on this information and the constraints above.

3   Simulation Results

We implemented the model presented above and evaluated three scenarios. All simulations used the following settings: 1000 agents ($n = 1000$), a vowel range of $0 \ldots 29 (q = 30)$, and a minimum difference between vowels of 4 ($d = 4$). To simulate social hierarchies we array the agents on a scale-free graph. We used the extended Barabasi-Albert scale free network generation process [11]. The parameters were $m_0 = 4, m = 2, p = q = 0.4$.

Experiment 1: Consensus to a Fully Solved Configuration. In the first experiment we investigated the emergence of a common language from random initial conditions. The population was initialized with randomly chosen assignments of vowel positions (that respect the ordering constraint but not the phonetic differentiation constraint).

Figure 2a shows time on the x-axis, and the average value of each vowel (over all agents in the population) on the y-axis. We see that the lines corresponding to the vowels become completely flat as the simulation progresses, and then stay that way. This demonstrates the emergence of a stable state. Further, the vowel positions are widely separated, which shows that the phonetic differentiation constraint is being satisfied.

Experiment 2: A New Subpopulation. For the second experiment we looked at the introduction of a new population of agents with a different language. This is similar to the sudden immigration of new language speakers into a country. Will the introduction of a new population cause a chain shift in the pronunciation of vowels?

In this experiment we replaced 30% of the population with a new population initialized with a different state. Initially, all agents had state $[0, 5, 10, 15, 20]$. The introduced population had state $[5, 10, 15, 20, 25]$, which is an overlapping but different state that satisfies the constraints. The entire vowel system is shifted by five positions with respect to the existing vowel system. The new population replaced the lowest degree 30% of nodes in the graph and was introduced at time step 1000. Figure 2b show the first 30K time steps.

Interactions between the new population and existing agents immediately start to cause a shift in the vowel positions. However, the vowels do not start shifting all together – the first vowels to shift are v_0 and v_4, which move in opposite directions. These are followed in turn by v_3 and v_2 (in the direction of v_4), while v_1 ends up staying more or less stable. Figure 2c shows the long run behavior of the same experiment. We see from this figure that eventually the entire population converges on a new stable state $[0, 5, 15, 20, 25]$, which is a combination of the vowel systems of the two populations. Further, the emergence of this new vowel system occurs through a chain shift, in two directions – v_0 moves down, while v_4, v_3, and v_2 move up, in that order.

Experiment 3: Varying the Social Network. In the final experiment we replicate Experiment 2, however we modify the social network to see the effect of different topologies on time to agreement. We run Experiment 2 on three different social networks, Complete, Scale-Free and Small-World.

A complete network is a network that contains all possible edges between nodes.

A Small-World network is a network that has high clustering yet small characteristic path lengths [12]. The networks used in this experiment were generated using the Watts-Strogatz algorithm described in [12] with parameters $p = .1$ and an average connectivity of 12.

Figure 3 summarizes the results of the experiments. For each network 10 simulations were run where the network was regenerated according to the parameters above and the populations were reinitialized.

In all cases the system converged to a language satisfying the constraints, although the time till convergence varied. Time till 90% agreement is a standard metric used in the literature [3]. It refers to the number of iterations till 90% of the population are in the same state.

The results of these experiments concur with results from previous work on diffusion, albeit under simpler settings. [13] show that small world networks took longer to reach 90% agreement than scale free or complete networks for the majority-rule model. [14] describes similar results for the voter model. The means were significantly heterogeneous (one-way ANOVA, $F_{2,27} = 1.663, P = .208$).

3.1   Discussion

As discussed earlier, the vowel space in figure 1 can be "unfolded" into a linear array, in which the a $\rightarrow$ æ$\rightarrow$ ɛ$\rightarrow$ e $\rightarrow$ i $\rightarrow$ əɪ$\rightarrow$ aɪ shift is a movement to the left, and the ɔ $\rightarrow$ o $\rightarrow$ u $\rightarrow$ əʊ$\rightarrow$ aʊ shift is a movement to the right. This matches,

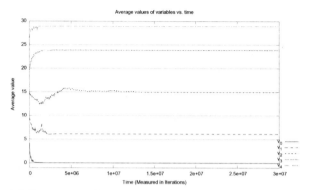

(a) Emergence of a consensus fully-solved configuration

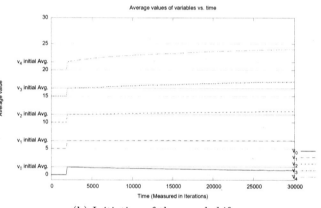

(b) Initiation of the vowel shift.

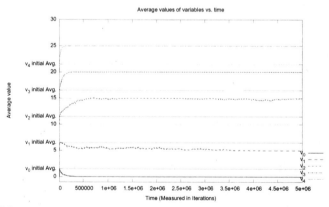

(c) Coordination to a new state with a new introduced population.

Fig. 2. Results from Experiment 1 & 2. Mean value of vowels vs. time

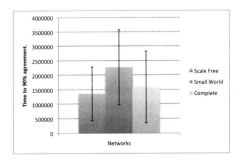

Fig. 3. Time till 90% convergence. Error bars indicate one standard deviation.

qualitatively, the movements observed in the experiments above, where some of the vowel positions shift up (to the left), and some shift down (to the right).

There are several competing explanations for the GEVS. One of the main contenders is that, after the Black Death there was a mass immigration into South-Eastern England, and the contact between the immigrants and locals led to a sudden change in vowel system. We have shown that this is a viable explanation, as it requires only the minimal assumptions that people tend to "accomodate" during interactions, i.e. they change their way of speaking to adapt to the other participant in the interaction, and that this accommodation is very limited, i.e. that they do not change their entire vowel system at once.

Taken together, the three experiments provide an interesting insight on norm emergence. Experiment 1 showed that a norm can emerge from a randomly initialized scenario.

Experiment 2 showed that our framework could capture complex behavior such as chain shifting. In addition, as mentioned above, these simulations provide some insight into the immigration explanation of the GEVS.

Experiment 3 illustrates the similarity of norm emergence in this model with other models by comparing the time to converge. As seen in the literature, the results show that norms take longer to emerge in small-world networks than in scale-free or complete networks.

4   Conclusions and Future Work

Norms are one avenue through which a society autonomously structures itself. The study of the emergence of norms in populations is an important aspect of reasoning about culture, and of social simulation in general. In this work we have presented a model to study linguistic change in populations as an example of norm emergence in complex scenarios. The study of language norms is a difficult undertaking, however, because language as a system consists of an intricate web of relationships and constraints between several variables. Through empirical simulation we have shown that a sudden influx of speakers of a different language can induce a drastic sound change.

Furthermore, we viewed the impact of three social network topologies on the time-to-convergence metric. We found that results were similar to other models; the small-world network took the longest time.

There is much work that can be done to extend the current work. Clearly, extensions to other aspects of language, such as syntax, remains to be developed. This approach can, in principle, also be applied to any situation where we have a system of constrained variables over which consensus must be achieved. Because of this, we are working to develop the theoretical underpinnings of our convergence plus constraints framework.

References

1. Burke, M.A., Young, H.P.: Social norms. In: Bisin, A., Benhabib, J., Jackson, M. (eds.) The Handbook of Social Economics. North-Holland (forthcoming)
2. Shoham, Y., Tennenholtz, M.: On the emergence of social conventions: modeling, analysis, and simulations. Artificial Intelligence 94(1-2), 139–166 (1997)
3. Delgado, J.: Emergence of social conventions in complex networks. Artificial Intelligence 141(1-2), 171–185 (2002)
4. Hock, H.H., Joseph, B.D.: Language History, Language Change, and Language Relationship: An Introduction to Historical and Comparitive Linguistics. Mouton de Gruyter (1996)
5. Lerer, S.: Inventing English: A Portable History of the Language. Columbia University Press (2007)
6. Leith, D.: A Social History of English, 2nd edn. Routledge (1997)
7. Perkins, J.: A sociolinguistic glance at the great vowel shift of english. Papers in Psycholinguistics and Sociolinguistics. Working Papers in Linguistics (22) (1977)
8. de Boer, B.: Self-organization in vowel systems. Journal of Phonetics 28, 441–465 (2000)
9. Ettlinger, M.: An exemplar-based model of chain shifts. In: Proceedings of the 16th International Congress of the Phonetic Science, pp. 685–688 (2007)
10. Steels, L.: Emergent adaptive lexicons. In: Proceedings of the Fourth International Conference on Simulating Adaptive Behavior (1996)
11. Albert, R., Barabàsi, A.L.: Topology of evolving networks: Local events and universality. Physical Review Letters 85(24), 5234–5237 (2000)
12. Watts, D.J., Strogatz, S.H.: Collective dynamics of 'small-world' networks. Nature 393(6684), 440–442 (1998)
13. Delgado, J., Pujol, J.M., Sanguesa, R.: Emergence of coordination in scale-free networks. Web Intelligence and Agent Systems 1(2) (2003)
14. Suchecki, K., Eguiluz, V.M., Miguel, M.S.: Voter model dynamics in complex networks: Role of dimensionality, disorder and degree distribution. ArXiv (2008); cond-mat/0504482v1

A Study of Informational Support Exchanges
in MedHelp Alcoholism Community

Katherine Y. Chuang and Christopher C. Yang

Department of Information Science and Technology,
Drexel University
{katychuang,chris.yang}@drexel.edu

Abstract. E-patients searching for online health information may seek support from the peers in health social media platforms especially when they cannot find the relevant information from authoritative Web sites. Many health social media sites have different 'architectural elements' to support the user communication. We seek to understand the relationship between social support and Computer-Mediated Communication (CMC) formats by comparing the social support types exchanged across multiple CMC formats (forums, journals, and notes) within the same community using descriptive content analysis on three months of data from MedHelp Alcoholism support community to find informational support (i.e. advice, opinions, and personal experiences, etc.). Forums are used for asking general questions related to Alcoholism. Notes are used for maintaining relationships rather than the main source for seeking information. Journal comments are similar to notes, which might indicate that journal readers consider the author as a friend. These descriptive results suggest that users may be initially attracted to the community forums for information seeking yet continue to engage in the online community due to relationships strengthened through journal or note formats.

Keywords: health informatics, social media, social support, computer-mediated communication.

1 Introduction

The Internet is a useful tool for finding health information and also for connecting people. People go online for communicative or social reasons; often, e-patients want to access user-generated or "just-in-time someone-like-me" health information [10,12,21]. In recent years, e-patients are increasingly sharing their health information with social networks by gathering information and seeking support as they face important decisions [10]. Such websites include community websites using software for newsgroups, blogs, social networking sites or micro blogging. These e-patients that seek, share, and sometimes create information about health and wellness from these sites can often benefit from sharing their experiences, discussing medical information, and exchanging social support [10]. Peer social support is beneficial for

S.J. Yang, A.M. Greenberg, and M. Endsley (Eds.): SBP 2012, LNCS 7227, pp. 9–17, 2012.

e-patients coping with difficult health conditions and increasing access to relevant information [15]. Previous research literature suggests that online communities become surrogate families of e-patients, where members share common problems, help each other toward mutual goals, and support each other through good times and bad [24]. Many users join online support groups for a sense of community with those who experienced similar situations because they are more likely to have the highly sought after compassion and experiential knowledge [24].

Online communities exist across many social web technologies (i.e. email lists, discussion boards, etc.), and most recently, social network sites (SNS). SNS enable users to find each other and build connections using profile pages and private and public communication tools to communicate [1, 3]. It is unique from previous text-based communication formats because of its emphasis on representing relationships between users and 'architectural elements' that encourage interpersonal relationships [18].

Studying the patterns of nurturing interactions within a support community will give useful insight into users' communication patterns. These results could contribute to improving the design of online intervention programs by suggesting new software features to promote a supportive environment. For example, the treatment for alcoholism often consists of participation in support groups for social support component (AA) or in recent cases, online interventions [5].

Previous research demonstrates e-patients' information needs beyond what their health professionals provide or technical support (i.e. medication reminders), most especially for social support in making healthcare decisions [24]. Previous findings about social support did not look specifically at the relationship between the CMC format and the behavior patterns of individuals participating within an online support group. We perceive that CMC enables interpersonal communication in a public environment; however users often conduct their conversations on this platform as if it were a private space [18,19,22].

Online support groups are a convenient place for health consumers to find conversations useful in guiding their healthcare decisions, by helping them find similar patients to talk with [13,24]. These self-help groups have a variety of social support types (informational, nurturant, instrumental). These sites are good sources for people such as alcoholics to gather information anonymously to avoid the stigma that comes from traditional face to face conversations. Social support exchanges can be thought of in economic terms. In a successful supportive interaction exchange, there will be a combination of someone offering and someone requesting support. People seek may ask for help when there is an information gap. With an architectural view, we see that the site design affects online transactions of social support exchanges [8,18]. Site design "promotes the development of particular culture or behaviors and identity presentation", which may be found through studying interactions between users [8]. Convenient features allow users to form and maintain online network "friends", where if one user invites another user to be friend and if accepted, a relationship is established on the website [1]. Friends can communicate through SNS in several ways, including private and public messaging systems [20].

The increasing socialization of online health information is a new phenomenon that could have untapped opportunities for future health services as people gather online to converse about health issues, especially in health support communities [10]. We investigate patterns of informational support exchanged across different social media communication formats. By conducting this study, we hope to gain a better understanding of the link between site design and communication for an online health community.

2   Methods

We extracted data from MedHelp Alcoholism Community (www.medhelp.com) using a web crawler and selected a 3-month period of discussion forums, user journals, and posts on users' profile pages called 'notes'. The messages in each sample were converted to spreadsheets for descriptive content analysis. Definitions for social support types were developed from reviewing examples in related literature and matching them with themes presented within our data [2,7,16,23]. Social support is generally the provision of psychological and tangible resources intended to benefit an individual's ability to cope with stress, such as information leading the subject to believe that he is cared [2,7,23]. Concepts and their definitions were drawn out and organized into three main categories (information, nurturant, instrumental), suggested by [7]. We chose these three categories because the categorization is most commonly used in related studies and also covers a wide range of support types. Only informational support and nurturant support were found in the data. The third type, instrumental support, is typically found in face to face interactions and not found in the data for this study. In this paper we only report findings of informational support.

MedHelp is a health-oriented SNS platform with peer support communities helping individuals connect with people and to information resources. It is open for any registered user to join. There are several interpersonal communication tools, including discussion forums, journals, and notes. The three CMC formats investigated in this study are available for any MedHelp community member to post content, where each varies in features . Users can post questions or polls to the forum. They are required to fill out a title, select a topic, describe their question, and are free to add tags. Posting to journals can optionally include title, entry, tags, photos, with selected privacy options per post. A journal thread must be initiated by the profile page owner. Posting notes on a user's profile includes type of note and the content in the note. A note can only posted by the users who have access to the profile page but the owner of the profile page cannot make a note to herself. If the user is not a friend, there is an option to befriend the user.

While the messages on each of these CMC formats might be displayed publicly depending on privacy level settings (public, friends, private), the literature review suggests that social interactions on each may have different kinds of support. All forum content created by users is set to publicly accessible. However, MedHelp allows journals and notes to be set to one of three options: 'Everyone', 'Only my friends', 'Only Me'. New posts to the forum can be viewed on the forum page, which is also known as the support community page. Updates to public journals (new posts or new comments) are listed on the support community page under 'recent activity' box. There is also a section that lists community members with links to their profile pages. Each profile page displays sections of the user's activity on the communication tools. Unlike the forum messages, journal and note messages can only be viewed on the individual profile pages. Privacy settings may affect what can be viewed on a user's profile page. If the setting for journal and notes are set to 'only my friends' then only users who are 'friended' may view these content. If setting is set to 'only me, only the user can see their own content when logged in. The content in each is organized chronologically.

Coding Scheme – Informational Support Types. Information support describes messages that convey instructions, including (a) advice, (b) referrals to other sources of information, (c) situation appraisal, (d) stories of personal experience, and (e) opinions. Messages coded as information support often appeared as an attempt to reduce uncertainty for the message recipient [2,7].

Support Type	Definition
Advice	Offers ideas and suggests actions for coping with challenges such as detailed information facts or news about the situation or skills needed to deal with the situation. i.e. *"....and i want to quit, but am not able to do so as my wife always gives me tensions , What should i do?"*
Referral	Referral to information is when recipient asks for information sources, or it could be efforts to link the recipient with a source of expertise. i.e. *"How can I get to the video link? Thanks so much."*
Facts or situation appraisal	Facts or situation appraisal is offered when someone reassesses the situation, often to provide a different way to look at things. i.e. *"I have talked bout this b4......PAWS....Post Acute Withdrawal Syndrome...comes from years of heavy drinking....takes a LONG time for the central nervous system to repair itself..."*
Personal Experiences	Stories about a person's experiences or incidents. It has a more story like form that is about self-disclosure and possibly personal information. i.e. *"i have 25 years sober/clean....every day u don't drink or use its a sober clean day!i"*
Opinions	Opinions are a form of feedback, which can be a view or judgment formed about something. It is not necessarily based on fact or knowledge. i.e. *"Antabuse is cheap and has two side effects men hate...onion breath and erectile dysfunction ..."*

3   Results

Our data contains three samples of user created messages from the discussion forums (n=493), the user journals (n=423), and from profile posts (n=1180). The messages in forums and journals were grouped into two types, *posts* (i.e. messages that start the thread) and *comments* to the post. There were 81 forum posts (FP) and 412 forum comments (FC). There were 88 journal posts (JP) and 335 journal comments (JC) and 1180 notes (N). We first identified five types of informational support in the samples that were both provided and requested. Some messages only offer support (i.e., *"Have you tried Naltrexone? It is supposed to help with the cravings there are other meds that can help with it too. If all else fails, make a picture of tea and pop some popcorn and hang out with him with your "drink"*), or only request support (i.e., "*Hi, is there a medicine to take to stop the craving for alcoholic drink?*").

The number of messages offering and requesting different types of informational support is given in Table 1 and Table 2, respectively.

Table 1. Percentage of messages with informational support types

	FP		FC		JP		JC		Notes	
	O (%)	R (%)	O (%)	R (%)	O (%)	R (%)	O (%)	R (%)	O (%)	R (%)
Advice	0.0	27.8	24.2	11.8	2.9	29.4	21.9	3.6	13.6	0.4
Referral	0.0	5.1	3.6	0.0	19.4	0	2.3	0	3.1	0.4
Fact	69.0	49.4	48.2	75.0	60.2	52.9	66.0	96.4	74.0	96.7
Personal	31.0	1.3	13.9	2.9	8.7	0	2.8	0	2.1	0.4
Opinion	0.0	16.5	10.1	10.3	8.7	17.6	7.0	0	7.2	2.1

Overall, fact was the most exchanged type of information across all samples. Notes and JC showed similar patterns of behaviors for both offered and requested. JP, FP, and FC showed similar patterns for requested informational support. The pattern appearing in notes was different from the other formats for the aspect of information sharing. In forum messages and journal posts, users were likely to request information types other than and in addition to fact; however, in the notes format users are more likely to exchange facts without mentioning stories or referrals. This is different in the longer messages of journals and forums, which contain the more stories, opinions, and advice.

There was a relationship between offered and requested support, for example, advice is offered in the comments, but not in posts. For all the samples, fact is exchanged the most. There were more offered than requested supports in notes, especially for exchanging fact, advice, and opinion. This could be an indication of using notes format for altruistic reasons. The high incidence of fact offered in JP suggests that users were documenting their thoughts. Perhaps they did not expect responses, unlike JC messages where requested fact is very high. JP requests opinion and advice along with fact, and JC offers these three types more than the other information types. In JP, personal experience and referrals not requested at all but referral and fact given in most messages, which can suggest that journals might be a place for sharing information. JP might also be a good place to seek advice, as comments offers advice. Perhaps users writing journal entries might have a close relationship where each party typically gives advice and opinions.

Although patterns in forum posts and forum comments seem correlated, it followed a slightly different relationship. FP messages offered personal stories, facts, or a combination of these two. The relationship between FP and FC messages can be characterized as polite and altruistic exchange, where more support is given than requested. Referral was given in some messages, possibly as a strategy to obtain advice, stories, and opinions, for example, because offering opinions may not be helpful in seeking advice from others. JP contains the highest percentage of messages of offering information referrals, despite forums having more messages. This might be because users are recording information they discover.

4   Discussion

Internet users join online health support communities (like those available on MedHelp.org) even while a plethora of alcohol and other health information is available on the Internet because they provide additional peer support. Because the members are not health professionals, the members are drawn to the a social place where participants share insights with each other as opposed to purely health information sites (e.g. WebMD), where users may experience difficulty in understanding the large quantity of information available. The added social components in support communities – where e-patients can have their questions answered, and hear other e-patients' experiences – provide more easily digestible information, for example advice about applying new lifestyle changes.

To the e-patient, the interactivity of an online community is different from perusing static information pages because of the added social component. Social media technology makes it easy to share and seek information from peers who have experienced similar situations and can offer targeted stories and practical advice [17,24]. Internet users may also want to use these websites to stay in touch with close friends and family [11,14]. A forum space is similar to a waiting room at the clinic, in that people know it is more public than the doctor's office. In terms of informational and emotional content exchanged in the community, users were selective in what they write and whom they interact with across the CMC formats. In the forums, it appeared that the space was used a Q&A forum, whereas on profile pages and journals the "personal nature" might explain their behavior for more emotional content. Environmental factors may play a role in shaping this behavior, however it is also possible that user perception of these environments alter their motives...

Our results slightly differ from related studies of the same type concerning the levels of support identified [2, 6,9,15]. First, this study collected data from different text-based communication formats (journal, notes) than previous studies (mailing lists, discussion boards). The architectural elements are different and can affect communication. Second, the members of the MedHelp community are allowed to and often communicate with each other across multiple CMC formats instead of just one (i.e. email lists). Features such as the profile page and journals are similar to providing rooms for people to talk about more specific things and have fewer interruptions, and this availability impacts the conversations on the communal areas to be more formal and the other areas to be less so.

The MedHelp communities have several communication formats, each used for a different purpose. Constructing an arena for people to talk (i.e. email list and bulletin boards) is good for group style but for more tailored communication between smaller groups (more focused topic) or between two individuals, the other formats are better suited. People will have different needs for participating in an online community, for example some members sought information, while others sought compassion and intimacy [4]. In addition, patients may go through waves of information needs [27]. Posting to the forum may be a different purpose than journal or notes, for example, one might disclose personal information as a strategy for finding tailored information or to document experiences.

A longitudinal view of social support exchange could help us better understand the why some users are more likely to offer support and others to request support. Users in online support groups go through a few phases of involvement (engagement, adoption, and diffusion) before they can become 'big brothers' to 'newbies [25]. Their involvement with the support group depends on amount of positive feedback received over time [26]. In fact, our findings through the lens of these social theories can advise the development and use of CMC for health care in specific instances where prescribing specific software design features for online intervention programs. For example, in initial stages of health treatment e-patients can be directed towards forum space for general information. Later on, if this e-patient is paired with in a buddy system such as those in Alcoholics Anonymous the e-patient can be direct to the notes type of CMC to enhance the relationship building. Further work specifically targeting the relationship formation in this community would help us better understand the evidence for the social theories.

We find the CMC format impacts communication behaviors, notes are similar to journal comments, forum comments are similar to journal comments, but forum posts stand out as having different pattern than other formats. Because privacy can be controlled through notes and journals, they are more personal than the public forum. It is possible that the users did not find it necessary to use privacy controls in the more personal areas because it seems more private. In a physical setting, it is easy for one to perceive the relative privacy of the space. However, in an online environment, the amount of privacy is not as transparent. In this case, perhaps the MedHelp users do not assess the online setting as they would a physical face to face setting. In light of the content observed through this community (i.e. blackouts, possible violent episodes, etc), the online setting diminishes amount of stigma that would be present in face to face support.

5   Conclusion

In this study we compared supportive interactions across different software features of a health social networking site. We found that there are different types of information exchanged as social support, and each CMC format has a different combination of patterns. While people can obtain social support from existing offline social networks, participation in online support groups have added benefits such as coping with chronic health conditions. We identified different types of informational support in the MedHelp alcoholism community across three text-based CMC formats. Each format was used differently. Forums were used for asking and sharing information with a wider audience. Journal comments were similar to notes with smaller groups of individuals interacting, which might be an indication that journal readers consider the author as a friend. Notes were not the main source for seeking information, but rather for maintaining relationships. Users joined the community seeking information however very likely remain active because of the community social connections presented.

References

1. Ahn, Y., Han, S., et al.: Analysis of topological characteristics of huge online social networking services. ACM (2007)
2. Bambina, A.D.: Online Social Support: The Interplay of Social Networks and Computer-Mediated Communication. Cambria Press (2007)
3. Boyd, D., Ellison, N.: Social network sites: Definition, history, and scholarship. JCMC 13(1) (2007)
4. Chuang, K.Y., Yang, C.C.: Helping you to help me: Exploring supportive interaction in online health community. In: Proceedings of ASIS&T, Pittsburgh, PA, October 22-27 (2010)
5. Cunningham, A.J., Humphreys, K., et al.: Formative Evaluation and Three-Month Follow-Up of an Online Personalized Assessment Feedback Intervention for Problem Drinkers. Journal of Medical Internet Research 8(2), e5 (2006)
6. Cunningham, J.A., van Mierlo, T., Fournier, R.: An online support group for problem drinkers: Alcohol Help Center.net. Patient Education and Counseling 70(2), 193–198 (2008)
7. Cutrona, C.E., Suhr, J.A.: Controllability of Stressful Events and Satisfaction With Spouse Support Behaviors. Communication Research 19(2), 154–174 (1992)
8. Donath, J.: Signals in social supernets. Journal of Computer-Mediated Communication 13(1), article 12 (2007),
   http://jcmc.indiana.edu/vol13/issue1/donath.html
9. Eichhorn, K.C.: Soliciting and Providing Social Support Over the Internet: An Investigation of Online Eating Disorder Support Groups. Journal of Computer-Mediated Communication 14(1), 67–78 (2008)
10. Fox, S., Jones, S.: The Social Life of Health Information Americans' pursuit of health takes place within a widening network of both online and offline sources (2009)
11. Gilbert, E., Karahalios, K.: Predicting tie strength with social media. In: Proceedings of the 27th International Conference on Human Factors in Computing Systems (2009),
    http://dx.doi.org/10.1145/1518701.1518736
12. Høybye, M.T., Johansen, C., Tjørnhøj-Thomsen, T.: Online interaction. Effects of storytelling in an internet breast cancer support group. Psycho-Oncology 14(3), 211–220 (2005)
13. Klaw, E., Dearmin Huebsch, P., et al.: Communication patterns in an on-line mutual help group for problem drinkers. Journal of Community Psychology 28(5), 535–546 (2000)
14. Kovic, I., Lulic, I., Brumini, G.: Examining the medical blogosphere: An online survey of medical bloggers. J. Med. Internet Res. 10(3), e28 (2008); PMID: 18812312
15. McCormack, A.: Individuals with eating disorders and the use of online support groups as a form of social support. Computers, Informatics, Nursing: CIN 28(1), 12–19 (2010)
16. McKenna, K.Y.A., Green, A.S., Gleason, M.E.J.: Relationship Formation on the Internet: What's the Big Attraction? Journal of Social Issues 58(1), 9–31 (2002)
17. Overberg, R., Otten, W., de Andries, M., Toussaint, P., Westenbrink, J., Zwetsloot-Schonk, B.: How breast cancer patients want to search for and retrieve information from stories of other patients on the internet: an online randomized controlled experiment. J. Med. Internet Res. 12(1), e7 (2021); PMID: 20215101
18. Papacharissi, Z.: The Virtual Geographies of Social Networks: A Comparative Analysis of Facebook, LinkedIn and ASmallWorld. New Media & Society 11(1-2), 199–220 (2009)
19. Smith, K.M.: Electronic Eavesdropping: The Ethical Issues Involved in Conducting a Virtual Ethnography. Online Social Research; Methods, Issues, & Ethics, 223–238 (2004)

20. Thelwall, M., Wilkinson, D.: Public dialogs in social network sites: What is their purpose? J. Am. Soc. Inf. Sci. Technol. 61(2), 392–404 (2010), doi:10.1002/asi.21241
21. Tufekci, Z.: Can You See Me Now? Audience and Disclosure Management in Online Social Network Sites. Bulletin of Science and Technology Studies 11(4), 544–564 (2008)
22. Walstrom, M.: Seeing and sensing in online interaction: An interpretive interactionist approach to USENET support group research Online social research: Methods, issues, & ethics, 81–97 (2004)
23. Winzelberg, A.: The analysis of an electronic support group for individuals with eating disorders. Computers in Human Behaviour 13, 393–407 (1997)
24. Wright, K.B., Bell, S.B.: Health-related Support Groups on the Internet: Linking Empirical Findings to Social Support and Computer-mediated Communication Theory. Journal of Health Psychology 8(1), 39–54 (2003), doi:10.1177/1359105303008001429
25. LaCoursiere, S.: A Theory of Online Social Support. Advances in Nursing Science 24(1), 60–77 (2001)
26. Taylor, D.A., Altman, I.: Self-Disclosure as a Function of Reward-Cost Outcomes. Sociometry 38(1), 18–31 (1975), http://www.jstor.org/stable/2786231

Real-World Behavior Analysis
through a Social Media Lens

Mohammad-Ali Abbasi[1], Sun-Ki Chai[2], Huan Liu[1], and Kiran Sagoo[2]

[1] Computer Science and Engineering, Arizona State University
[2] Department of Sociology, University of Hawai'i
{Ali.abbasi,Huan.liu}@asu.edu,
{Sunki,sagoo}@hawaii.edu

Abstract. The advent of participatory web has enabled information consumers to become information producers via social media. This phenomenon has attracted researchers of different disciplines including social scientists, political parties, and market researchers to study social media as a source of data to explain human behavior in the physical world. Could the traditional approaches of studying social behaviors such as surveys be complemented by computational studies that use massive user-generated data in social media? In this paper, using a large amount of data collected from Twitter, the blogosphere, social networks, and news sources, we perform preliminary research to investigate if human behavior in the real world can be understood by analyzing social media data. The goals of this research is twofold: (1) determining the relative effectiveness of a social media lens in analyzing and predicting real-world collective behavior, and (2) exploring the domains and situations under which social media can be a predictor for real-world's behavior. We develop a four-step model: community selection, data collection, online behavior analysis, and behavior prediction. The results of this study show that in most cases social media is a good tool for estimating attitudes and further research is needed for predicting social behavior.

1 Introduction

The advent of participatory web has created user-generated data [1], that leave massive amounts of online "clues" that can be examined to infer the attributes of the individuals who produced data. As it becomes easier and easier to create content in the virtual world, more and more data is generated in various aspects of life for studying user attitudes and behaviors. Sam Gosling in [7] reveals how his team gathers a large amount of information about people without asking any questions but only by examining the work and living places of their subjects. As we can understand people by studying their physical space and belongings, we are now able to investigate users by studying their online activities, postings, and behavior in a virtual space. This method can be a replacement for traditional data collection methods.

Among traditional social science data collection techniques, surveys or experiments are structured and active, and generating new data is an important

S.J. Yang, A.M. Greenberg, and M. Endsley (Eds.): SBP 2012, LNCS 7227, pp. 18–26, 2012.
© Springer-Verlag Berlin Heidelberg 2012

part of the process. The researcher defines what s/he needs, designs question-naires or experimental treatments, and collects the data based on the results of administering them. The results provide a greater degree of control over the measurement but is expensive, time consuming, and may even be dangerous sometimes. On the other hand, studying social media, as an alternative to sur-veys or experiments, can be considered as an extension of passive methods of traditional social research such as field research and content analysis to observe people's attitude.

Real-World Behavior Prediction. Attitudes[1] among individuals in a pop-ulation can be determined in social sciences. More specifically, attitudes may be measured using established data collection techniques, such as surveys, ex-periments, field research, and content analysis. Alternatively, attitudes may be determined without direct measurement through models that allow them to be predicted from individual or collective structural position and/or past actions and experiences. Such approaches are often described as exogenous and endoge-nous analysis, respectively [5].

Online Behavior Prediction. The use of World Wide Web content to predict the attitudes and behavior of individuals or groups is an issue that is increasingly tantalizing and frustrating to social and computer scientists [13]. Conversely, information available online seems to offer a gold mine of useful data - it is copious, usually publicly accessible, can be located with the aid of search engines, and often has built-in annotations in the form of meta-tags and link information. Furthermore, because such information, when public, can be downloaded by virtually anyone with an Internet connection, regardless of location, its collection generally incurs less time, expense, intrusiveness, and danger (depending on the population being studied) than traditional primary social science data collection techniques such as surveys, experiments, and field research.

On the other hand, there is not a straightforward relationship between online content and attitudes/behavior in the real-world. Although the Web is growing exceedingly fast, some sectors of populations are more likely to use it than others [9]. Furthermore, among the many interesting research issues, novel techniques are needed to explain the relationship between web content and attitudes of those producing the content, and the relationship between these attitudes and actions on the ground by the groups that the content producers represent. Nei-ther relationship is simple to determine, and both require the implementation of innovative methodologies in order to provide the impetus to making productive use of online information as a predictor of collective behavior.

Prediction and understanding of the attitudes and behaviors of individuals and groups based on the sentiment expressed within online virtual communities is a natural area of research in the Internet era. Ginsberg et al [6] used Google search engine query data to measure concern about influenza, which in turn was

---

[1] An attitude is a hypothetical construct that represents an individual's degree of like or dislike for something.

used to predict influenza epidemics. They used the idea that when many from specific area are searching for influenza or topics related to it, this is a sign that there is an epidemic in that place. O'Connor et al. [10] analyzed sentiment polarity of a huge number of tweets and found a correlation of 80% with results from public opinion polls. Bollen et al [3] used Twitter data to predict trends in the stock market. They showed that one can predict general stock market trends from the overall mood expressed in a large number of tweets. In other research, Asure and Huberman [2] used Twitter data to forecast box-office revenues for movies. They showed that there is strong correlation between the amount of attention a movie has and its future revenue.

2   Methodology

The prediction of human behavior using social media has been active for years. Most of the work in this field can be classified into two categories: extraction of attitudes and prediction of behaviors. To extract attitudes, researchers mine archived or online data and map specific data patterns to specific attitudes based on the frequent patterns. Patterns are then used to predict actions and outcomes. In order to predict real-world behavior based on online behavior analysis, we follow a 4-step procedure below.

1. Select real-world communities, and find corresponding community or communities in social media;
2. Collect attitudes and online behavior from social media;
3. Analyze online behavior; and
4. Predict real-world behavior based on observed online behavior and attitudes.

Online Community Selection. The initial step is selecting an online population that in some sense represents the same group as that of a real-world community. One way to do so is base selection of both the subject population and the virtual community on a particular ascriptive characteristic [11], i.e. a characteristic that is for practical purposes static, or at least difficult to change, within individuals and groups. Such characteristics include race, ancestral religion, primary language, and country/region of origin. Contemporary cross-national social science theories of ethnicity generally accept that ethnic groups are defined by one or more ascriptive boundaries [4]. Selecting populations based on ethnic characteristics is helpful because of the stability of what individuals possess them. This in turn allows the researcher to ensure that any group comprising those who share those characteristics will be a relatively stable one whose members are relatively easy to identify and are not constantly moving in and out of the group. Our ethnic group in this study is countries involving in Arab Spring revolutions.

Data Collection. We used Twitter to collect 35 million tweets related to Arab Spring event almost for all of the countries involved in the revolutions. In addition we collected more than one million articles and blogposts from popular Middle Eastern social tagging websites. Moreover we crawled 135,000 popular Facebook pages to collect data on posts, comments and like behavior on Facebook. The data on real-world events has been collected from Reuters.com website, which contains an archive of all published articles dating back several years.

Text Processing and Online Behavior Analysis. We performed preliminary text mining techniques including removing stop words, stemming, and extracting the most frequent words and phrases. Then we translated the most repeated words into English. Detecting events from newswire stories is a form of event detection from text streams, something that has been an active research field in recent computational studies of the Internet [8], [12]. In order to examine the relationship between social media artifacts and real world events, we ordered the events chronologically and performed multiple forms of statistical analysis against time-stamped results from online content analysis. For our main analysis, we used weekly word frequency data drawn from the blogosphere and compared this against a measure of the magnitude of collection action events, also weekly, drawn from the Reuters database.

Real-World Behavior Prediction. We used correlational analysis to identify word categories whose frequency of mention was most significantly related statistically to the magnitude of social action during the same period, then used multivariate regression analysis to assign coefficients to them for predictive analysis. We found that our ability to predict events was preserved even when independent variables were lagged to a period of up to two weeks. For zero, one, and two week lags, we could identify at least two word categories whose coefficients were significant at the $p < 0.5$ level in a two-tailed test. Lagging independent variables has the added advantage of demonstrating that large social action events can be predicted well before their actual occurrence. Indeed, there were two categories with significant coefficients even with a lag period of 1.5 months. The number of word categories was deliberately chosen to be relatively small in number in order to ensure that significance would not be the result of spurious results due to excessive statistical degrees of freedom.

3   Observations

Event Prediction. During social movements, social media has been used for either to organize future events or to report past events. For the former case we are able to find online conversation on events, even weeks before the real-world event but for the later one, as there is no or few information, prediction is not possible. We extracted frequent keywords for each event before and after that event which shows amount of online discussion for each event. Figures 1a through 1d show the amount of related discussion in social media on each topic. The peaks align

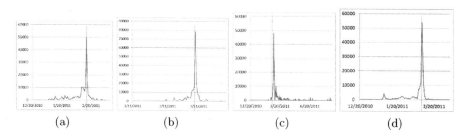

Fig. 1. (a, b, c, and d) The correlation between online behavior and real events. The vertical red line shows the day that real-world event is happened. Each graph is related to an event. x-axis is the date and y-axis is frequency of keywords related to the event. (a) Shows lots of online conversation almost 5 weeks before and two weeks after the event. This kind of events are easily predictable. (b) Almost the same as (a) but with few conversation after the real-world event. (c) shows little conversation before the event (this is the pattern for unexpected events) and much more after the events even for three months. (d) shows an event with only few days conversation after the event.

very closely with the dates that real-world action activity peaked as well. This synchronization supports the hypothesis that online dialog is closely related to events on the ground. For some (usually unexpected) events, there is nothing or few conversation before the events but much more after that (Figure 1c). However there are much more tweet or blogpost after the event. The prediction results usually does not contain information about details of corresponding real-world event such as time, location, magnitude, length, effects, and participants. Also we are not able to observe or predict small events (usually those that are limited to small portion of the target society). In some events, in spite of large amount of online conversation, the real-world event was not as large as expected and vice versa (A large real-world event with negligible online conversation).

Attitude Extraction. By mining the frequent patterns we are able to extract attitudes. For example Arab Spring tweets show that Yemenis were more concerned about Security whereas Egyptians were more concerned about Revolution and Freedom (Figures 2b and 2c). More interesting, we are able to track change of attitudes during a social movement by using a time series of tweets and blogposts.

Key People Detection. By employing the same method as we used for attitude extraction, we are able to find Key people. Our definition of Key people is those whom most mentioned in social media during an event. These people usually play an important rule and knowing them in social movements is very important. The data extracted using this method is highly correlated with the data from real-world.

Mood Analysis. We used method introduced by [14] to evaluate sentiment orientation of words, sentences and documents. By using this method we are able to measure the overall sentiment of the society before, during and after events. Results show that usually, people are much happier before than after events. They also are more exited when event is happening in the real-world.

4   Discussion

This section provides analyis on the results of previous section. Analysis are given for all parts of the predifined method including, Community selection, Data collection, Online behavior analysis and Prediction. In summary, the observations show that in many cases by using social media data we are able to extract attitudes, mood, events, and even key figures. In general as we can see social media data can be used for mining purpose. But for prediction, the results are not satisfactory. In some cases resutls match with real-world data and in some cased does not. In this section we provide more details about the prediction process using social media data.

Community Selection. The first step of a good prediction is attitude extraction from desired group. As we want to use social media data to extract attitudes, we should find an online substitute for our real-world group. But in most cases finding the same population both in the real-world and social media is very challenging or even impossible. This problem leads us to select a community with lower similarity with real-world group. As we can see in opinion polling, the main challenge is selecting a sample of elements from a target population. As an example, analyzing Arab Spring tweets show that roughly 75 percent of the 1 million clicks on Libya-related tweets and 89 percent of the 3 million clicks for Egypt-related Tweets came from outside of the Arab world[2]. Since almost 90% of this data is generated outside of the Arab world, how it can represent Arabs' attitudes? And in what extent the prediction based on this data is accurate?

Another example is analyzing the relation between number of followers on Twitter or number of page likes on Facebook and number of supporter in real-world. What is the relation between candidates' number of followers in Twitter and their chance to win the next presidential election? Barack Obama, Mitt Romney, and Rick Perry have 11,045,000 and 161,000 and 103,000 followers on Twitter respectively. This data is not valid for prediction because voters are not this society of Twitter user.

Data Collection. Most of Twitter users who tweeted from inside of Arab world, have missing information in their profiles. Usually they don't use Twitter's geotag feature when they tweet. They reveal little information about themselves and usually use nicknames instead of their real name. This is very common in non-democrat countries. Lack of information about the source of data, make it

---

[2] http://www.stripes.com/blogs/stripes-central/stripes-central-1.8040/
researchers-skeptical-dod-can-use-social-media-to-predict-future-
conflict-1.155296

(a)   (b)   (c)

Fig. 2. (a) The day after Steve Jobs' death his Facebook page had one million likes three times more than the day before. At the same time Apple's stock was experiencing a drop in real-world. (b, c) Tag-clouds show that Yemenis were more concerned about Security whereas Revolution for Egyptians.

unreliable and we can't use them in our analysis with confident. More seriously, people in non-democratic countries do not reveal their real believes and intension. Unrelated and spam data is another challenge in social media. Paid bloggers (or Twitterers) are another problem that misleads the prediction results. In most of the events we observed many accounts actively tweet fan of the government. Statistical methods are very sensitive to this kind of tweets.

Online Behavior Analysis. Extracting attitudes from plain text due to language complexity is not a straight forward process. In most cases researchers are not able to extract complex patterns from the text. We also need to infer collective behavior by observing bunch of individuals' online behavior that generates more complexity. In some cases though there is connection between attitudes and behaviors, but it is hard to be discovered by using automatic methods.

Real-World Behavior Prediction. Prediction is hard even if we have enough sources of data. In many cases there is not a straight forward relation between online and Real-world behavior and even by using sophisticated methods to extract online behaviors, we will not be able to predict the real-world behavior. An example of this case is events related to Steve jobs death. Before the announcement of his death, his Facebook page had less than 300,000 likes. But few days after that, more that a million of his fans liked his page (Figure 2a). People used social media and weblogs to write comments about him and his company, Apple. People everywhere praised him and his company. By using the method proposed by Bollen et al [3], we should have a huge rise for Apple's stock but in real world Apple's stock was experiencing a drop. This example and many others show that even we collect enough data from social media and run sophisticated algorithm to analyze social media data, we would not be able to predict the real-world's outcome.

5   Conclusions and Future Work

In this paper we report a comprehensive research on using social media data to analyze online behavior and predict real-world behavior. The prediction task has been divided into four sub tasks. These tasks are community selection, data collection, online behavior analysis, and real-world behavior prediction. The main challenges of this process are task one and task four which are how to identify key individuals or groups, as well as how to predict real-world behavior by interpreting the attributes of these individuals or groups. Once these two issues are addressed, analyzing the data generated by people in social media can help to understand their attitudes and to predict future real-world activities. In some cases we are not able to find a match for the real-world population, this problem causes a set of non relevant attitude and therefore misleading to wrong behavior prediction. Insufficient data in some cases is another source of failure. Compexity of language also leads to misinterpretation of online behavior. And the last item it compexity of behaviors. Sometimes information available in social media is not enough for prediction. As we see in the case of Steve Jobs' death, information available in social media is not enough to predict Apple's stock so we need more data and need to use more comlex algorithms for prediction. We also observed that we are not able to predict minor events or details about major events.

Future research would look at comparison between these methods and more traditional social science methods, including surveys, and methods such as prediction through change in social structural variables (political and economic). Comparisons with hybrid methods such as social science attitudinal content analysis of online social media would also yield additional insights into the relationship between activity in virtual communities and on the ground.

Acknowledgments. This research is sponsored, in part, by Air Force Office of Scientific Research.

References

1. Agarwal, N., Liu, H., Tang, L., Yu, P.: Identifying the influential bloggers in a community. In: Proceedings of the International Conference on Web Search and Web Data Mining, pp. 207–218. ACM (2008)
2. Asur, S., Huberman, B.A.: Predicting the future with social media. In: 2010 IEEE/WIC/ACM International Conference on Web Intelligence and Intelligent Agent Technology, pp. 492–499. IEEE (2010)
3. Bollen, J., Mao, H., Zeng, X.: Twitter mood predicts the stock market. Journal of Computational Science (2011)
4. Chai, S.: A theory of ethnic group boundaries. Nations and Nationalism 2(2), 281–307 (1996)
5. Chai, S.: Choosing an identity: A general model of preference and belief formation. Univ. of Michigan Pr. (2001)
6. Ginsberg, J., Mohebbi, M., Patel, R., Brammer, L., Smolinski, M., Brilliant, L.: Detecting influenza epidemics using search engine query data. Nature 457(7232), 1012–1014 (2008)

7. Gosling, S., Drummond, D., NetLibrary, I.: Snoop: What your stuff says about you. BBC Audiobooks America (2008)
8. Kleinberg, J.: Bursty and hierarchical structure in streams. Data Mining and Knowledge Discovery 7(4), 373–397 (2003)
9. Norris, P.: Digital divide: Civic engagement, information poverty, and the Internet worldwide. Cambridge Univ. Pr. (2001)
10. OConnor, B., Balasubramanyan, R., Routledge, B., Smith, N.: From Tweets to polls: Linking text sentiment to public opinion time series. In: Proceedings of the International AAAI Conference on Weblogs and Social Media, pp. 122–129 (2010)
11. Parsons, T., Shils, E., Smelser, N.J.: Toward a general theory of action: Theoretical foundations for the social sciences. Transaction Pub. (2001)
12. Rattenbury, T., Good, N., Naaman, M.: Towards automatic extraction of event and place semantics from flickr tags. In: Proceedings of the 30th Annual International ACM SIGIR Conference on Research and Development in Information Retrieva (2007)
13. Smith, M., Kollock, P.: Communities in cyberspace. Psychology Press (1999)
14. Turney, P.: Thumbs up or thumbs down? semantic orientation applied to unsupervised classification of reviews. In: Proceedings of the 40th Annual Meeting of the Association for Computational Linguistics, ACL 2002 (2002)

The Mythology of Game Theory

Mathew D. McCubbins[1], Mark Turner[2], and Nicholas Weller[3]

[1] University of Southern California, Marshall School of Business, Gould School of Law
and Department of Political Science
[2] Case Western Reserve University, Department of Cognitive Science
[3] University of Southern California, Department of Political Science

Abstract. Non-cooperative game theory is at its heart a theory of cognition, specifically a theory of how decisions are made. Game theory's leverage is that we can design different payoffs, settings, player arrays, action possibilities, and information structures, and that these differences lead to different strategies, outcomes, and equilibria. It is well-known that, in experimental settings, people do not adopt the predicted strategies, outcomes, and equilibria. The standard response to this mismatch of prediction and observation is to add various psychological axioms to the game-theoretic framework. Regardless of the differing specific proposals and results, game theory uniformly makes certain cognitive assumptions that seem rarely to be acknowledged, much less interrogated. Indeed, it is not widely understood that game theory is essentially a cognitive theory. Here, we interrogate those cognitive assumptions. We do more than reject specific predictions from specific games. More broadly, we reject the underlying cognitive model implicitly assumed by game theory.

Keywords: game theory, human behavior, Nash equilibrium, economics, Trust game, prediction markets.

1 The Mythology of Non-cooperative Games

Game theory is essentially cognitive: it assumes that people know the actions available to them, that they have preferences over all possible outcomes, and that they have beliefs about how other players will choose. It assumes that this structure of knowledge, preferences, and beliefs is causal for their choices. Game theory dictates the direction and the patterns of relationships across preferences, beliefs, and choices. Human actions, according to game theory, leap forth fully-formed from our preferences, beliefs, and knowledge of the structure of the game, according to the patterns of execution assumed by game theory.

Researchers have already found that behavior in experimental settings does not accord with these predictions derived from game theory [1]. To explain these discrepancies between theory and behavior, scholars have pointed to (1) cognitive biases and dysfunctions in the decision-making of players [2, 3]; (2) mismatches between a game's payoffs and an individual's utility [4, 5]; and (3) the effects of uncertainty, bounded search ability, or limitations in thinking about others' likely

S.J. Yang, A.M. Greenberg, and M. Endsley (Eds.): SBP 2012, LNCS 7227, pp. 27–34, 2012.

behavior [6, 7, 8,9].To be sure, people regularly make choices that do not comport with Nash equilibrium strategies. But this does not necessarily indicate dysfunction: humans solve many tasks that are cognitively quite difficult, and they show great flexibility in their approaches to these tasks [10, 11, 12]. To predict human behavior we must begin with how we actually reason[13], which must be discovered, not assumed. As game-theoretic models are increasingly used across many domains to address important problems [14, 15], it is important to correct these assumptions. These assumptions are pointed out in [16], quoted below. (See also [17]).

> "[T]he Nash equilibrium (NE) concept . . . entails the assumption that all players think in a very similar manner when assessing one another's strategies. In a NE, all players in a game base their strategies not only on knowledge of the game's structure but also on *identical conjectures about what all other players will do.* The NE criterion pertains to whether each player is choosing a strategy that is a best response to a shared conjecture about the strategies of all players. A set of strategies satisfies the criterion when all player strategies are best responses to the shared conjecture. In many widely used refinements of the NE concept, such as subgame perfection and perfect Bayesian, the inferential criteria also require players to have shared, or at least very similar, conjectures." [16: 103-104].

The mythology of game theory is that these identical conjectures spring forward automatically and reliably in all situations, across all settings, consistently for all actors. This is the core that is protected in game theory.

Yet, there is prior work on subjects' beliefs in experimental settings suggesting that subjects in fact do not hold the conjectures and beliefs assumed for them in game theoretic mythology [18, 19, 20]. In what follows, using a within-subjects design across a large battery of common experimental games, we investigate choices, beliefs, and their relationship, if any. It is to a description of our experiments that we turn.

2   Experimental Design

When subjects show up, they are divided into two rooms of 10 subjects each, and seated behind dividers so they cannot see or communicate with each other. Each subject is randomly paired with a subject in the other room. They complete the tasks using pen and paper. Subjects were recruited using flyers and email messages distributed across a large public California university and were not compelled to participate in the experiment, although they were given $5 in cash when they showed up. A total of 180 subjects participated in this experiment. The experiment lasted approximately two hours, and subjects received on average about $41 in cash.

We report on a portion of our battery of experimental tasks derived from the canonical Trust game involving two players [21]. For the Trust game, each player begins with a $5 endowment. The first player chooses how many dollars, if any, to pass to an anonymous second player. The first player keeps any money he does not pass. The money that is passed is tripled in value and the second player receives the tripled amount. The second player then has the initial $5 plus three times the amount the first player passed, and decides how much, if any, of that total amount to return the first player. This is common knowledge for the subjects and they know that their choices are private and anonymous with respect to the other player and to the experimenters. The subgame perfect Nash equilibrium (SPNE) is that Player 1 will send $0 and Player 2 will return $0. This is also a dominant strategy equilibrium.

Equilibrium strategies derive from assumed beliefs: the assumption is that all players maximize economic payoffs and believe that all other players do the same. In the trust game, for example, a Player 1 with these beliefs concludes that Player 2 will return nothing and so, as a maximizer, sends nothing. The beliefs that players hold about other players lead to the belief, at every level of recursion, that all players will send $0, will guess that others will send $0, will guess that others will predict that everyone will send $0, and so on ad infinitum.

But what if a subject who does have these assumed Nash beliefs finds himself off the equilibrium path? In the Trust game, only Player 2 could be required to make a choice after finding himself presented with an off-the-equilibrium-path choice. If Player 2 receives any money, the SPNE strategy is still to send $0 back.

A novel feature of our experimental battery borrows from the idea of a prediction market [22]. We add elements to the basic Trust game in order to tap into subjects' beliefs and conjectures. We ask subjects to "guess" other subjects' choices, or to guess other subjects' "predictions." We do not ask subjects to report their expectations or beliefs, because asking for a report might have normative implications. In general, we try to provide little or no framing of the experimental tasks offered to our subjects. After Player 1 makes his choice about how much to pass, we ask him to guess how much Player 2 will return and to guess how much Player 2 predicted that Player 1 would transfer. Before Player 2 learns Player 1's choice, we ask Player 2 to guess how much money Player 1 passed. We also ask Player 2 to guess how much Player 1 predicted she would transfer. After Player 2 learns Player 1's choice, we ask Player 2 to guess how much Player 1 predicted she would return. All players know that all players earn $3 for each correct guess and earn nothing for a wrong guess.

The questions we ask vary for each task, but as an example, here is an exact question we ask Player 2: "How much money do you guess the other person transferred to you? If you guess correctly, you will earn $3. If not, you will neither earn nor lose money." Players do not learn whether their predictions were right or wrong and subjects never have any information about other subjects' guesses.

Players in the Trust game know they are randomly paired with another subject in a different room. Later in the experiment, all subjects also make choices as Player 2, randomly assigned to a player in the other room who was Player 1. All subjects first make choices as Player 1 and then, roughly 90 minutes later, make choices as Player 2. Subjects never learn the consequences of their actions as Player 1, but of course, when the subject is Player 2, the subject can infer the consequences of her choice.

Subjects also make decisions in a variety of other games, including a Dictator game and what we call the Donation game. In both these games, each subject is randomly rematched with another subject in another room. In the Dictator game, there are two players: the Dictator and the Receiver. It is arranged that the Dictator has the same endowment he or she has in the role of Player 2 in Trust, and that the Receiver has the same endowment he or she has in the role of Player 1 in Trust. These endowments are common knowledge. Accordingly, the Dictator game is identical to the second half of the Trust game. In effect, each subject plays the second half of the Trust game twice, but only once with the reciprocity frame. The SPNE is for the Dictator to send $0 to the Receiver.

The Donation game has two players, a Donor and a Receiver. It is identical to the Dictator game, except that both players start with a $5 endowment, and any money sent by the Donor is quadrupled before it is given to the Receiver. This places the Donor in the same strategic setting as Player 1 in Trust, because the obvious dominant strategy for Player 2 in Trust is to return $0. The SPNE is for the Donor to send $0.

At the end of the battery, we present the subjects with those few tasks that would allow them to learn something about the choices made by subjects in the other room. Subjects are asked to make their choice as Player 2 in the Trust game as one of these final tasks. But in no case do subjects get feedback on their choices.

In what follows, we show that subjects do not behave according to NE, that even in deviating they do not deviate consistently and that they do not hold beliefs that are consistent across similar tasks. We do not see identical conjectures across subjects or anything remotely in that vein.

3   Inconsistent Behavior within and Across Games

The standard approaches to explaining departures from NE strategies (other-regarding preferences, cognitive constraints, or decision-making biases) implicitly assume that players deviate from game-theoretical expectations in consistent ways. For example, if players prefer to reduce inequality, that preference should be stable across all manner of economic games. If players cannot detect and reject dominated strategies or cannot perform iterated deletion of dominated strategies and then reassess, then they cannot reach a dominant-strategy equilibrium. If they cannot perform backward induction, then they also cannot reach SPNE. Such handicaps should operate in all game environments of equal difficulty. We will show that subjects' behavior in Trust, Dictator, and Donation are strongly inconsistent. We do so in three steps.

First, examining play within the Trust game, we find that 56% of subjects as Player 1send money. On average, they send $1.43 (s.d. $1.70). On average, in the role of Player 2, they return $1.23 (s.d. $2.29). Such results are well-documented in the literature [3]. Our emphasis is not on the well-known deviance from SPNE, but rather on the large variance in behavior both across subjects in a specific task and by the identical subject across different tasks. This variance casts doubt on the prospects for a single, simple explanation for people's behavior.

Second, of the 100 subjects who as Player 2 receive money, only 62 of them return any money. The average returned is $2.22, again with a large variance (s.d. $2.71). Let's follow the 62 who return money after receiving money. We might expect them to be consistent in sending money when SPNE dictates that they not. But of those 62, only 40 send money when they are in the role of Dictator, and of those 40, only 29 send money when they are in the role of Donor.

Third, there are 60 subjects who behave consistently with SPNE in both roles in the Trust game. Since Dictator and Donation lack the reciprocity frame of Trust, we might expect them all to play SPNE in Dictator and Donation. Indeed, of these 60, 57 send $0 in Dictator, and of these 57, 48 send $0 in the Donation game. In short, 20% of the subjects who play SPNE in both roles in Trust deviate from SPNE within Dictator and Donation. Overall fewer than 27% of our 180 subjects consistently play SPNE across these four tasks.

Alternatively, we might expect that a subject's pattern of deviation from SPNE to be consistent. There are 42 subjects who deviate from SPNE in both roles in Trust. Of these 42, only 33 send money in Donation, and of these 33, only 26 send money in Dictator. We see that fewer than 15% of our subjects consistently deviate from SPNE across these four tasks.

Only 41% of our subjects either consistently follow SPNE in these four tasks or consistently deviate from SPNE in these four tasks. Our subjects do not rigidly follow or deviate from SPNE strategies.

4   Are Beliefs and Behavior Consistent?

Cognitive science gives us considerable reason to doubt that players will behave or hold beliefs identically across different environments, because changes in environments lead to changes in mental activation, which affect behavior and beliefs. As Sherrington famously wrote, the state of the brain is always shifting, "a dissolving pattern, always a meaningful pattern, though never an abiding one" [23]. If the particular tasks induce different states of mental activation, then belief and behavior may well vary accordingly. We have just shown that most subjects are not consistent in their choices. We now show that they do not hold consistent beliefs.

Although to our knowledge it has not previously been done, it is easy to take the strategy that is predicted by NE and see whether players believe that other players will follow the NE strategy. In the Trust game, subjects make guesses as Player 1 about the behavior of Player 2 and, likewise, as Player 2 about the behavior of Player 1. As Player 1, subjects guessed what Player 2 would return, and as Player 2, they guessed what Player 1 would send. Only 38 of 180 guessed both times that the other player would send $0. In other words, only 21% of our subjects have NE beliefs inside just the Trust game. We find that there are 54 subjects who possess NE beliefs as Player 1 but not as Player 2 and 30 subjects who possess NE beliefs as Player 2 but not as Player 1. The overwhelming majority of subjects deviate from NE beliefs during even this single experimental game.

We can compare subjects' beliefs about others in one part of the Trust game with their choices in that same part of the Trust game. For example, we can examine the difference between what a subject chooses to do as Player 1 in the Trust game and what as Player 2 they believe Player 1 will do. The modal category is subjects who believe that other subjects will play like them: 109 of the 180 subjects guess that the choice of the Player 1 with whom they are randomly matched will be the same as their own choice when they were Player 1. Perhaps most surprising, there is a large variance, with 71 subjects (39%) making guesses that differ from their own choices. Their conjectures about what others will do and believe in a situation do not match what they do and believe in the same situation. It is difficult to see how a notion of shared, identical conjectures across players can withstand such a result.

We now return to the 60 subjects who chose $0 as both Player 1 and Player 2 in Trust. We will call them "fully Nash actors" in Trust. We examine here whether their beliefs are also "fully Nash" in the Trust game and whether their actions are "fully Nash" in the related Donation and Dictator games. Of these 60 subjects, 56 of them guess as Player 1 that Player 2 will return nothing, which is consistent with SPNE. We also ask Player 1 to guess how much he or she believes the other player (Player 2) will guess that Player 1 is sending to them. In this task, only 40 of the 60 (66%) subjects guessed that the other player would predict that $0 would be sent. In addition, we ask Player 1 to guess how much Player 2 will predict that Player 1 guesses Player 2 will return and in this task 49 of the 60 subjects (81%) have beliefs consistent with SPNE. Even among these 60 subjects, the percent that have SPNE-consistent beliefs varies across questions, further demonstrating that subjects do not have rigid beliefs.

We now turn to a different part of the mythology of game theory having to do with beliefs and behavior. The mythology of game theory assumes that people enter every setting with preferences over outcomes and with beliefs and conjectures about how other players will act and what they will believe. In this mythology, it cannot make a difference whether one asks them to choose an action in the game before making a prediction or the reverse. To interrogate this mythology, we report [24] on results from a unanimous Public Goods game. In this game, each subject is paired with 9 other subjects. Each player has a $5 endowment, and is asked whether they wish to contribute this $5 to a pot or withhold it. The contributors lose that $5 and the non-contributors keep it. If they all contribute, each receives $15. If fewer than ten contribute, then each receives nothing.

In some cases, we ask the subjects first to choose whether to contribute and second to guess the number of other subjects who will contribute. In other cases, we present the tasks in the reverse order. Our experiments show that subjects who choose first guess on average that 3.3 other players will contribute, while subjects who guess first guess on average that 4.6 other players will contribute ($p=0.03$ in a Kolmogorov-Smirnov equality of distributions test), with 80 subjects in each group. Further, in an equality of proportions tests, 25% of subject choose to contribute when making their choice before their prediction, whereas 43% choose to put their money in the pot when prompted about their beliefs before they made their choice ($p<.03$). This result suggests that changing the order of belief elicitation and choice significantly affects subjects' beliefs. This simple change in task order does not accord with Nash equilibrium expectations. Are the subjects who guess after they choose simply winging it first and rationalizing later, or are the others simply winging their guesses first and then choosing according to something else later?

5   Discussion

Our results show, as is usually shown, that subjects deviate from NE predictions. They also show that these deviations are not simple, consistent, or easily explained: they depend on the specific setting and task. Our results also demonstrate that there are not shared beliefs about game strategy. Individuals' beliefs seem to be specific to particular settings and not generalizable from one setting to the next. Indeed, it may be misleading to refer to these patterns of action and belief as "deviations" at all.

The assumptions about human cognition that are part of the mythology of game theory, Nash equilibrium, and its refinements are at odds with what we know about actual human cognition. This is not a surprise, because the equilibrium concepts were not constructed based on how actual humans think, reason, or make decisions. Models that use false assumptions may not be problematic if our goal is to predict, rather than to understand, outcomes. However, Nash equilibrium and related models fail to predict behavior, which means we cannot resort to predictive success to justify the use of false assumptions.

We have shown that the protected core of game theory—the unrecognized cognitive model of non-cooperative game theory—fails repeatedly in hypothesis testing. We are not the first to say so, and the results are not shocking. Human beings for tens of thousands of years have been adept at moving through different settings and roles, exactly because they can adjust their demeanor, preferences, beliefs, and actions to suit their situations. Relative to all other species, they can turn on a dime, and this has made them astonishingly successful at inhabiting and constructing different forms of life. That people are not inflexibly fixed in their strategies does not mean that they are arbitrary and random. Rather than trying to fit people to a poor mythology, we should construct models that fit the reality of human behavior. We have not yet found a source—in neurobiology, computer science, evolution, or economics—from which this model can spring, like Athena from the head of Zeus, fully formed in convincing simplicity and power.

Acknowledgments. McCubbins acknowledges the support of the National Science Foundation under Grant Number 0905645. Any opinions, findings, and conclusions or recommendations expressed in this material are those of the author(s) and do not necessarily reflect the views of the National Science Foundation. Turner acknowledges the support of the Centre for Advanced Study at the Norwegian Academy of Science and Letters.

References

1. Camerer, C.: Behavioral Game Theory. Russell Sage Foundation. Princeton University Press, Princeton, New York (2003)
2. Kahneman, D., Amos, T.: Prospect Theory: An Analysis of Decision under Risk. Econometrica 47(2), 263–292 (1979)
3. Rabin, M., Richard, H.T.: Anomalies: Risk Aversion. Journal of Economic Perspectives (1), 219–232 (2001)

4. Elizabeth, H., Kevin, M., Keith, S., Vernon, S.: Preferences, Property Rights, and Anonymity in Bargaining Games. Games and Economic Behavior 7(3), 346–380 (1994)
5. Rabin, M.: Incorporating Fairness into Game Theory and Economics. American Economic Review LXXXIII, 1281–1302 (1993)
6. Costa-Gomes, M.A., Crawford, V.P.: Cognition and Behavior in Two-Person Guessing Games: An Experimental Study. American Economic Review 96(5), 1737–1768 (2006)
7. Gigerenzer, G., Selten, R.: Reinhard Bounded Rationality: The Adaptive Toolbox. MIT Press (2002)
8. Simon, H.: A Behavioral Model of Rational Choice. In: Models of Man, Social and Rational: Mathematical Essays on Rational Human Behavior in a Social Setting. Wiley, New York (1957)
9. Stahl, D., Wilson, P.: Experimental Evidence on Players' Models of Other Players. Journal of Economic Behavior and Organization 25, 309–327 (1994)
10. Gigerenzer, G.: Adaptive thinking: Rationality in the real world. Oxford University Press, New York (2000)
11. Gigerenzer, G.: Rationality for mortals: How people cope with uncertainty. Oxford University Press, New York (2008)
12. Turner, M.: The Scope of Human Thought (2009),
    http://onthehuman.org/humannature/
13. McCubbins, M.D., Turner, M.: Going Cognitive: Tools for Rebuilding the Social Sciences. In: Sun, R. (ed.) Grounding Social Sciences in Cognitive Sciences, ch. 14. MIT Press, Cambridge (in press)
14. Fudenberg, D., Tirole, J.: Game Theory. MIT Press (1991)
15. Nisan, N., Roughgarden, T., Tardos, E., Vazirani, V.V.: Algorithmic Game Theory. Cambridge University Press (2007)
16. Lupia, A., Levine, A.S., Zharinova, N.: Should Political Scientists Use the Self Confirming Equilibrium Concept? Benefits, Costs and an Application to the Jury Theorem. Political Analysis 18, 103–123 (2010)
17. Aumann, R., Brandenberger, A.: Epistemic conditions for Nash equilibrium. Econometrica 63, 1161–1180 (1995)
18. Croson, R.: Theories of Commitment, Altruism And Reciprocity: Evidence From Linear Public Goods Games. Economic Inquiry 45(2), 199–216 (2007)
19. Kuhlman, D.M., Wimberley, D.L.: Expectations of choice behavior held by cooperators, competitors, and individualists across four classes of experimental game. Journal of Personality and Social Psychology 34, 69–81 (1976)
20. McKenzie, C.R.M., Mikkelsen, L.A.: A Bayesian view of covariation assessment. Cognitive Psychology 54(2007), 33–61 (2007)
21. Berg, J.E., Dickhaut, J., McCabe, K.: Trust, Reciprocity, and Social History. Games and Economic Behavior 10, 122–142 (1995)
22. Wolfers, J., Zietzwitz, E.: Prediction Markets. Journal of Economic Perspectives. American Economic Association 18(2), 107–126 (2004)
23. Sherrington, C.S.: Man on his Nature. New American Library, New York (1964)
24. McCubbins, M.D., Turner, M., Weller, N.: The Challenge of Flexible Intelligence for Models of Human Behavior. Technical Report of the Association for Advancement of Artificial Intelligence Spring Symposium on Game Theory for Security, Sustainability and Health (2012)

Crowdsourced Cyber Defense:
Lessons from a Large-Scale, Game-Based Approach
to Threat Identification on a Live Network

Barton Paulhamus, Alison Ebaugh, C.C. Boylls, Nathan Bos,
Sandy Hider, and Stephen Giguere

Johns Hopkins University, Applied Physics Laboratory,
11100 Johns Hopkins Rd., Laurel, Maryland, 20723, United States
{Bart.Paulhamus,Alison.Ebaugh,Curt.Boylls,Nathan.Bos,
Sandy.Hider,Stephen.Giguere}@jhuapl.edu

Abstract. Today, the responsibility for U.S. cyber defense is divided asymmetrically between a large population of cyber-naïve end-users and a small cadre of cyber-savvy security experts in government and the private sector. We foresee the rise of "Cyber Civil Defense" driven by the perception of vulnerabilities in our present over-reliance on professionals and propelled by two additional factors: crowdsourced cyber *offense* and crowdsourced innovation. To explore crowdsourcing cyber defense, we developed an online game called Flux Hunter and deployed the game on a large-scale live network at APL, attracting over 700 players. In this paper, we discuss the concept of crowdsourced cyber defense, describe our online game, present our results, and analyze the performance and behaviors of players individually and collectively, looking for the "wisdom of crowds".

Keywords: crowdsourcing, cyber defense, games with a purpose, cyber militia, wisdom of crowds, fast-flux.

1   Introduction

Today, the responsibility for U.S. cyber defense is divided asymmetrically between a large population of cyber-naïve end-users and a small cadre of cyber-savvy security experts in government and the private sector. Since end-users typically lack the knowledge (or interest) to defend beyond their own endpoints, the experts are entrusted with watching over community assets. This arrangement resembles the classical, physical-space use of police to defend neighborhoods, or the military to defend borders and critical infrastructure. It is a model that is vulnerable to attacks that exploit either the paucity of professional defenders or the ignorance and apathy of end-users. In physical communities, that realization has led to the formation of nonprofessional bands of civilian watchdogs—Neighborhood Watch, Civil Defense—who simultaneously amplify the sensing and reach of professional defenders, while raising threat-awareness within the populace. We foresee an analogous evolution in

S.J. Yang, A.M. Greenberg, and M. Endsley (Eds.): SBP 2012, LNCS 7227, pp. 35–42, 2012.

cyber communities—forms of "Cyber Civil Defense" driven by the perception of vulnerabilities in our present over-reliance on professionals, and by:

1. Crowdsourced cyber *offense*. From Russia's cyberattacks against Georgia and Estonia [1], to cyber flashmobs mobilized in support of Wikileaks [2], we are witnessing the recruiting of nonprofessional cyber-attack "militias." Certain nation-states are incorporating such militias into their cyberwar doctrine [3], betting that a million amateur attackers who know one thing can overwhelm one expert defender who knows a million things. While one could also achieve numerical advantage with bots, humans are readily instructed, can self-organize, recognize situations, and exercise judgment. And social media provide a ready channel for real-time, decentralized command and control. Why would these arguments not also apply to cyberdefense?

2. Crowdsourced innovation. In addition to mounting attacks, a cyber militia can also be tapped for innovation in the attack itself, either through a competition among individual ideas (cf. Kaggle [4]) or through idea-aggregation and resulting synthesis of an approach owned by no one, but better than anyone (cf. "wisdom of crowds" [5]). For game-changing ideas on defending U.S. cyberspace, we presently ask the experts. But should we also be asking "the crowd"?

We thus launched the present study to answer the following questions:

- In a given threat situation, can we actually extract "cyberdefense wisdom" from a diverse crowd of non-expert strangers?
- Will the crowd's insights and actions improve upon those of professional defenders?
- What are realistic ways to recruit, instruct, and motivate the crowd to provide its contributions in time to make an incisive difference?
- How do we work with a crowd using public networks, without encryption, yet without revealing our defensive interests or capabilities to adversaries?

2   FLUXHUNTER: A Game for Detecting Fast-Flux Domains

We elected to use a game, Flux Hunter, to map a cyber threat problem—defense against fast-flux domains—into a task that would attract a "crowd" of employees at our institution (APL). The goal of Flux Hunter is to use the crowd to identify when an APL computer is attempting to access an external fast-flux domain (for more information on fast-flux domains see [6], on "games with a purpose" see [7][8]).

To play Flux Hunter, players log onto a website where they are presented with a series of four images that represent a single web domain. The images are derived from network-sensor information (e.g., observed IP addresses), but players are told nothing about this, or what the images "mean". Players select one of six options corresponding to the degree of threat that they perceive. Fig. 1A shows an example of the four images and the six voting options. Once a player casts a vote, we provide ground-truth feedback ("correct", "wrong", or "unknown") followed by the next set of images. In addition, we give each player an overall performance score and provide a scoreboard with the names and scores of the top ten players. The score is based on a combination of quantity of votes and accuracy. Before players begin the game, they are given a small amount of training about what to look for in the images (Fig. 1B).

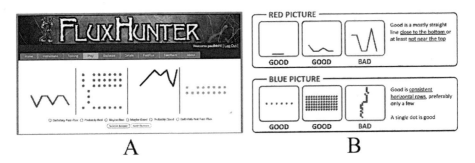

Fig. 1. A) Screenshot of the Flux Hunter online game. B) The provided training for interpreting the Red and Blue pictures. Similar training was provided for Green and Magenta pictures.

The user interface of Flux Hunter was influenced by two design constraints: First, per agreement with APL's network security team, we were not allowed to display domain names or IP addresses. This was to prevent APL staff from visiting these potentially hostile websites and possibly furthering the damage. The second constraint arose from our desire to convince non-experts that they too can participate in cyber defense. To this end, we tried to eliminate cyber jargon and technical data from the game. Flux Hunter operates on live APL network traffic. When sensors monitoring DNS traffic detect a potentially malicious website, network data about the domain is sent to Flux Hunter. Flux Hunter converts the data into the four simple images described above and stores the images in an internal database for later presentation. The Flux Hunter game was released to APL staff on an internal network for nine business days in late May 2011. Over that time, 330 domains were sent to Flux Hunter for evaluation.

3   Results

3.1   Data Collection

Our population is limited to the staff at APL. APL has approximately 5200 staff; a majority (72%) has technical degrees. Technical staff fall into three broad classifications, Associate, Senior, and Principal, that roughly align to experience levels. Of the 330 domains presented, we had ground truth on 43 domains. 11 were actual fast-flux, and 32 were innocuous. The ground truth domains were sprinkled

Table 1. Levels of participation from the Flux Hunter game. Data is presented as raw counts followed by percentages of the total in parentheses. Population is APL Staff and values have been approximated.

	Population	Full Data Players	Full Data Votes	Subset Players	Subset Votes
Total	5200 (100)	785 (100)	31658 (100)	200 (100)	4200 (100)
Technical	3750 (72)	668 (85)	25545 (81)	159 (80)	3339 (80)
Associate	700 (13)	201 (25)	8505 (27)	53 (27)	1113 (27)
Senior	2050 (39)	385 (49)	13729 (43)	88 (44)	1848 (44)
Principal	1000 (19)	82 (10)	3311 (10)	18 (9)	378 (9)
Non-technical	1450 (28)	117 (15)	6113 (19)	41 (21)	861 (21)

throughout the game to give players "random" instances of feedback. All domains were presented to players in the same order regardless of when they signed on to play the game. When players logged off and then logged back in, the game resumed where they left off. We offered a small daily prize given to one player (chosen at random) who played on that day.

Because participation in Flux Hunter varied from individuals who cast only a few votes to players who voted on every domain, we identified a subset of data that jointly maximized the number of known domains (21) and the number of players who voted on all of those domains (200). Of the 21 domains, 8 were fast-flux and 13 were not fast-flux. Player and vote distributions in this sample are statistically consistent with the full data set. Table 1 summarizes Flux Hunter participation.

785 staff cast at least 1 vote within Flux Hunter. Over 30,000 total votes were collected. Participation levels variedly slightly from the APL population with the largest discrepancies being the number of Associate staff and Non-technical staff who played Flux Hunter. Although Associate Staff only make up 13% of APL Staff, they made up 25% of game players. In general, the younger Associate Staff probably had more free time and more interest in online games than their more senior counterparts. One of our goals was to attract non-technical staff who might otherwise feel intimidated by contributing to cyber defense. Non-technical staff accounted for 15% of players. Although this is lower than expected from overall APL demographics, non-technical staff participation exceeded our expectations. Many non-technical staff do not work at computers all day. The non-technical staff who did play were some of the most active players, ultimately contributing 19% of total votes. Furthermore, only 11% of overall participants came from APL's Applied Information Sciences Department where most of APL's more cyber-savvy staff reside (the 11% is consistent with APL's department breakdown demographic).

3.2   Fast-Flux Detection Performance: Individuals

For this study, we measure the performance of individuals by comparing their votes to the ground truth domains. If the ground truth is "not fast-flux" and the player voted one of the three not fast-flux votes ("maybe good", "probably good", or "definitely not fast-flux"), then the vote was deemed correct. The same rules applied to the "fast-flux" ground truth domains. The unknown ground truth domains were not used. An individual's performance was the percentage of ground truth domains voted on correctly. Figure 2 shows the distribution of performance of the individuals in the subset data. The mean performance was 72.2% and the top performer achieved 90.5%. The worst performer was a clear outlier and only scored on 19.1%. This suggests the player was not trying to vote correctly or intentionally voted incorrectly.

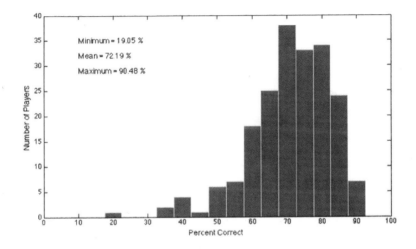

Fig. 2. Distribution of player performance from the subset data

3.3   Fast-Flux Detection Performance: The Crowd

Eliciting "the wisdom of crowds" depends in part upon how individual opinions are aggregated into a crowd decision or estimate. For Flux Hunter, our initial aggregation method was simple majority voting: if more than 50% of the crowd voted that a domain was not fast-flux, then that was deemed to be the crowd decision. Under this regime, for the 21 domains and 200 players in our data subset, we found that the crowd decision about a domain was correct in only 71% of cases, and that 49% of individuals in the crowd actually made better decisions than the crowd overall. Fig. 3 hints at the reason for these unpromising results. Here, we first compare the ground truth about each domain with the crowd's decisions. Then we "X-ray" the crowd, showing how each player voted. The players have been sorted by individual performance (see Fig. 2). One can readily see that, in terms of vote unanimity, the crowd is much more "decisive"—and accurate—about fast-flux domains versus non-fast-flux, even when the crowd makes an error. Crowd decisions about good domains are more equivocal and error-prone. We discuss possible reasons for this disparity in later sections. However, the results suggest that the accuracy of crowd decisions could be significantly improved by reducing the proportion of non-fast-flux votes required

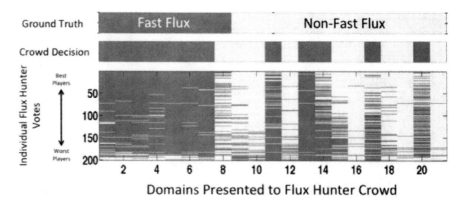

Fig. 3. Anatomy of crowd decisions made by Flux Hunter players. Domains presented to the crowd are plotted horizontally, with the ground-truth for each indicated by the shading in the topmost bar. The middle bar shows the corresponding crowd decision on the domain, as determined by the majority vote of the crowd. The lower bar documents the vote of each player for each domain. Voting records are sorted so that the best and worst individual performances are at the top and bottom of the bar, respectively.

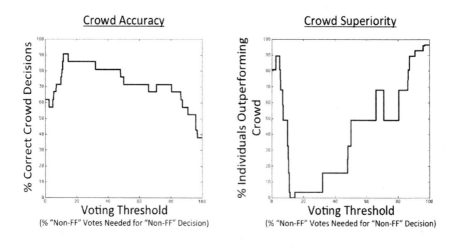

Fig. 4. Effect of shifting the percentage of votes required to post a "non-fast-flux" (non-FF) decision by the Flux Hunter crowd. See text for details.

to label a domain "non-fast-flux." In Fig. 4, we illustrate the effect of shifting this "voting threshold" from 0.1% through 99.9%, examining both the accuracy of crowd decisions about domains, and also the degree to which the crowd outperforms the individuals within it—the "crowd superiority." As anticipated, crowd accuracy reaches a peak of 90.5% when the plurality of votes needed for a non-fast-flux decision falls between 11.5% and 14.4%. Furthermore, at that point, crowd performance equals or exceeds that of any individual, consistent with "wisdom of crowd" expectations. This optimum, however, is fragile with respect to the voting threshold and, in practice, would likely require empirical discovery. This suggests that replacing simple voting schemes with more sophisticated classifiers might yield more robust crowd decisions (and possibly higher accuracy; see Section 4).

4   Conclusions and Future Work

Although proposals exist for "volunteer cyber defense militias" [9], [10], we believe ours to be the first trial of the "wisdom of crowds" in actual cyber threat detection. In this very limited experiment, we see that non-experts will indeed volunteer to participate in cyber defense. We also see that, even when experts (APL technical and principal staff) volunteer, their performance as a crowd differs little from rank amateurs (associate and non-technical staff), assuming one assesses crowd opinion using simple majority voting. Of course, had we used more sophisticated methods of vote aggregation—such as a classifier that weights and blends voters based upon their individual performance [11]—differences might have emerged and crowd performance might have improved. We also intentionally withheld fast-flux technical information from game players, even when they requested it, so we don't know what improvement might result from "instructing the crowd." Both vote-aggregation and crowd instruction are areas for future research.

Another topic for the future lies in ascertaining and compensating for the crowd's biases or "priors." For example, we note that the crowd was more accurate and "decisive" (based on voting pluralities) in identifying actual fast-flux domains versus non-fast-flux. This may reflect a player bias toward reducing missed-detections of fast-flux at the expense of additional false-alarms. All players knew that Flux Hunter was about hunting for fast-flux. However, it is also possible that our results are explained by properties of the game's visual display and the associated instructions to players. Both were designed around phenomena observed, more or less uniquely, in fast-flux. We attempted to gauge this effect by examining reaction times. We found that the crowd got faster as the game progressed without sacrificing performance. We did not see a correlation between spikes in longer reaction times and poor performance (indicating more difficult visual displays). Visual analysis of pictures for domains associated with poor performance or longer reaction times is an area for future research.

Given the nature of the Flux Hunter game, we would not anticipate deploying Flux Hunter as a permanent, long-term solution to APL network security. After only nine days, participation levels dwindled to about 1% of the first day. However, given the high levels of performance from the Flux Hunter crowd, we are currently analyzing the Flux Hunter data to learn how to build an effective machine-based solution. This is where we see the most value from a crowdsourced, cyber defense game like Flux Hunter—as a way to learn about data and threats before designing and implementing automated solutions or as a quick reaction defense to a new, previously unseen cyber threat/attack by standing up cyberdefense flash mobs. The Flux Hunter framework has been integrated within the APL network infrastructure in such a way that we can more quickly deploy similar games in the future (Flux Hunter took about four weeks to deploy). We look forward to deploying similar types of games based on what we learned from Flux Hunter.

Acknowledgments. The authors would like to thank our colleagues Kim Glasgow and Clay Fink for their useful technical discussions.

References

1. Owens, W.A., Dam, K.W., Lin, H.S. (eds.): Technology, Policy, Law, and Ethics Regarding U.S. Acquisition and Use of Cyberattack Capabilities. The National Academies Press, Washington, D.C (2009)
2. Clayton, M.: How pro-WikiLeaks hackers wage cyberwar without hijacking your computer, http://www.csmonitor.com/USA/2010/1209/How-pro-WikiLeaks-hackers-wage-cyberwar-without-hijacking-your-computer
3. Military and security developments involving the People's Republic of China, Office of the Secretary of Defense Annual Report to Congress (2010)
4. Carpenter, J.: May the best analyst win. Science 331, 698–699 (2011)
5. Surowiecki, J.: The Wisdom of Crowds: Why the Many Are Smarter than the Few and How Collective Wisdom Shapes Business, Economies, Societies, and Nations. Doubleday Books, New York (2004)
6. Honeynet Project, Know Your Enemy: Fast-Flux Service Networks, http://www.honeynet.org/papers/ff/
7. von Ahn, L.: Games with a purpose. Computer 39, 92–94 (2006)
8. Khatib, F., Cooper, S., Tyka, M.D., Xu, K., Makedon, I., Popovic, Z., Baker, D., Players, F.: Algorithm discovery by protein folding game players. PNAS (2011), http://www.pnas.org/content/early/2011/11/02/1115898108
9. Fink, E., Sharifi, M., Carbonell, J.G.: Application of machine learning and crowdsourcing to detection of cybersecurity threats, white paper. Carnegie-Mellon University (2011), http://www.cs.cmu.edu/afs/.cs.cmu.edu/Web/People/eugene/research/full/ml-detection.pdf
10. Hollis, D., Hollis, K.: Cyber defense: U.S. cybersecurity must-do's. Armed Forces J. 148, 16–19 (2011)
11. Chen, K.Y., Fine, L.R., Huberman, B.A.: Predicting the future. Inform. Sys. Frontiers 5, 47–61 (2003)

How Many Makes a Crowd? On the Evolution of Learning as a Factor of Community Coverage

Yaniv Altshuler[1], Michael Fire[2], Nadav Aharony[1], Yuval Elovici[2],
and Alex (Sandy) Pentland[1]

[1] MIT Media Lab, Cambridge
{yanival,nadav,sandy}@media.mit.edu
[2] Deutsche Telekom Laboratories at Ben-Gurion University of the Negev, Israel
{mickyfi,elovici}@bgu.ac.il

Abstract. As truly ubiquitous wearable computers, mobile phones are quickly becoming the primary source for social, behavioral and environmental sensing and data collection. Today's smartphones are equipped with increasingly more sensors and accessible data types that enable the collection of literally dozens of signals related to the phone, its user, and its environment. A great deal of research effort in academia and industry is put into mining this raw data for higher level sense-making, such as understanding user context, inferring social networks, learning individual features, and so on. In many cases, this analysis work is the result of exploratory forays and trial-and-error. In this work we investigate the properties of learning and inferences of real world data collected via mobile phones for different sizes of analyzed networks. In particular, we examine how the ability to predict individual features and social links is incrementally enhanced with the accumulation of additional data. To accomplish this, we use the *Friends and Family* dataset, which contains rich data signals gathered from the smartphones of 130 adult members of a young-family residential community over the course of a year and consequently has become one of the most comprehensive mobile phone datasets gathered in academia to date. Our results show that features such as ethnicity, age and marital status can be detected by analyzing social and behavioral signals. We then investigate how the prediction accuracy is increased when the users sample set grows. Finally, we propose a method for advanced prediction of the maximal learning accuracy possible for the learning task at hand, based on an initial set of measurements. These predictions have practical implications, such as influencing the design of mobile data collection campaigns or evaluating analysis strategies.

Keywords: Sampling size, Social network, Mobile sensing, Inferring attributes.

1 Introduction

Mobile phones are increasingly used as social and behavioral data collection instruments. Their increased pervasiveness makes them an ideal wearable sensing platform for location, proximity, communications and context. Eagle and Pentland[1]

S.J. Yang, A.M. Greenberg, and M. Endsley (Eds.): SBP 2012, LNCS 7227, pp. 43–52, 2012.
© Springer-Verlag Berlin Heidelberg 2012

coined the term *"Reality Mining"* to describe the collection of sensor data pertaining to human social behavior. As the field of computational social science matures, a need for more structured tools and methodologies is further required. In particular, we are interested in tools that would assist the researcher or practitioner in designing data collection campaigns, understanding the potential of collected datasets and estimating the accuracy limits of current analysis strategy versus alternative ones.

Conducting mobile-phone based field studies is challenging and costly. Some of these costs might include subject compensation, technical system development and maintenance, ongoing support for subject's phone hardware and software and mobile phone plans. There is also the added resource and manpower cost related to constant communication with subject populations, recruitment and minimizing attrition of subjects as well as researchers. On the other hand, such studies give us an unprecedented window into the lives of individuals and entire communities. As demonstrated in [2], conducting user-centric data collection provides a wide range of signals on participants and allows us to directly gather ground-truth and ancillary information such as demographic information, physiological and psychological information, self-perceptions and attitudes and a variety of additional data types gathered via surveys, interviews, and interactions with the subject population. These types of information are usually not available for datasets "donated" to researchers by mobile service providers and other commercial entities.

Mobile-phone studies span along a wide range of studies. Some studies are all encompassing and broadly defined living laboratory or "social observatory" types of studies. These include the Reality Mining study [7], the Social Evolution study [10], and the Friends and Family Study [2]. There are also more targeted studies, focusing on specific research questions or data types, like physical activity levels or environmental impacts.

In these and similar initiatives, there is a continuing and conflictive tension. On one hand, there is an idealistic desire to make the best use of the invested time and resources; collection of as an encompassing a dataset as possible which includes a maximal amount of subject from the target populations. Real-world data is often noisy and challenging to work with. When attempting to establish inferences or generalizations, we wish to increase our confidence in the data by assuring that the dataset is as complete as possible. Indeed, there exist practical considerations of the time and cost of additional resources as well as the fact that due to many reasons, it is never really possible to recruit all members of a target community to any study. However, we ask the question, are all target members really necessary? And how many, exactly, is "enough"?

In this work, we aim to gain a better understanding of the evolution of the process of learning personal features and behavioral properties based on mobile-phone sensed data as we increase the size of the sample group. We investigate issues related to the "coverage" of a given community, i.e., the number of subjects we have access to with respect to the actual size of the measured sub-network. For this analysis, we are less concerned about the specific learned models and their generalizability, and more about using them to study and benchmark the evolution of learning accuracy. Understanding this process is of significant importance to researchers in a variety of fields as it would provide an approximation for the needed level of coverage (sample size vs. community or network size) in order to "learn" specified features for some given accuracy. Alternatively, it could give the investigators an idea of the expected level of accuracy that can be obtained for a given socially-relevant data collection initiative.

To carry out our research we use the *Friends and Family* dataset which contains rich data signals gathered from the smartphones of 140 adult members of a young-family residential community and collected over the course of a year [2], as well as self-reported personal and social-tie information. We build voting classifiers for predicting personal properties such as nationality and religion, based on the participants' SMS messages graph's topology. We demonstrate the characteristics of incremental learning of multiple social and individual properties from raw sensing data collected from mobile phones while the information is accumulated from a multitude of individuals. We observe similar behavior among different learning processes as sample size increases. We also observe a limit, or "saturation", where additional community coverage does not increase the accuracy of prediction, or in some cases, increases it only marginally. Furthermore, we propose a method for advanced prediction of the maximal learning accuracy possible for the learning task at hand using just the first few measurements. This information can be useful in many ways, including:

- Informing real-time resource allocation for data collection for an ongoing study.
- Estimating the monitoring time needed for desired accuracy levels of a given method.
- Early evaluation of modeling and learning strategies.

The paper is organized as follows: Related work is presented in Section 2. Our analysis is described in Section 3 and concluding remarks in Section 4.

2   Scientific Background

In recent years the social sciences have been undergoing a digital revolution heralded by the emerging field of "Computational Social Science". Lazer, Pentland, et al. [3], describe the potential of computational social science to increase our knowledge of individuals, groups, and societies, with an unprecedented breadth, depth, and scale. Computational social science combines the leading techniques from network science[4-6] with new machine learning and pattern recognition tools aimed for the understanding of people's behavior and social interactions [7].

2.1   Mobile Phones as Social Sensors

The pervasiveness of mobile phones the world over has made them a premier data collection tool of choice and they are increasingly used as social and behavioral sensors of location, proximity, communications and context. Eagle and Pentland[1] coined the term "Reality Mining" to describe the collection of sensor data pertaining to human social behavior. They show that by using call records, cellular-tower IDs, and Bluetooth proximity logs collected via mobile phones at the individual level, the subjects' regular patterns in daily activity can be accurately detected[1, 7]. Furthermore, mobile phone records from telecommunications companies have proven to be quite valuable for uncovering human level insights. For example, Gonzales et al. show that cell-tower location information can be used to characterize human mobility and that humans follow simple reproducible mobility patterns[8]. This approach has already expanded beyond academia, as companies like Sense Networks [9] are putting such tools to use in the commercial world to understand customer churn, enhance targeted advertisements, and offer improved personalization and other services.

2.2   Individual Based Data Collection

Data gathered through service providers include information on very large numbers of subjects. However, this information is constrained to a specific domain (email messages, financial transactions, etc.) and there is very little, if any, contextual information on the subjects themselves. The alternative data gathering approach at the individual level allows collection of many more dimensions related to the end user which are many times not available at the operator level.

Madan et al.[10] expand Eagle and Pentland's work[1] and show that mobile social sensing can be used for measuring and predicting the health status of individuals based on mobility and communication patterns, as well as the spread of political opinion within the community[11]. Other examples of using mobile phones for individual-based social sensing are Montoliu et al.[12], Lu et al.[13], and projects coming from CENS center, e.g. Campaignr by Joki et al.[14], as well as additional works described in [15]. Finally, the Friends and Family study, which our paper uses as its data source, is probably the richest mobile phone data collection initiative to date, in relation to the number of signals collected, study duration and the number of subjects. In addition to mobile phones, there have been other types of wearable sensor-based social data collection initiatives. A notable example is the *Sociometric Badge*, which captures human activity and socialization patterns via a wearable device and is mostly used for data collection in organizational settings [16]. Our results are applicable to these types of studies as well.

2.3   Learning and Prediction of Social and Individual Information

A great deal of studies involving predicting individual traits and social ties were conducted in the recent years in the general context of social networking. Example include works by Liben-Nowell and Kleinberg [17], Mislove[18] and Rokach et. al. [19], combining machine learning algorithms with social network data in order to build classifiers. In computational learning theory, "Probably Approximately Correct" (PAC) learning tries to solve similar problems of finding efficient classifier functions that depends on the train set sample size [31].

3   Analysis and Results

The goal of this work is to study and analyze the evolution of the learning process of personal features and behavioral properties with a constant increase in sampling group size. Our analysis focuses less on specific learned models and their generalizability, and is more concerned by their use as benchmark for the learning process as data accumulates. Understanding this process is of significant importance to researchers in a variety of fields as it provides an approximation for the amount of data-samples needed in order to "learn" these features for some given accuracy, or alternatively, revealing the level of accuracy that can be obtained for a given community.

We constructed the SMS messages social network which was created after 65 weeks of the study, using more than 97,000 SMS messages sent between March 1st 2010 and May 31th, 2011. An edge represents users that had at least two common friends in the network (see Figure 4). We then attempted to predict each participants' personal information by using their friends personal information. We divided the

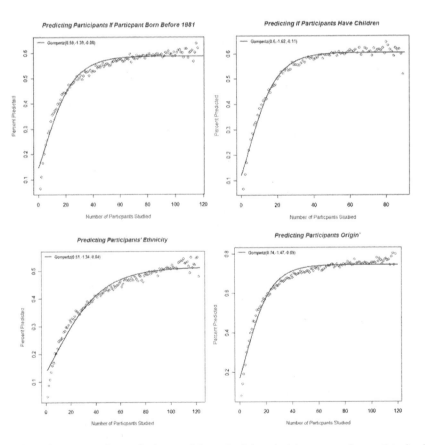

Fig. 1. Learning process for predicting participants' origin, ethnicity, age, and parenthood, with data regarding groups of growing sizes. For each group's size, 100 random groups were generated and their performance values averaged. Red lines represent the *Gompertz* regression.

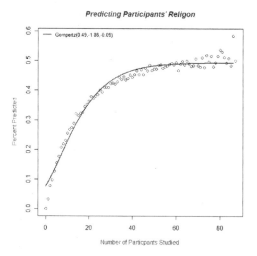

Fig. 2. Learning process for predicting participants' religion with data regarding groups of growing sizes. For each group's size, 100 random groups were generated and their performance values averaged. The red line represents the *Gompertz* regression.

social network graph to communities using the *Louvain* algorithm for community detection [23]. We look at the community which the participant belongs to and predict that attribute *a* would be equal to the attribute *a* majority in the community. We then used the aforementioned classifiers on 100 random training groups, each containing a single user (and test groups that contained the rest of the community), using the average performance. We then took 100 random groups of 2 users each, and so on, until testing 100 groups of 90% of the users. Figures 1 and 2 present the results of 5 of the classifiers we have run.

We used the *Gompertz function* in order to model the evolution of this process :
$$f = ae^{be^{ct}}$$
This well known model is flexible enough to fit various social learning mechanisms while providing the following important features:

(a) Sigmoidal advancement; a monotonous increase in accuracy with increase in group size.

(b) The rate at which information is produced is smallest at the start and end of the process.

(c) Asymmetry of the asymptotes, as for any value of *t*, the amount of information gathered in the first *t* time steps is greater than the amount gathered at the last *t* time steps.

The *Gompertz function* is frequently used for modeling a great variety of processes due to its flexibly, such as mobile phone uptake [25], population in a confined space [26] and growth of tumors [27]. In addition, the applicability of the *Gompertz function* for modeling progress of behavior patterns prediction of mobile and social users was demonstrated in [28].

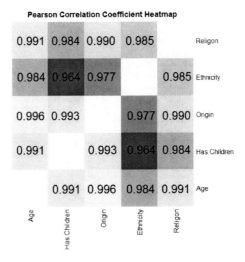

Fig. 3. The correlation matrix between the prediction vectors of the 5 classifiers

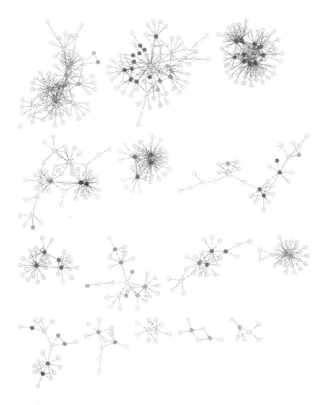

Fig. 4. SMS Social Network Graph created over 65 weeks (each unknown node connected to at least two known nodes). Different colors represent different religions.

The regression yielded surprisingly accurate results:

(a) **Ethnicity :** Residual standard error: 0.02026, Achieved convergence tolerance: 3.167e-06
(b) **Religion :** Residual standard error: 0.02115, Achieved convergence tolerance: 8.747e-06
(c) **Children :** Residual standard error: 0.01901, Achieved convergence tolerance: 2.464e-06
(d) **Origin :** Residual standard error: 0.02515, Achieved convergence tolerance: 8.127e-06
(e) **Age :** Residual standard error: 0.02056, Achieved convergence tolerance: 6.189e-06

4   Conclusion

In this paper we have studied the effects of the amount of sensors information on the ability to predict personal features of community's members. We have shown that the dynamics of this process can be modeled using the *Gompertz* function and hence can be further extrapolated in order to predict bounds on the overall ability to learn specific features. This information can be used to inform and design the data collection for an ongoing or future data collection experiment or initiative. This can be done by extrapolating the *Gompertz* function with the regression values found in several initial small sampling sets. Based on this extrapolation, an approximation for the maximal amount of information (or accuracy) that can be achieved with large sample sets, as well as the accuracy that a given group size would result in, can be produced.

In addition, correlations between the evolutions of the different learning processes, as depicted in Figure 3, may imply an underlying correlation between the raw data itself, and can hence be used as further validation for correlated features and observations, such as the suggestion that people are more likely to marry within their own ethnic group, as observed in [29,30].

In conclusion, it is also interesting to compare how does the dynamics of the learning process as a function of the community's size correlates with the dynamics of the learning process over time. We intend to conduct this study on the Friends and Family dataset in the coming months.

References

1. Eagle, N., Pentland, A.: Reality Mining: Sensing Complex Social Systems. Personal and Ubiquitous Computing 10, 255–268 (2006)
2. Aharony, N., et al.: Social fMRI: Investigating and shaping social mechanisms in the real world. Pervasive and Mobile Computing (2011)
3. Lazer, D., et al.: Life in the network: the coming age of computational social science. Science 323, 721 (2009)

4. Barabasi, A.-L., Albert, R.: Emergence of scaling in random networks. Science (1999)
5. Newman, M.E.J.: The structure and function of complex networks
6. Watts, D.J., Strogatz, S.H.: Collective dynamics of 'small-world' networks. Nature (1998)
7. Eagle, N., Pentland, A., Lazer, D.: From the Cover: Inferring friendship network structure by using mobile phone data. Proceedings of The National Academy of Sciences 106(36), 15274–15278 (2009)
8. Gonzalez, M.C., Hidalgo, A., Barabasi, A.-L.: Understanding individual human mobility patterns. Nature (2008)
9. Networks, S.: http://www.sensenetworks.com/
10. Madan, A., et al.: Social sensing for epidemiological behavior change. In: Ubiquitous Computing/Handheld and Ubiquitous Computing, pp. 291–300 (2010)
11. Madan, A., Farrahi, K., Gatica-Perez, D.: Pervasive Sensing to Model Political Opinions in Face-to-Face Networks (2011)
12. Montoliu, R., Gatica-Perez, D.: Discovering human places of interest from multimodal mobile phone data, 1–10 (2010)
13. Lu, H., et al.: The Jigsaw continuous sensing engine for mobile phone applications. In: Conference on Embedded Networked Sensor Systems, pp. 71–84 (2010)
14. Joki, A., Burke, J.A., Estrin, D.: Campaignr: A Framework for Participatory Data Collection on Mobile Phones (2007)
15. Abdelzaher, T.F., et al.: Mobiscopes for Human Spaces. IEEE Pervasive Computing 6(2), 20–29 (2007)
16. Olguín, D.O., et al.: Sensible Organizations: Technology and Methodology for Automatically Measuring Organizational Behavior. IEEE Transactions on Systems, Man, and Cybernetics 39(1), 43–55 (2009)
17. Liben-Nowell, D., Kleinberg, J.: The link-prediction problem for social networks. Journal of the American Society for Information Science and Technology 58(7), 1019–1031 (2007)
18. Mislove, A., et al.: You are who you know: inferring user profiles in online social networks. In: Web Search and Data Mining, pp. 251–260 (2010)
19. Rokach, L., et al.: Who is going to win the next Association for the Advancement of Artificial Intelligence Fellowship Award? Evaluating researchers by mining bibliographic data. Journal of the American Society for Information Science and Technology (2011)
20. Funf. Funf Project, http://funf.media.mit.edu
21. Hagberg, A.A., Schult, D.A., Swart, P.J.: Exploring Network Structure, Dynamics, and Function using NetworkX (2008)
22. Hall, M., et al.: The WEKA data mining software: an update. Sigkdd Explorations 11(1), 10–18 (2009)
23. Blondel, V.D., et al.: Fast unfolding of communities in large networks. Journal of Statistical Mechanics: Theory and Experiment 10 (2008)
24. Xie, J., Szymanski, B.K.: Community Detection Using A Neighborhood Strength Driven Label Propagation Algorithm. Computing Research Repository (2011)
25. Rouvinen, P.: Diffusion of digital mobile telephony: Are developing countries different? Telecommunications Policy 30(1), 46–63 (2006)
26. Erickson, G.M.: Tyrannosaur Life Tables: An Example of Nonavian Dinosaur Population Biology. Science 313(5784), 213–217 (2006)
27. Donofrio, A.: A general framework for modeling tumor-immune system competition and immunotherapy: Mathematical analysis and biomedical inferences. Physica D-nonlinear Phenomena 208(3-4), 220–235 (2005)

28. Pan, W., Aharony, N., Pentland, A.: Composite Social Network for Predicting Mobile Apps Installation. In: Intelligence, AAAI 2011, San Francisco, CA (2011)
29. Kalmijn, M.: Intermarriage and Homogamy: Causes, Patterns, Trends. Annual Review of Sociology 24(1), 395–421 (1998)
30. McPherson, M., Smith-Lovin, L., Cook, J.M.: Birds of a Feather: Homophily in Social Networks. Annual Review of Sociology 27(1), 415–444 (2001)
31. Haussler, D.: Part 1: Overview of the Probably Approximately Correct (PAC) learningframework (1995)

Cultural Consensus Theory: Aggregating Signed Graphs under a Balance Constraint

Kalin Agrawal and William H. Batchelder

Institute for Mathematical Behavioral Sciences, School of Social Sciences,
University of California, Irvine, CA 92697, USA
{kagrawal,whbatche}@uci.edu

Abstract. Cultural Consensus Theory (CCT) consists of cognitive models for aggregating the responses of experts to test questions about some domain of their shared cultural knowledge. This paper proposes a new CCT model for a situation where experts judge the ties in a complete signed graph. New to CCT is that the model imposes a side constraint on the aggregation process that requires that the consensus signed graph satisfy the social network property of structural balance. Balanced signed graphs require that the nodes can be partitioned into two sets with positive ties between nodes in the same set and negative ties between nodes in different sets. While the balance constraint is imposed on the consensus aggregation, it is not assumed that each expert's responses satisfy balance because they may be error-prone or biased. The model is presented in terms of signal detection assumptions that allow heterogeneity in expert ability and item difficulty. Bayesian inference of the model is developed using a specially designed Markov Chain Monte Carlo sampler. It is shown that the sampler can recover parameters from simulated data, and then the model is applied to interpret experimental data. Of particular interest is that the model aggregation reveals a single consensus balanced signed graph with a high posteriori probability despite the fact that none of the experts' responses satisfy the balance constraint.

Keywords: Signed Graphs, Balance, Bayesian Inference, Information Aggregation, Graph Aggregation, MCMC.

1 Introduction

Cultural Consensus Theory (CCT) is a popular statistical modeling approach to information aggregation (pooling, fusion) [1,2,3,4,5]. It is assumed that a researcher has access to the responses of several knowledgeable experts to a set of questions about some domain of their shared knowledge or beliefs. Aggregation of the responses is desired because the researcher does not know the answers to the questions that represent the consensus shared by the experts. Questions are in some format, e.g. true/false, ordered categories, probability judgments, and CCT consists of cognitive models for different questionnaire formats, where each model includes parameters for the competence level and response bias characteristics of each expert as well as the difficulty level and the consensus answers

S.J. Yang, A.M. Greenberg, and M. Endsley (Eds.): SBP 2012, LNCS 7227, pp. 53–60, 2012.
© Springer-Verlag Berlin Heidelberg 2012

to the questions. If a CCT model adequately fits the responses of the experts, the estimates of the consensus answers constitute the aggregation process. This paper introduces a new CCT model for the case where experts provide positive or negative ties concerning some affective relationship between pairs of actors in a social network.

The new CCT model is designed to aggregate information from experts concerning the ties in a *complete signed graph*. A complete signed graph $S = \langle A, V, \sigma \rangle$ consists of a finite set $A = \{a_1, ..., a_N\}$ of nodes, the set V of all two element subsets of A, and a function σ from V into the set $\{0, 1\}$, where 0 codes a negative tie and 1 a positive tie. Signed graphs were introduced by Cartwright and Harary [6] to represent a network of social actors, where symmetric relationships among pairs of actors exhibit a positive or negative affect on a relationship such as friendship, agreement, or cooperation; however, signed graphs also play a role in other areas of applied mathematics [7]. The authors were particularly interested in a property of complete signed graphs called *balance*. Formally, a complete signed graph is balanced in case the set of nodes can be partitioned into two sets $\{A_1, A_2\}$, one of which may be empty, such that all within-set ties are positive and all between-set ties are negative. It is easily established [6] that the global property of balance holds for a complete signed graph if and only if for every distinct triple of nodes, the product of the signs of their three ties is positive (using the ordinary rules of arithmetic). A complete signed graph is also balanced if and only if the product of tie signs along every distinct cycle is positive. In general if the graph has N nodes, there are $N(N-1)/2$ distinct pairs of nodes, consequently there are $2^{N(N-1)/2}$ possible complete signed graphs. However, under the balance constraint, only 2^{N-1} of these graphs satisfy balance because this is the number of distinct binary partitions of a set of N nodes.

The property of balance was invented to describe a common situation where a group of social actors evolves into two opposing factions, where members within each faction have positive ties and members in different factions have negative ties. Such situations approximately characterize many situations in the social world, e.g. Republicans and Democrats concerning the size of government, Catholics and Protestants in Ireland, or Sunnis and Shiites in Iraq. The model we present is designed for a situation where a researcher hypothesizes that the experts' consensus knowledge about the graph satisfies balance even though the responses of individual experts may not satisfy balance because they may be error-prone, lack complete knowledge, or may render knowledge on only some of the ties. The major new feature of our model is that the balance constraint is imposed on the aggregation process, so it requires that the aggregation process deliver only balanced graphs even though the responses of the experts may not satisfy this property. As we will see, the balance constraint places non-trivial conditions on the aggregation process.

2   A CCT Model for Balanced Signed Graphs

Let $S = \langle A, V, \sigma \rangle$ be a complete signed graph on N nodes. If the graph satisfies balance, let $\{A_1, A_2\}$ denote the corresponding binary partition of A, and define the partition vector $\mathbf{W} = \langle W_j \rangle_{1 \times N}$ such that for all $a_j \in A$, $W_j = 1$ if $a_j \in A_1$, otherwise $W_j = 0$. The partition vector is a useful representation of S since

$$\forall a_j, a_k \in A, \sigma(\{a_j, a_k\}) = 1 - (W_j - W_k)^2 . \tag{1}$$

Of course for any complete signed graph, the partition vector has a dual, $\mathbf{W}^* = \langle W^*_j \rangle$ where $W^*_j = 1$ iff $W_j = 0$, that generates the same graph. In the aggregation of the experts' responses we will obtain a posterior distribution over the space of all 2^{N-1} such pairs of partition vectors.

To apply the model, we require judgments from $M > 1$ expert sources of the signed ties between pairs from a fixed set A of N nodes. Let the random variable $X_{i,jk}$ with space $\{0, 1\}$ denote the signed response of expert i to tie jk, $1 \le j < k \le N$. Then $X_i = \{X_{i,jk} | 1 \le j < k \le N\}$ represents the responses of expert i, and $\mathbf{X} = \langle X_i \rangle_{1 \times N}$ the responses from all the experts. The response model assumes that each tie judgment by each expert is either detected directly from the consensus graph or it is a guess. This assumption is typical of other CCT models, and it is based on the so-called two high threshold model (2HT) of signal detection theory [8]. The basic idea of the 2HT assumption is that detection thresholds for positive and negative ties are both sufficiently high so that detections only lead to the consensus tie; however responses that depart from the consensus graph may occur when detection fails and a guess is made. The model specifies parameters for expert detection probabilities and guessing probabilities as follows. Let $D_{i,jk} \in [0, 1]$ be the probability that expert i detects the sign of tie jk, and $G_i \in [0, 1]$ the probability expert i guesses '+' to any tie that they do not detect. We collect these expert parameters into arrays $\mathbf{D} = \langle D_{i,jk} \rangle$ and $\mathbf{G} = \langle G_i \rangle$. The model is stated below in four axioms.

Axiom 1 (Common Truth). *There is a fixed partition vector* $\mathbf{W} = \langle W_j \rangle_{1 \times N}$ *applicable to all experts.*

Axiom 2 (Conditional Independence). *The responses from all the experts are conditionally independent given the partition vector, namely for all realizations* $\mathbf{x}$ *of* $\mathbf{X}$,

$$\Pr\left(\mathbf{X} = \mathbf{x} | \mathbf{W} = \langle W_1, ..., W_N \rangle, \mathbf{D}, \mathbf{G}\right)$$
$$= \prod_{i=1}^{M} \prod_{j=1}^{N-1} \prod_{k>j}^{N} \Pr\left(X_{i,jk} = x_{i,jk} | \langle W_j, W_k \rangle, D_{i,jk}, G_i\right) . \tag{2}$$

Axiom 3 (Marginal Edge Responses). *The component response probabilities in (2) are given by*

$$\Pr\left(X_{i,jk} = x_{i,jk} | \langle W_j, W_k \rangle, D_{i,jk}, G_i\right)$$
$$= [D_{i,jk} + (1 - D_{i,jk})G_i]^{x_{i,jk}T_{jk}} [(1 - D_{i,jk})(1 - G_i)]^{(1-x_{i,jk})T_{jk}} \tag{3}$$
$$[(1 - D_{i,jk})G_i]^{x_{i,jk}(1-T_{jk})} [D_{i,jk} + (1 - D_{i,jk})(1 - G_i)]^{(1-x_{i,jk})(1-T_{jk})} ,$$

where for simplicity $T_{jk} = 1 - (W_j - W_k)^2$.

Axiom 4 (Rasch Knowledge). *Each expert has an ability parameter, $\alpha_i \in Re$, and each tie has its own difficulty parameter, $\beta_{jk} \in Re$, such that*

$$\forall 1 \leq i \leq M, \forall 1 \leq j < k \leq N, D_{i,jk} = [1 + exp(-(\alpha_i - \beta_{jk}))]^{-1} . \qquad (4)$$

The first axiom states that there is a single balanced complete signed graph that represents the consensus knowledge of all the experts. Because of this consensus graph, one would expect that there would be a lot of dependencies between the responses from the experts even though they respond without collaboration. The second axiom assures that these dependencies disappear when the responses are conditioned on the consensus partition and the other parameters. A conditional independence assumption like Eq. (2) is typical of response models in psychometric test theory [9] as well as many other parametric models. Axiom 3 expresses the marginal response probabilities defined in Eq. (2) as a function of the relevant parameters. Eq. (3) consists of four exponentiated terms. The terms correspond, respectively, to the usual signal detection terms of 'hits', 'misses', 'false alarms', and 'correct rejections'. For example if $T_{jk} = x_{i,jk} = 1$, then the tie is positive and the expert correctly responds positive. This is a hit and it occurs in the 2HT model if the expert detects the positive tie, with probability $D_{i,jk}$, or fails to detect it and guesses 'positive', with probability $(1 - D_{i,jk})G_i$, as reflected in the first term in Eq. (3).

Notice that in Axiom 3 there are knowledge and detection parameters for every combination of expert and pair, so some further restriction in parameters is necessary to make the model workable. Axiom 4 addresses this problem by applying a standard psychometric modeling idea in Eq. (4) to the $D_{i,jk}$ called the Rasch model [9]. In test theory, the Rasch model is applied to the probability that respondent i is correct to item j; however, in our case we apply it to a latent parameter that is also indexed by respondents i and items jk. In essence the Rasch model implies that the effects of respondent ability and item difficulty on the probability of detecting the signal are additive as can be seen by computing

$$Logit(D_{i,jk}) = log\left[D_{i,jk}/(1 - D_{i,jk})\right] = \alpha_i - \beta_{jk} . \qquad (5)$$

It is well known that the Rasch model is non-identified as one can add a positive constant to both the abilities and difficulties in Eq. (5) without changing its value. This problem is overcome in inference by fixing some property of the ability parameters as described later.

The model comprised by Axioms 1-4 is similar to the General Condorcet Model (GCM) in [2] and [3] with two important differences. First in [3] the items were N separate dichotomous true/false questions rather than dichotomous questions about the ties between nodes in a network. Second, and most important, in [3] there were no constraints on the 2^N possible consensus answer patterns. In our case there are $2^{N(N-1)/2}$ possible response patterns over the ties; however, from Axiom 1 only a small fraction, namely 2^{N-1} of them, satisfy the constraint of balance.

3   Bayesian Inference for the Model

In Bayesian inference one desires the posterior distribution of the model parameters given the data. In our case, the posterior distribution is $\Pr(\mathbf{W}, \mathbf{G}, \boldsymbol{\alpha}, \boldsymbol{\beta}|\mathbf{X})$, and by applying Bayes' theorem, it is well known that it is proportional to the likelihood function times the prior distribution [10]. The result is

$$\Pr(\mathbf{W}, \mathbf{G}, \boldsymbol{\alpha}, \boldsymbol{\beta}|\mathbf{X}) \propto L(\mathbf{X}|\mathbf{W}, \mathbf{G}, \boldsymbol{\alpha}, \boldsymbol{\beta})\pi(\mathbf{W}, \mathbf{G}, \boldsymbol{\alpha}, \boldsymbol{\beta}) , \qquad (6)$$

where $L(\mathbf{X}|\mathbf{W}, \mathbf{G}, \boldsymbol{\alpha}, \boldsymbol{\beta})$ is the likelihood function given by inserting Eq. (3) into Eq. (2), and $\pi(\mathbf{W}, \mathbf{G}, \boldsymbol{\alpha}, \boldsymbol{\beta})$ is the prior distribution of the parameters.

To complete the Bayesian model, we need to specify prior distributions. All priors are assumed to be independent. We assume bias is $G_i \sim$ Uniform$(0, 1)$. To anchor the Rasch parameters we centered both α_i and β_{jk} at mean zero, where $\alpha_i \sim$ Normal$(0, 1)$ and $\beta_{jk} \sim$ Normal$(0, 2)$. We assume that each node is assigned to A_1 or A_2 with equal probability, i.e. $W_j \sim$ Bernoulli$(1/2)$. Therefore, the prior on the set of all binary partitions is "flat".

We employ a Gibbs sampler with Metropolis-Hastings (M-H) steps to estimate model parameters. These MCMC algorithms are described well in texts on Bayesian modeling, e.g. [10] and [11]. MCMC samplers for other applications of the Rasch models already exist, such as for population size estimation [12] and for the GCM [3]. Inference follows that of [3], but with a different parameterization of the Rasch model and a constrained answer key. For simplicity, ability parameters and item difficulty parameters were each sampled using Normal$(0, 1)$ candidate generators without tuning, and re-sampled in the case of numeric underflow/overflow. Bias candidates were sampled from the Uniform$(0, 1)$ distribution. The partition was sampled as a block, with candidates drawn by first randomly choosing a cell and, if non-empty in the previous iteration, then a random node from within that cell is moved to the other cell, else the same partition is used again.

4   Performance of the Model

4.1   Simulated Data

We applied the model to simulated data to establish that our MCMC sampler was working. In particular we selected values of the expert and item parameters in Eq. (6) and generated observed values of $\mathbf{X}$ from the model equations. Then we ran our MCMC sampler to see if the means of the posterior distributions of the parameters correlated highly with the generating parameters. We were satisfied with the results in parameter recovery. Work with simulated data suggests that, with 10 nodes and numbers of experts ranging from 12 to 40, discarding at least the first 1000 samples was necessary to assure convergence in multiple sampler chains.

4.2   Real Data

Pairwise judgments of 45 items consisting of all pairs of 10 nodes were elicited from 19 UC Irvine students. The nodes consisted of 5 names from among "most valuable players" from earlier years in the professional sports of basketball and baseball. Together, these comprised a ground truth partition known objectively to the researchers, but only accessible by memory for the students. Students responded to each item with either "same sport" or "different sport" and scored their confidence on a 3-point scale.

An important experimental design issue concerns the order that one presents the questions about pairs of nodes. For example if the three pairs $\{a_1, a_2\}$, $\{a_2, a_3\}$, $\{a_1, a_3\}$ are presented consecutively an expert might remember their responses to the first two pairs and be tempted to make sure the response to the third pair is consistent. Such triples form a cycle in the graph, and recall that balance requires that the product of the signs of pairs on any cycle be positive. To avoid possibilities like this, an algorithm was constructed that sequenced all the pairs to minimize the possibility that some responses would be inferred logically from short term memory of earlier responses. The algorithm creates three stages of question presentations. The first stage presents pairs where no inferences from balance are possible, the second stage involves pairs that do complete cycles but the longer cycles are completed before shorter ones, and finally the third stage presents the remaining pairs. At all stages of the algorithm one avoids presenting pairs with a common node consecutively.

Of primary importance in using a CCT model to aggregate experts' responses is its ability to estimate the consensus truth accurately. After a sufficient burnin of 1000 iterations, the modal posterior partition corresponded with the known ground truth over 0.999 of the time. This strong, single peaked distribution is evidence for the assumption in Axiom 1 that there is a single fixed consensus partition of the nodes for the experimental situation. This narrow distribution for the balanced consensus graph comes despite the experts imperfect balance in their responses, uncertainty in the ability, guessing and difficulty parameter estimates, and lack of self-reported confidence. None of the experts reported a fully balanced graph, though a general tendency towards balance is discussed below. Based on the known partition, the simple rates of correct response correlated very highly with the median estimated ability parameters (0.97). It is interesting to note that fully 516 of 855 item responses were coupled with a reported confidence of "don't know".

It may be that the experts are not responding according to an underlying partition, which can be detected as follows. For each of the 19 experts we calculated the proportion of positive ties P_i in their responses. Then we calculated the expected number of unbalanced triples E_i assuming ties are assigned independently with probability P_i. Then we calculated the observed number of imbalanced triples O_i and calculated chi-square values. On the independence assumption this should have a chi-square distribution with 19 df, and a $p = 0.95$ critical value of 30.14. The observed value was 71.03 which is considerably larger, and this establishes that the experts' responses exhibited a significant tendency

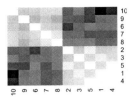

Fig. 1. Mean marginal post-predictive tie values. Negative is darker, positive is brighter.

toward balance even though none of the experts' responses exactly satisfied the balance constraint.

Visualization of posterior-predictive mean marginal tie values is provided in Figure 1, where rows and columns have been permuted to show the two-set structure.

5 Conclusion

CCT models and estimation methods have provided cognitively-based means to calibrate expert judgments for variety of tasks. This paper aims to further this work by adding the plausible assumption of a consensus graph constrained by balance. Though estimation for constrained models is not straightforward due to the dependencies among expert judgments, we have demonstrated the use of MCMC methods in recovering a sort of "wisdom of the crowds", where experts can have different abilities and biases. We have also highlighted a complimentary issue in survey methodology: when judgments are not independent, it may be useful to mitigate experts relying on logic and calculation.

There are two restrictions of the CCT model of balance in this paper that are useful to remove in future research. The first is the assumption in Axiom 1 that the consensus signed graph is complete. One can easily imagine situations where there may not be a tie of either sign between some pairs of nodes. In this case there are three possible values for a tie, $\{-1, 0, +1\}$, where '0' codes a no-tie. Then balance is achieved as before with a partition of the graph into two sets, where positive ties occur only between nodes in the same set and negative ties only between nodes in different sets. In this generalization, the criterion for balance is also that the product of the signs of the ties along every distinct cycle of nodes is positive, but cycles with no-ties are excluded. A second restriction of the current model is that it only handles the case where the set of nodes splits into two cliques. Davis [13] defines the concept of *clustering* to refer to the case where the nodes can be partitioned into two or more sets, where positive ties occur only between nodes in the same set and negative ties only between nodes in different sets. In future work we plan to extend the MCMC algorithm to place this more general constraint on the consensus signed graph.

Acknowledgments. Support is gratefully acknowledged from Gregory Alexander and from grants to William Batchelder from NSF, the Air Force Office of Scientific Research, the Army Research Office, and IARPA.

References

1. Batchelder, W.H.: Cultural consensus theory: Aggregating expert judgments about ties in a social network. In: Liu, H., Salerno, J.J., Young, M.J. (eds.) Social Computing and Behavioral Modeling, pp. 1–9. Springer, Boston (2009)
2. Batchelder, W.H., Romney, A.K.: Test theory without an answer key. Psychometrika 53(1), 71–92 (1988)
3. Karabatsos, G., Batchelder, W.H.: Markov chain estimation for test theory without an answer key. Psychometrika 68(3), 373–389 (2003)
4. Romney, A.K., Weller, S.C., Batchelder, W.H.: Culture as consensus: A theory of culture and informant accuracy. American Anthropologist 88(2), 313–338 (1986)
5. Romney, A.K., Batchelder, W.: Cultural consensus theory. In: Wilson, R., Keil, F.C. (eds.) The MIT Encyclopedia of the Cognitive Sciences, pp. 208–209. The MIT Press, Cambridge (1999)
6. Cartwright, D., Harary, F.: Structural balance: a generalization of Heider's theory. Psychological Review 63(5), 277–293 (1956)
7. Zaslavsky, T.: A mathematical bibliography of signed and gain graphs and allied areas. Electronic Journal of Combinatorics Dynamic Surveys 8, 151 (1999)
8. Macmillan, N.A., Creelman, C.D.: Detection Theory: A User's Guide, 2nd edn. Cambridge University Press, New York (2004)
9. Fischer, G.H., Molenaar, I.W. (eds.): Rasch Models: Foundations, Recent Developments, and Applications. Springer, Heidelberg (1995)
10. Gelman, A.: Bayesian Data Analysis, 2nd edn. Chapman & Hall/CRC, Boca Raton (2004)
11. Jackman, S.: Bayesian Analysis for the Social Sciences. Wiley (2009)
12. Fienberg, S.E., Johnson, M.S., Junker, B.W.: Classical multilevel and bayesian approaches to population size estimation using multiple lists. Journal of the Royal Statistical Society. Series A (Statistics in Society) 162(3), 383–405 (1999)
13. Davis, J.A.: Clustering and structural balance in graphs. Human Relations 20(2), 181 (1967)

Using Organizational Similarity to Identify Statistical Interactions for Improving Situational Awareness of CBRN Activities

David Melamed[1], Eric Schoon[1], Ronald Breiger[1], Victor Asal[2],
and R. Karl Rethemeyer[2]

[1] University of Arizona
[2] University at Albany, SUNY

Abstract. Distinctive combinations of attributes and behaviors lead us to improve on existing representations of terrorist organizations within a space of group properties. We review and extend a four-step procedure that discovers (statistical) interaction effects among relevant variables based on clusters of organizations derived from group properties. Application of this procedure to 395 terrorist groups in the period 1998-2005 identifies distinctive patterns of unconventional weapons activity and improves prediction of the groups that use or pursue chemical, biological, radiological, or nuclear (CBRN) weapons.

Keywords: CBRN use or pursuit, terrorism, regression modeling, profiling.

1 Introduction

How do we determine which non-state organizations are threats because they are likely to use or pursue chemical, biological, radiological or nuclear (CBRN) weapons? Various methods have been leveraged with respect to this question. No single profile fits all non-state organizations that seek or employ CBRN weapons [1]. As a result, there is an ever increasing need for analytic methods that afford greater accuracy.

In this paper we generalize a recently proposed algorithm for detecting statistical interactions in data organized in a cases × variables format. In our application the cases are 395 terrorist groups worldwide, and the variables pertain to CBRN activity (data of Asal and Rethemeyer, Project on Violent Conflict, University at Albany). The usual regression models pertain to relations among the columns of the data matrix (i.e., relations among variables). By contrast, we use the variables to learn about the cases. We thus shift the usual emphasis from factors to actors in addressing situational awareness.

We begin by reviewing the statistical interaction-finding algorithm developed by Melamed, Breiger and Schoon [2]. The algorithm is based on linear models, but our application uses a binary outcome (CBRN use and pursuit). As such, we generalize the algorithm to account for binary outcomes. We then describe the data, and demonstrate the application of this generalized approach using the Big Allied and

S.J. Yang, A.M. Greenberg, and M. Endsley (Eds.): SBP 2012, LNCS 7227, pp. 61–68, 2012.
© Springer-Verlag Berlin Heidelberg 2012

Dangerous (BAAD1) database. Our results suggest a modification to the hypothesis (see [3] for a literature review) that organizations with an extreme religious ideology are most likely to use or pursue CBRN weapons. In fact, we find that religiously-based organizations are statistically more likely to use or pursue CBRN weapons only when their ideology is rooted in a shared ethnicity. We conclude by discussing the implications of our findings for an economic club theory of terrorism, as outlined by Berman and colleagues [4-5].

In brief, the motivation for our procedure is improved prediction of CBRN activity by incorporating appropriate statistical interactions into our model. In order to identify these interactions among variables, we first identify sets of cases that exhibit similarity in their profiles across the variables. We exploit the idea that interactions among variables imply clusters of cases within which statistical effects differ [2].

2   Background

Building on the foundational work of Breiger and colleagues [1][6], Melamed, Breiger and Schoon [2] developed a four-step procedure for discovering statistical interaction effects. The algorithm is based on the idea that clusters of cases can be productively understood as sets of cases in which statistical effects differ. That is, while interactions are typically conceived only as relations among the columns of the data matrix (figure 1a), the approach outlined below is based on the idea that clusters of cases co-constitute statistical interaction effects. We examine how membership in a given cluster of cases is related to predicting the outcome (figure 1b).

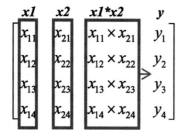

Fig. 1a. Standard conceptualization of statistical interactions.

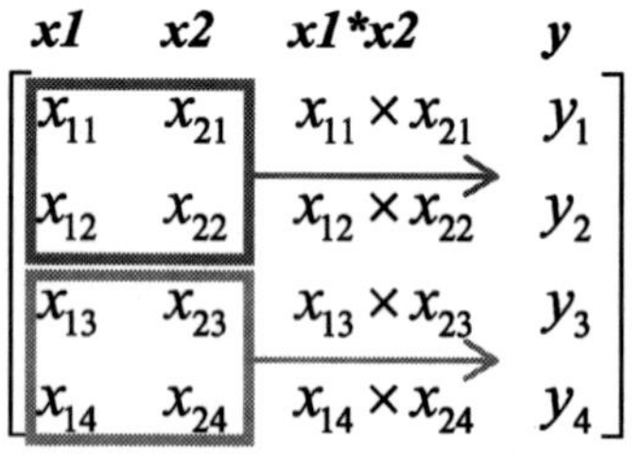

Fig. 1b. Melamed et al.'s [2] conceptualization of statistical interactions.

The first step in this procedure is to identify a parsimonious clustering of cases based on a series of attributes. Let $\mathbf{X}$ denote an $n \times p$ matrix where n refers to the number of cases and p refers to the number of variables or attributes, including a column of 1's representing the constant term. Using the Singular Value Decomposition (SVD)[1] of $\mathbf{X}$, we generate a matrix of orthogonal row scores ($\mathbf{U}$), a matrix of orthogonal column scores ($\mathbf{V}$), and a diagonal matrix of weights ($\mathbf{S}$). Geometric representation of each case (a row of $\mathbf{X}$) is usually given by the

---

[1] Belsley, Kuh, and Welsch [7] provide an introduction to the SVD decomposition, according to which $\mathbf{X} = \mathbf{U} \mathbf{S} \mathbf{V}^{\mathrm{T}}$.

Mahalanobis distance of this vector from the origin, but we note that this is identical to the Euclidean distance of the corresponding row of $\mathbf{U}$ from the origin. With linear models, it is more straightforward to cluster rows of $\mathbf{U}$ rather than rows of the data matrix $\mathbf{X}$. Using k-means clustering, we calculate the within sums of squared errors resulting from nine different clustering solutions, representing two through ten clusters. We then plot the sums of squared errors to identify a clustering solution that balances parsimony (fewer clusters) and model fit (lower squared error). The resulting plot resembles a scree plot in its appearance and interpretation.

The second step is to include each cluster's cases as a variable in the regression analysis. Membership in a cluster is treated as a dummy variable, and these are included along with the original variables comprising $\mathbf{X}$. Statistically significant cluster membership indicates a non-linearity in the data with respect to the outcome.

The third step in the procedure is to compute each of the cluster's contributions to each regression coefficient. In the linear regression case, the solution to the estimated regression coefficients ($\mathbf{b}$) is provided in equation 1, and the SVD analog (obtained by substituting the definition of SVD; see footnote 1) is provided in equation 2. From equations 1 or 2, linear regression implies $\mathbf{Xb} = \hat{\mathbf{y}}$, where $\hat{\mathbf{y}}$ refers to a vector of predicted values on the outcome variable. To compute the clusters' contributions to the regression coefficients using (2), the analyst simply multiplies each row of $\mathbf{VS^{-1}U^T}$ (a variables × cases matrix) by the outcome vector and then sums the columns according to cluster membership. Denote this matrix as $\mathbf{A}$. Matrix $\mathbf{A}$ is of size p by c, where p refers to the number of parameters and c refers to the number of clusters. Importantly, summing across the columns of $\mathbf{A}$ yields the regression coefficients for the full model (see discussion of Table 2). Examining the coefficients for each of the clusters allows one to identify potentially important interaction effects.

$$\mathbf{b} = (\mathbf{X^TX})^{-1}\mathbf{X^Ty} \, . \tag{1}$$

$$\mathbf{b} = \mathbf{VS^{-1}U^Ty} \, . \tag{2}$$

The final step in the procedure is to include the interaction term as identified in step three into a regression model that also includes the attributes contained in $\mathbf{X}$. Due to the dual relationship between clusters of cases and statistical interactions among variables, if cluster identification is significant, Melamed, et al. [2] argue that the interaction effect will also be significant.

We aim to implement the procedure detailed above with respect to the use or pursuit of CBRN weapons among 395 terrorist organizations during the time period 1998-2005. However, CBRN use or pursuit is a binary outcome amenable to logistic regression. We therefore turn to a generalization of steps three and four of this procedure in order to account for binary outcomes.

3   Generalized Interaction Detection

In the case of Ordinary Least Squares (OLS) regression, equations 2 and 3 can be used to obtain the regression coefficients. That is, there are several ways to obtain the closed form solution to equations 1 and 2. In the case of logistic regression, there is no closed form solution to the regression coefficients; hence an iterative procedure is required. Maximum likelihood estimation of the parameters is most common, but other algorithms, such as iteratively reweighted least squares, also provide solutions for the coefficients.

Once a solution to the logistic regression parameters is obtained, one can compute the logistic regression analog of equation 2. Equation 5 presents the linear solution to the logistic regression parameters[2] and equations 3 and 4 presents the components of equation 5. In equations 3 and 4, $\mathbf{W}$ is a diagonal matrix with elements $w_{ii} = \hat{p}_i(1-\hat{p}_i)$ that can be thought of as comprising a matrix of weights.[3] The right-hand side of equation 3 is obtained by computing the SVD of the product of $\mathbf{W}^{1/2}\mathbf{X}$. In equation 4, $\mathbf{z}$ is a vector of pseudovalues that stands for the observed binary outcome, $\mathbf{y}$ [8].[4]

$$\mathbf{W}^{1/2}\mathbf{X} = \mathbf{U}^*\mathbf{S}^*\mathbf{V}^{*T}. \tag{3}$$

$$\mathbf{y}^* = \mathbf{W}^{1/2}\mathbf{z}. \tag{4}$$

$$\mathbf{b} = \mathbf{V}^*\mathbf{S}^{*-1}\mathbf{U}^{*T}\mathbf{y}^*. \tag{5}$$

Multiplying each row of $\mathbf{V}^*\mathbf{S}^{*-1}\mathbf{U}^{*T}$ by the logistic regression outcome analog (i.e., $\mathbf{y}^*$) and summing the columns by cluster membership yields each cluster's contribution to the estimated logistic regression coefficients. That is, equation 5 allows us to compute a logistic regression analog to matrix $\mathbf{A}$. Let $\mathbf{A}^*$ denote this matrix. $\mathbf{A}^*$ is of size parameters × clusters (see, e.g., Table 2 below). $\mathbf{A}^*$ can be used to identify potential interaction effects to include within a logistic regression model at step four of the interaction-finding procedure. In summary, step one remains unchanged when extending our approach to logistic regression. In step two, we include cluster membership as a series of dummy variables into a logistic regression model predicting the binary outcome. Step three is completed by examining matrix $\mathbf{A}^*$ as a means to identify interaction effects. The identified interaction from step three can then be included into a logistic regression model predicting the binary outcome.

4   Data and Baseline Model

To demonstrate the value added by our approach, we analyze data on CBRN activity from the BAAD1 database of Asal and Rethemeyer (see also [3]). Our data include 395 terrorist organizations worldwide, of which 23, or 5.82%, are coded as having used or pursued CBRN weapons between 1998 and 2005. The organization is the unit of analysis, and we leverage five organizational-level variables to explain CBRN use or pursuit. Organizational membership is a four-category order-of-magnitude measure of group size ranging from under 100 members to over 10,000 members.

---

[2] Note that the iterated parameter values are required to estimate components of both $\mathbf{W}$ and $\mathbf{z}$.

[3] $\hat{p}$ refers to the predicted value on the binary outcome.

[4] The pseudovalues are defined as: $\mathbf{z} = \mathbf{X}\beta + \mathbf{W}^{-1}(\mathbf{y} - \hat{p})$.

Table 1. Summary of a logistic regression model predicting CBRN use or pursuit

Factor	Model 1	Model 2	Model 3
Constant	-4.306***	-5.896***	-3.866
Organizational Age	.450*	.508*	.508*
Eigenvector Centrality	.928***	1.070***	1.070***
Organization Size	.630*	.664*	.664*
Religious Ideology	.151	.796	-1.231
Ethnic Ideology	.746	1.778*	-.251
Cluster 1		2.030*	
Cluster 2		Omitted	
Religious*Ethnic			2.030*

Note: *p < .05, ** p < .01, ***p < .001. Standard errors (not shown) are adjusted for clusters.

Organizational age is the number of years since the group's inception to 2005. Organizational connectedness is measured as each organization's eigenvector centrality, computed by Asal and Rethemeyer from the "related groups" module of the Terrorist Knowledge Base [3]. This measure of centrality accounts for the relative incidence of each group's ties, while also adjusting for the centralities of the groups to which the group is connected. We also include two ideological indicators: whether or not the group is based on a religious orientation, and whether or not the group is based on an ethnic orientation. (In our coding, neither orientation is necessarily exclusive of the other.) Together these five organizational variables constitute our predictors of CBRN use and pursuit.

Of the 395 terrorist organizations, 261 are coded as having fewer than 100 members, 77 are coded as having between 101 and 1,000, 45 are coded as having between 1,001 and 10,000, and 12 are coded as having more than 10,000 members. The average organizational age is 11.27 years, with a standard deviation of 12.16. The average eigenvector centrality score is 2.13, with a standard deviation of 6.80. Finally, 116 of the groups are coded as having a religious orientation, and 153 of them are coded as being based on an ethnic orientation in their ideology.

Model 1 of Table 1 presents the results of our baseline model. All three of the continuous variables (age, size, and centrality) have been standardized, and are standardized in subsequent models. The results suggest that size, age, and centrality are significantly related to CBRN use and pursuit. Specifically, for each standard unit increase in size there is a corresponding 87.7% increase in the odds of use or pursuit of CBRN; for each standard unit increase in age there is a corresponding 56.8% increase in the odds of use or pursuit of CBRN; and, for each standard unit increase in centrality there is a corresponding 152.9% increase in the odds of use or pursuit. Moreover, the coefficients for the two ideology variables, religion and ethnicity, are not significant. This is surprising in light of the fact that prior research has shown that

Table 2. Logistic regression coefficients decomposed by clusters

Factor	Cluster 1	Cluster 2	Cluster 3	Net Effect
Constant	-2.623	-1.028	-.655	-4.306
Organizational Age	.312	.141	-.003	.450
Eigenvector	.227	-.077	.779	.928
Organization Size	.290	.024	.316	.630
Religious Ideology	1.067	1.819	-2.736	.151
Ethnic Ideology	1.328	-1.243	.661	.746

Note: Net effect in column 4 is a sum across rows of this table, and is identical to Model 1 of Table 1.

religious terrorist organizations are more likely to inflict casualties than non-religious organizations [4-5]. With model 1 as our baseline, we now turn to an application of our generalized algorithm.

5   Interaction Detection: Application

Using k-means clustering on the row space matrix, **U**, we cluster the 395 terrorist organizations based on the five attributes used in model 1. Following our procedure as outlined above, we select a three-cluster solution, which yields a cluster (#1) with 188 groups, a cluster (#2) with 91 groups, and a cluster (#3) with 116 groups. Cluster #1 turns out to consist of the 188 groups coded as being neither ethnic nor religious. Cluster #2 contains the 91 groups that are coded as being ethnic but not religious. Cluster #3 includes the other 116 groups that are coded as being religious; there is, however, variability on ethnicity in the third cluster. This variability in the third cluster will provide us with the necessary "traction" to estimate cluster effects in a logistic regression context.

Model 2 of Table 1 presents the results of our model when we include these clusters as dummy variables in our logistic regression. The first thing to notice about model 2 is that cluster #2 has been omitted. This is because there is no variability on religion within this cluster that can differentiate it from the main effect of religion. Consequently, the parameter for cluster #1 reflects significant differences between cluster #1 and the other two categories (since cluster #2 was omitted and cluster #3 was left out as the reference category). As a result of the significant differences between clusters, we proceed to step three of the interaction identification algorithm.

As noted above, step three of our procedure consists of multiplying each row of $\mathbf{V}^*\mathbf{S}^{*-1}\mathbf{U}^{*T}$ by the logistic regression outcome analog (i.e., $\mathbf{y}^*$) and summing the columns by cluster membership, which yields each cluster's contribution to the estimated logistic regression coefficients. Table 2 presents the results of these computations. We see that summing the cluster's contributions to the logistic regression coefficients in table 2 yields the estimated coefficients from model 1 in

Table 1 (e.g., for the constant: -2.623 + -1.028 + -.655 = -4.306). In general, Melamed, et al. [2] advise looking for systematic differences in coefficients across the clusters, coupled with relatively large magnitude shifts in coefficients. Based on the decomposition of the effects presented in Table 2, it appears that there may be an interaction between religion and ethnicity. In cluster #1, both terms (religious and ethnic ideology) are relatively large and positive. In cluster #2, both terms are relatively large, but they have opposite signs. And in cluster #3 the effects again have different signs. In light of this patterning of effects by cluster, we deduce an ethnicity by religious interaction effect.

Model 3 of Table 1 presents the results of adding the discovered interaction effect to the original logistic regression model. Indeed, the discovered statistical interaction effect is significant. Substantively, this result suggests that organizations that are based on both religion and ethnicity are 661.4% more likely to use or pursue CBRN weapons than the reference category (groups that are not religious and not ethnic), net of the other factors in the model. Of the 62 groups that are both religious and ethnic, 10 of them (16.13%) are coded as using or pursuing CBRN weapons between 1998 and 2005—this is a conditional probability that is 2.8 times the baseline.

The results in Table 1 also illustrate Melamed, et al.'s [2] point that clusters of cases co-constitute statistical interactions among variables. Indeed, the dummy variable representing cluster 1 has the exact same coefficient and standard error as the interaction effect. This is due to that categorical nature of the data (i.e., religion and ethnicity are binary variables), so in this case, there is an exact one-to-one correspondence between the cluster coefficient and the statistical interaction coefficient.

6   Discussion

Several authors have demonstrated that a terrorist group's ideology can be predictive of their lethality [4][5] and of their use or pursuit of CBRN [3]. However, these studies have either focused exclusively on religious ideologies [4][5] or have found insufficient evidence to support ethnicity as a predictive factor. Our analysis corresponds with existing research on lethality [4], suicide bombing [5], and CBRN [3], finding that ethnicity is not significant for predicting CBRN use or pursuit. However, by applying the new analytic approach outlined above, we find that groups with ideologies that combine religion and ethnicity are more than 600% more likely to use or pursue CBRN than those groups without either ideological basis. These results add important and previously overlooked nuance to our understanding of the role of ideology in predicting terrorist activities.

Berman and colleagues' [4][5] application of an economic club model to the study of terrorism is one excellent example of a productive theoretical framework which examines the functional role of ideological affiliations for terrorist organizations. These authors argue that religious sects can be understood as economic clubs that provide valuable resources to those who are willing to demonstrate their commitment to the group. Other social scientists (e.g., Coleman [9]; Tajfel [10]) also write that shared in-group membership based on religion promotes solidarity and trust, qualities which are necessary for groups that use unconventional violent tactics [4]. Comparing our results to Berman and colleague's research on lethality [4][5], we see that the organizations consistently used as cases to demonstrate the club model—Hezbolah, Hamas, the Mahdi Army, and the Taliban—are all found in cluster #3 of our analysis. Of these organizations, our data show that ethnicity and religion are central ideological components for the former three, while the fourth consists largely of a single ethnic group, the Pashtuns. This concurrence between our research and the work by Berman and colleagues lends further support to the assertion that ethnicity significantly amplifies, but does not analytically overshadow, the effect of religion.

Our results provide important new information for analysts seeking to create more accurate profiles of terrorist threats. We have generalized an analytic procedure for discovering interaction effects. In the ways indicated, this innovative approach can facilitate research aiming to move beyond existing knowledge.

Acknowledgement. This work was supported by the Defense Threat Reduction Agency, Basic Research Award # HDTRA1-10-1-0017, to the University of Arizona (R.L. Breiger, Principal Investigator). Eric Schoon was supported by a National Science Foundation Graduate Research Fellowship. For helpful comments we thank Elisa Bienenstock, Patrick Doreian, Sean Everton, H. Brinton Milward, James Moody, and John Tidd.

References

1. Breiger, R.L., Ackerman, G., Asal, V.H., Melamed, D., Milward, H.B., Rethemeyer, R.K., Schoon, E.: Application of a Profile Similarity Methodology for Identifying Terrorist Groups that Use or Pursue CBRN Weapons. In: Salerno, J., Yang, S., Nau, D., Chai, S. (eds.) Social Computing and Behavioral Cultural Modeling and Prediction, pp. 26–33. Springer, New York (2011)
2. Melamed, D., Bregier, R.L., Schoon, E.: The Duality of Clusters and Stastical Interactions. Sociological Methods & Research 41 (2012) (in press)
3. Asal, V.H., Rethemeyer, R.K.: Islamist use and pursuit of CBRN terrorism. In: Ackerman, G., Tamsett, J. (eds.) Jihadists and Weapons of Mass Destruction, pp. 335–358. CRC Press, Boca Raton (2009)
4. Berman, E.: Radical, Religious, and Violent: The New Economics of Terrorism. The MIT Press, Cambridge (2009)
5. Berman, E., Laitin, D.D.: Religion, Terrorism and Public Goods: Testing the Club Model. Journal of Public Economics 92 (2008)
6. Breiger, R.L., Melamed, D., Schoon, E.: Report on a Profile Similarity Methodology for Turning Terrorist Attributes into Network Connections. Unpublished technical report
7. Belsley, D.A., Kuh, E., Welsch, R.E.: Regression Diagnostics: Identifying Influential Data and Sources of Collinearity. John Wiley, Hoboken (2004)
8. Hosmer, D., Lameshow, S.: Applied Logistic Regression, 2nd edn. Wiley, Hoboken (2006)
9. Coleman, J.: Foundations of Social Theory. Harvard University Press, Cambridge (1990)
10. Tajfel, H.: Human Groups and Social Categories. Cambridge University Press, Cambridge (1981)

Opinion Dynamics in Gendered Social Networks: An Examination of Female Engagement Teams in Afghanistan

Thomas W. Moore, Patrick D. Finley, Ryan J. Hammer, and Robert J. Glass

Sandia National Laboratories, Albuquerque, New Mexico
{tmoore,pdfinle,rhammer,rjglass}@sandia.gov

Abstract. International forces in Afghanistan have experienced difficulties in developing constructive engagements with the Afghan population, an experience familiar to a wide range of international agencies working in underdeveloped and developing nations around the world. Recently, forces have begun deploying Female Engagement Teams, female military units who engage directly with women in occupied communities, resulting in more positive relationships with those communities as a whole. In this paper, we explore the hypothesis that the structure of community-based social networks strongly contributes to the effectiveness of the Female Engagement Team strategy, specifically considering gender-based differences in network community structure. We find that the ability to address both female and male network components provides a superior ability to affect opinions in the network, and can provide an effective means of counteracting influences from opposition forces.

1   Introduction

Success in Afghanistan depends on developing an understanding of, and avoiding a hostile reaction from, the communities that comprise the Afghan population: failure to do so has resulted in fractionalization and destabilization several times in the history of the region [1]. Despite realization of the importance of community engagement, U.S. attempts to effect regional stabilization through local community interventions have met with, at best, mixed success [2].

In 2009, the United States and coalition forces initiated the deployment of Female Engagement Teams (FETs) [3]. Derived from the military employment of females to conduct personal searches of female civilians in Iraq(dubbed Lioness teams)[4],FETs are composed of female military personnel who expand the Lioness mission to include humanitarian engagement and provision of medical care [3].

Although anecdotal accounts exist regarding the successful operations of FETs, as yet there exists no study of the theoretical foundations that might account for the successes. In this investigation, we employ an opinion dynamics model on idealized social networks characterized by strong gender assortativity to analyze how gendered networks contribute to opinion formation, and how FETs and other groups might act

S.J. Yang, A.M. Greenberg, and M. Endsley (Eds.): SBP 2012, LNCS 7227, pp. 69–77, 2012.
© Springer-Verlag Berlin Heidelberg 2012

to influence opinions and behaviors in those networks. A successful model would allow the investigation of elements that determine whether a given community might respond well to an FET-based intervention strategy, would allow for the generalization of engagement strategies based on gendered and otherwise highly assortative networks outside the immediate context of Afghanistan, and would permit the application of those principles in contexts other than counter-insurgency operations.

Our focus includes modeling how opinions related to international forces evolve in a community, and how extra-community agents (including FETs, traditional male-only engagement teams, and local opposition forces), interacting with gendered networks, can influence opinions in the community.

2   Origin of Structure in Gendered Social Networks

The topologies that characterize social networks emerge from the relationships between and among individuals [5]. The kinds of relationships that contribute to individual studies can vary widely. Social networks have been investigated in a wide variety of contexts, including corporations [6], schools [7], and electronically mediated communities, such as Internet sites [8].

In addition to enabling and mediating relationships among individuals, social contexts can act to constrain the kinds of relationships available. These constraints can originate implicitly, determined by existing community characteristics. For example, if two individuals are unlikely to encounter one another in a social context, they are unlikely to form a relationship [9]. Communities that constrain interactions between individuals of different classes can thus exhibit segregation within their social networks.

In addition, individuals tend to form friendships with friends of their friends. This property, called transitive closure, can act to increase the density of ties in social cliques by forming triangles (connected triads) of individuals [10]. It has also been observed that individuals tend to form relationships with people who resemble themselves according to some sets of socio-demographic characteristics, including ethnicity, age, class, and gender [11]. This may be due to the contributions of proximity and contact frequency, or due to opinion propagation of affective evaluations, as described in structural balance theory [12].

Gender assortativity, the tendency for like-gendered individuals to form relationships, has been observed in many types of networks, including childhood and elementary schools [13] and secondary schools [10] in the United States, in workplaces and entrepreneurial endeavors [14, 15], and in immigrant populations [16].

Gender can also influence the qualities associated with relationships within social networks. An examination of the National Longitudinal Study of Adolescent Health ("Add Health") data set indicated that the social relationships formed by female adolescents were more likely to result in triadic closures than those established by males [10].

Some studies have found that gender differences in perceived relationship intimacy exist, generally characterizing female relationships as more often incorporating mutual emotional support and male relationships as more often incorporating shared activities and instrumental support [18]. Cross-cultural studies have identified possible interacting roles for both biological explanations and socialization constraints [19]. Female-female social relationships in non-human primates have also been studied, and have supported the hypothesis that both ecological/environmental and biological factors influence gender-specific characteristics of social interactions[20].

3 The Existence of Gendered Networks in Afghan Communities

Sociological and anthropological studies of Afghan communities indicate several common factors relevant to this study. Afghanistan is multi-ethnic and multi-linguistic, with Pashtun and Tajik ethnicities and Afghan Persian and Pashto languages in the majority, and is almost exclusively Islamic, with 80% of the population being Sunni and 19% of the population being Shia [21].

Especially among the Pashtun, the conceptualization of tribal values of honor and shame lead to the general seclusion of women as a class both through the employment of a full veil and through walled family compounds [22]. In a 2002 study of female narratives in Afghanistan, one participant observed that she "talk[s] to every woman, and some men who are relatives" [23]. The strong limitations of interaction between the sexes are collectively termed *purdah*, and are generally common (with cultural variations) in many conservativeIslamic societies and throughout South Asia [24].

Although women in Afghanistan are often seen as subservient and in some contexts areconsidered to be property [22], some researchers studying the region have cautioned againstthe dismissal of women as a powerful and influential force in the community, citing mother-son and wife-husband relationships, as well as relationships among the women themselves [3, 4, 25].

Women interact with each other in extended households and in labor contexts [25]. Solidarity among women plays an important role and can mark a counterpoint to the patriarchal culture that outwardly characterizes Afghan society [26, 25]. This dynamic has the potential to create strong social ties among women in Afghan communities, although rivalries can exist both within and between households.

Men in Pashtun culture hold significantly to a warrior-oriented value system, endorsing individual-oriented bravery and, in principle, a largely egalitarian social structure [22]. The concept of *Pashtunwali*defines a set of core values of freedom, honor, revenge, and chivalry [27]. Conflict resolution is multi-level, with familial components (in the person of the father or grandfather), and local, regional/tribal, and national *jirga*s, or egalitarian authoritative bodies [28]. These individualistic concepts of honor and familial loyalties translate, in this model, to generally lower levels of affective association and opinion propagation between males than between females, resulting in relatively sparser networks of intimate relationships and influence.

4   The Opinion Dynamics Model

The bounded confidence opinion dynamics model used in this study is an extension of a model initially proposed to study the diffusion of environmental agricultural practices among European producers [29]. It is one of several related techniques derived from statistical physics Ising spin models [30].

The Deffuant-Weisbuch (DW) approach used here incorporates a community of individuals who have opinions modeled on the range [0,1]. Each individual is then potentially influenced by the people with whom she interacts, subject to the constraints of bounded confidence, according to the discrete time equation:

$$x_i(t+1) = x_i(t) + \frac{1}{|S_i|}\sum_{j \in S_i} \mu_{ij}[x_j(t) - x_i(t)] \tag{1}$$

where $x_i(t+1)$ is the opinion of the ith individual at the next time step, $x_i(t)$ is the current opinion of the ith individual, S_i is the set of neighbors of the ith individual and $|S_i|$ is the cardinality of the set, μ_{ij} is the plasticity value between the ith individual and her jth neighbor, and $x_j(t)$ is the opinion value of her jth neighbor. When applied to a directed social network, S_i consists of the out-degree neighbors of ith individual, and μ_{ij} is a weighting value characterizing the relative strength of the relationship between i and j with respect to i's opinion formation about the subject, and is analogous to an edge weight in a social network. This averaging approach is distinct from that employed in the original DW model, and is similar to that proposed by Hegselmann and Krause [31]. This approach is also similar in its theoretical basis to that proposed by DeGroot [32] and used extensively in modeling consensus formation in economics and political science.

Application of the opinion update function in Equation 1 is subject to the constraints of bounded confidence:

$$|x_i(t) - x_j(t)| < \varepsilon_i \tag{2}$$

where $x_i(t)$ and $x_j(t)$ are as described above, $|x_i(t) - x_j(t)|$ is the absolute value of the difference in opinion values between the ith individual and her jth neighbor, and ε_i is the tolerance value, or confidence bound, of the ith individual. The tolerance value further restricts the individuals with whom the ith individual interacts to those that hold an opinion already "close enough" to the opinion held by i. This value represents the tendency of individuals to reject opinions seen as being too different from the ones they currently hold.

Opinion dynamics models can be seen as algorithmic realizations of structural balance theory as proposed by Cartwright and Harary [12] in which individuals having a positive relationship with each other will tend to develop over time the same opinion toward a third individual or idea. A related model proposed by French described how different patterns of influence between individuals connected by directed relationships could be seen to affect opinion formation [33]. In these two models, social power, affective ties, and attempts to minimize cognitive dissonance result in individuals altering their opinions to achieve a level of agreement with their neighbors in social space. Additional supporting research came from the seminal work of Asch, which demonstrated that perceived social pressure could influence test subjects to alter their answers to simple questions of observation [34].

5   Network Structures

One topic of current interest in social network research is the elicitation of community structure from network topological data. In this restricted sense of the term, a community within a social network can be broadly defined as a group of individuals who have more connections to each other than they have to other individuals or groups [35].

We construct the graph for a social network representing an Afghan community by generating two independent graphs with a uniform probability of vertex connection, based on the class of random graphs proposed by Erdos and Renyi [36]. These algorithms generate graphs with a Poisson distribution of vertex degree, creating a graph whose vertices exhibit an average level of connectivity that then characterizes the network. These networks exhibit the formation of a single, giant connected component around the phase transition point $P = \frac{\ln(N)}{N}$ where P is the probability of a connection existing between any two given nodes, and N is the number of nodes in the network [36]. We generate the networks representing females and males separately, with a higher P value for the female network representing the greater degree of relationships exhibited by females in these communities. We then create links between the male and female networks, representing the relationships within households between husbands and wives, and mothers and sons.

In our opinion dynamics model, networks represent patterns of social influence relative to a specific idea or set of ideas, rather than simple definitions of categorical relationships, such as friendships or family ties. Idealized graph topologies, including the Poisson distribution generating topologies of Erdos and Renyi [36], the small world networks of Watts and Strogatz [37], and the power law networks of Barabasi and Albert [38], are approximations of real world networks capturing some aspects of certain types of relationships. For this study, the Poisson distribution networks were selected to approximate the idealized egalitarian structure of Pashtun community and tribal organizations [39].

6   Experimental Results

This analysis considers the ability of units including Female Engagement Teams to influence opinions in a population by examining average opinion values in the male and female communities under various scenarios. In this model, opinion is used to measure the opinion about international forces, with higher opinion values indicating a higher opinion about international forces. A lower opinion is considered to indicate more support for opposition forces.

We simulate an Afghan community using a highly abstracted network model. The communities consist of random networks of 50 females and 50 males. In keeping with the sociological literature cited above, more female-to-female links exist in the community than male-to-male links (as illustrated in Figure 1). Initial opinions are assigned to community members following a truncated normal distribution centered at 0.5. The OD algorithm is allowed to run to equilibrium on the network, resulting in clusters of nodes in the community sharing similar opinions, and potentially differing sharply from other individuals in the community.

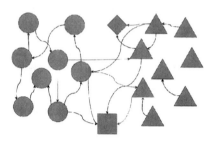

Fig. 1. Notional illustration of sample network topology. Circles represent females; triangles represent males; square represents FET; diamond represents opposition influence. Note the higher density of edges among females, the connections between the FET and both gender-based subnetworks, and the male-only connections of the opposition influence.

We model the effect of outside influences by introducing interventions in which an actor (allied FET or opposition group) introduces an opinion into a subset of the community (Table 1). The community social network is allowed to return to equilibrium and the mean population opinion values indicate the effectiveness of the intervention. The results shown in Table 1 represent the mean of 1,000 simulations over stochastically derived social networks.

Table 1. Community Opinions Resulting from Interventions

Intervention				Mean Results		
Code	**Actors**	**Population**	**Opinion**	**Female**	**Male**	**All**
NONE	N/A	N/A	N/A	0.500	0.500	0.500
FEM	FET	FEMALE	0.7	0.640	0.584	0.613
MALE	FET	MALE	0.7	0.532	0.556	0.544
MIX	FET	ALL	0.7	0.615	0.585	0.600
OPPO	OPPO	MALE	0.3	0.480	0.441	0.451
FEMOPPO	OPPO / FET	MALE/FEMALE	0.3/0.7	0.617	0.517	0.567
MIXOPPO	OPPO / FET	MALE/ALL	0.3/0.7	0.583	0.519	0.551
MALEOPPO	OPPO / FET	MALE/MALE	0.3/0.7	0.497	0.499	0.498

Experimental results support the hypothesis that FETs, by extending contact to the female community in a population, can bring about a greater shift in opinion than engagement teams who interact with the male community alone (Figure 2). In addition, units which include FETs can shift the opinions of populations in a favorable direction despite the influence of opposition forces due to the opposition's inability to successfully engage the female community. Under this model, simultaneous engagement of female and male communities (MIX) provides very good performance. Engaging either the female or the male networks singly (FEM or MALE)also provides an improvement over non-engagement (NONE).FET engagements are capable of counteracting opposition influences in communities (Table 1 and Figure 2).Importantly, FET's interacting with female population (FEMOPPO) or with an integrated population (MIXOPPO) are significantly more effective at countering opposition influence than an allied team interacting with an exclusively male population (MALEOPPO) at a 95% confidence level.

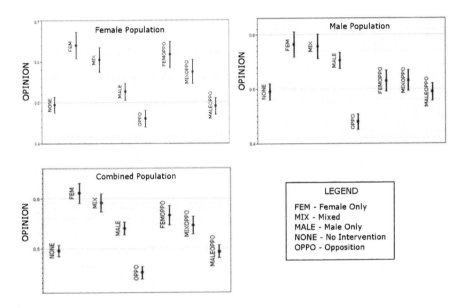

Fig. 2. Effect of Interventions on Community Opinions. Data points represent the mean opinion value of each population obtained from 100 stochastic simulation runs. The error bars represent 95% CI of the mean.

7   Discussion

Female Engagement Teams can significantly enhance the ability of international forces to effect changes in opinion in populations that exhibit strong assortativity in gender with associated community structure in the resulting social network. By engaging with both male and female populations in a given community, FETs allow for more efficient dissemination of information and opinions.

Opposition forces, if compelled by social rules to engage only the male population in a given community, can still significantly affect community opinion in the absence of counter-messaging by international forces. However, by engaging both male and female community members, FETs provide an effective countermeasure.

Our modeling indicates that FETs acting on female and male social networks with topologies consistent with the literature can generate positive opinion shifts, which agrees with anecdotal accounts from units in the field.

These results were obtained using a simplified model with a highly aggregated opinion value. Real world opinions are more complex, and can consist of many different sets of concepts. In addition, this study considered a limited subset of network topologies. Future investigations should expand the number of topologies investigated, including scale-free networks and networks with more complex community structures within the gendered subdivisions. Future investigations should also introduce more complex structures representing individual opinions and behavioral decisions.

Acknowledgements. The authors wish to thank the participants of the International Data Farming Workshop 23/NATO MSG-088 Meeting 5,including Gary Horne and Steven Anderson, and the associated Social Network Analysis group for their assistance and support in the foundations that underlie this work. Ted Meyer of the Naval Postgraduate School contributed to the initial modeling efforts, and Narelle Silwood of the New Zealand Defense Technology Agency contributed invaluable insight into the underlying social phenomena.

References

Note: Due to space limitations, the full bibliography of this paper is available on our website:http://www.sandia.gov/CasosEngineering/SBP2012_FETReferences.html

1. Biddle, S., Christia, F., Their, A.: Defining success in Afghanistan. Foreign Affairs 89, 48–51 (2010)
2. Peck, S.R.: PRTs: Improving or undermining the security for NGOs and PVOs in Afghanistan. DTIC Document (2004)
3. Pottinger, M., Jilani, H., Russo, C.: Half-Hearted: Trying to Win Afghanistan without Afghan Women. Small Wars Journal (2010)
4. Mehra, S.: Equal Opportunity Counterinsurgency: The Importance of Afghan Women in U.S. Counterinsurgency Operations (2010)
5. Wasserman, S., Faust, K.: Social Network Analysis: Methods and Applications. Cambridge University Press, Cambridge (1994)
6. Krackhardt, D., Kilduff, M.: Friendship Patterns and Culture: The Control of Organizational Diversity. American Anthropologist 92, 142–154 (1990)
7. Valente, T.W.: Social network influences on adolescent substance use: An introduction. Connections 25, 11–16 (2003)
8. Lampe, C., Ellison, N., Steinfield, C.: A Face (book) in the crowd: Social searching vs. social browsing. In: Proceedings of the 2006 20th Anniversary Conference on Computer Supported Cooperative Work, pp. 167–170 (2006)
9. Moody, J.: Race, School Integration, and Friendship Segregation in America. American Journal of Sociology 107, 679–716 (2001)

10. Goodreau, S.M., Kitts, J.A., Morris, M.: Birds of a feather, or friend of a friend? using exponential random graph models to investigate adolescent social networks. Demography 46, 103–125 (2009)
11. McPherson, M., Smith-Lovin, L., Cook, J.M.: Birds of a feather: Homophily in social networks. Annual Review of Sociology, 415–444 (2001)
12. Cartwright, D., Harary, F.: Structural Balance: A Generalization of Heider's Theory. Psychological Review 63 (1956)
13. Belle, D.: Gender Differences in Childern's Social Networks and Supports. Children's social networks and social supports. John Wiley and Sons (1989)
14. Lincoln, J.R., Miller, J.: Work and Friendship Ties in Organizations: A Comparative Analysis of Relational Networks. Administrative Science Quarterly 24, 181–199 (1979)

Socio-cultural Evolution of Opinion Dynamics in Networked Societies

Subhadeep Chakraborty[1] and Matthew M. Mench[1,2]

[1] Department of Mechanical, Aerospace, and Biomedical Engineering,
University of Tennessee, Knoxville, Tennessee
[2] Energy and Transportation Science Division, Oak Ridge National Laboratory,
Oak Ridge, Tennessee 37831, United States

Abstract. This paper introduces a modeling paradigm based on a language theoretic framework for stochastic simulation of decision-making in a social setting, where choices and decisions by individuals are increasingly being influenced by a person's online social interactions. In this paper, the dynamics of opinion formation in a networked society have been studied with a joint model that bridges micro-level decisions based on reward maximization and the corresponding social influences which alter the estimate of these reward values. The effect of long term government policies on the stability and dynamics of the population opinion and the effect of including an influencing agent group has been studied. Simulated results on a sample society demonstrate the major impact of a relatively small but sharply opinionated influencing group toward pushing the society toward a desired outcome.

Keywords: Socio Cultural Decision Modeling, Effective Influence, Behavioral Dynamics.

1 Introduction

The dynamics of opinion formation in a networked society has developed into an active field of study in the last few decades. In the context of econometrics, individual decision-making has been largely modeled as an optimal policy estimation problem, aimed towards maximization of some utility function defined over a discrete set of choices [1]. But at the same time, it has been argued [2] that the human decision mechanism is never entirely rational and independent of external (social, political, economic, fashion trends) influences.

This paper is an attempt to introduce a modeling paradigm based on a language-theoretic framework for stochastic simulation of a person's decision-making process in a social setting, where the said individuals are influenced by their social interactions, including online social networking, and global (political) events. It has been conjectured that individuals in a population, exhibiting such collective behavior as a result of making decisions largely influenced by the actions of others, lead to rapid unstable fluctuations in the society. These cultural, political and social *snowballing* effects are called *information cascades* [2].

S.J. Yang, A.M. Greenberg, and M. Endsley (Eds.): SBP 2012, LNCS 7227, pp. 78–86, 2012.

The concept of *information cascades*, based on *observational learning theory* was formally introduced in a 1992 article by Bikhchandani S. et. al. [2]. Watts D.J. [3] has studied the origin of rare cascades in terms of a sparse, random network of interacting agents using *generating functions*. Statistical mechanics tools, such as the Ising model [4] as well as non-equilibrium statistical models [5] has been used extensively to model the spread of influence in a networked society. *Influence maximization*, deals with finding the optimal set of people in a society to start an information cascade. Approximate solutions to this problem have been studied by Kempe [6] using *submodular functions*. However, in all these models, individual decision logic is largely overlooked.

Econometric theory, in contrast, contains a vast array of analyses tools for discrete outcomes [1]. Here, emphasis is on the logic behind discrete choices, specifically on selecting an optimal action policy which will lead to the largest discounted average future reward, but, interactions and influences are neglected.

To the best of the authors' knowledge, there have been relatively few attempts to merge these two areas into a more realistic model of opinion evolution in networked societies. In this paper, both complex microscopic logic, as well as social interactions that affect and are affected by these micro-choices are jointly studied to determine the advent of macroscopic social patterns which are observable, metrizable and potentially controllable.

2   The Framework for Discrete Choice Modeling (DCM)

Assumption 1. *Finite set of discrete choices*

At each instant, every individual in the network is faced with the same set of finite discrete choices – for example, to vote for political candidate 'A' or 'B' or not to vote at all. In Markov Decision Modeling [1] as well as in the current framework, the problem is posed as finding the optimal choice policy for maximizing the rewards gained as a result of one's own choices. However, in a network, the rewards are also functions of choices by every other individual in the community. This feedback dynamics add a further layer of intricacy.

Assumption 2. *Equivalent normative (rational) perspective*

The term *social norm* refers to rules or expectations for behavior that are shared by members of a group or society. For the sake of simplicity it is assumed in the model that majority of the society subscribe to the same rational consensus about right or wrong as a basic fact of organized social life. By no means does this imply that each individual makes decisions or reacts to stimulus identically, but simply that a rational structure may be imposed on the behavior of individuals.

Assumption 3. *Probabilistic individual decisions*

It is assumed that, even when presented with the exact same choices with the exact same pay-offs, different individuals, and possibly even the same individual, may probabilistically make alternate decisions; the only restriction being that the decision cannot be *deviant* from the *normative perspective* (Assumption 2).

Assumption 4. *Two kinds (external and internal (ε)) of events*

It is assumed that the set of events may be classified as *external/global* and *internal/local*. External events affect everyone simultaneously, according to the logic dictated by normative perspective. However, an individual's personal choice influences his own decisions, and possibly his networked neighbors to an extent. Some external events can lead to uncontrollable transitions, as explained next.

2.1   Normative Perspective Modeled as a Probabilistic Finite State Automata (PFSA)

The assumption of *normative perspective* allows rational behavior, to be encoded as a PFSA. In this example, each individual faces the *internal* decision of supporting the existing government, the opposing group, or remain in a state of indecision. Additionally, the individual can reach a state of political advantage, or disadvantage, but the uncontrollable transition to these two states can only occur through an *external* event, namely, the success or failure of the revolution.

Table 1. List of PFSA States and Events

States	Description	Events	Description
I	State of indecision	g	A popular act by the government
R	State of supporting the revolution	$\tilde{g}$	An unpopular government act
G	State of supporting the government	ε	An internal decision
A	State of political advantage	s	Success of the revolution
D	State of political disadvantage	f	Failure of the revolution

Figure 1 gives a schematic of the assumed normative perspective encoded as a PFSA. It may be noticed that transitions such as $g : G \to R$ or $g : I \to R$ are unauthorized, since it is assumed that a favorable act by the government should not make anyone decide to join the opposing group. Also, the same event can cause alternate transitions from the same state; the actual transition will depend probabilistically on the measure of attractiveness of the possible target states.

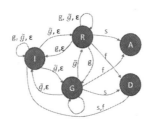

Fig. 1. Schematic of normative perspective coded as a PFSA

2.2   Rewards, Transition Costs and Probabilities

This section derives the logic behind probabilistic micro level choices by individuals. The probability of transitioning to a different state is dependent on the reachability of that state from the current state, the current event (external or internal), and also the relative degree of attractiveness of the target state. The state attractiveness measure is calculated using the concept of positive real measure attributed to a string of events [7], which is explained briefly here.

Let the discrete choice behavior be modeled as a PFSA as:

$$G_i \equiv (Q, \Sigma, \delta, q_i, Q_m) \tag{1}$$

where $Q = \{I, R, G, A, D\}$ is the finite set of choices with $|Q| = 5$ and the initial state $q_i \in Q = I$. The distribution of states may be represented as a coordinate vector of the form $\bar{v}_i$, defined as the $1 \times N$ vector $[v_1^i, v_2^i, ..., v_N^i]$, given by

$$v_j^i = \begin{cases} 1 \text{ if } i = j, \\ 0 \text{ if } i \neq j. \end{cases} \tag{2}$$

$\Sigma = \{\varepsilon, g, \tilde{g}, s, f\}$ is the (finite) alphabet of events with $|\Sigma| = 5$; the Kleene closure of Σ is denoted as Σ^*; the (possibly partial) function $\delta : Q \times \Sigma \times Q \to [0, 1]$ represents probabilities of state transitions and $\delta^* : Q \times \Sigma^* \times Q \to [0, 1]$ is an extension of δ; and $Q_m \subseteq Q$ is the set of marked (i.e., accepted) states.

Definition 1. *The reward from each state* $\chi : Q \to [0, \infty)$ *is defined as a characteristic function that assigns a positive real weight to each state* q_i, *such that*

$$\chi(q_j) \in \begin{cases} [0, \infty) & \text{if } q_j \in Q_m, \\ \{0\} & \text{if } q_j \notin Q_m. \end{cases} \tag{3}$$

Definition 2. *The event cost, conditioned on a PFSA state at which the event is generated, is defined as* $\tilde{\pi} : \Sigma^* \times Q \to [0, 1]$ *such that* $\forall q_j \in Q, \forall \sigma_k \in \Sigma, \forall s \in \Sigma^*$,

(1) $\tilde{\pi}[\sigma_k, q_j] \equiv \tilde{\pi}_{jk} \in [0, 1);$ $\sum_k \tilde{\pi}_{jk} < 1;$
(2) $\tilde{\pi}[\sigma, q_j] = 0$ *if* $\delta(q_j, \sigma, q_k) = 0 \forall k;$ $\tilde{\pi}[\epsilon, q_j] = 1;$

The event cost matrix, ($\tilde{\Pi}$-matrix), is defined as: $\tilde{\Pi} = \begin{bmatrix} \tilde{\pi}_{11} & \tilde{\pi}_{12} & \cdots & \tilde{\pi}_{1m} \\ \tilde{\pi}_{21} & \tilde{\pi}_{22} & \cdots & \tilde{\pi}_{2n} \\ \vdots & \vdots & \ddots & \vdots \\ \tilde{\pi}_{n1} & \tilde{\pi}_{n2} & \cdots & \tilde{\pi}_{nm} \end{bmatrix}$

The characteristic vector $\bar{\chi}$ is chosen based on the individual state's impact. For example, if the states represent various job choices, the remuneration from these jobs can serve as the characteristic vector. The event cost is an intrinsic property of the nominal perspective. The event cost is conceptually similar to the state-based conditional probability of Markov Chains, except $\sum_k \tilde{\pi}_{jk} = 1$ is not allowed to be satisfied. The condition $\sum_k \tilde{\pi}_{jk} < 1$ provides a sufficient condition for the existence of the real signed measure, as discussed in [7].

Definition 3. *The state transition function of the PFSA is defined as a function* $\pi : Q \times Q \to [0, 1)$ *such that* $\forall q_j$, *and* $q_k \in Q$,

(1) $\pi(q_j, q_k) = \sum_{\sigma \in \Sigma : \delta(q_j, \sigma, q_k) \neq 0} \tilde{\pi}(\sigma, q_j) \equiv \pi_{jk}$

(2) and $\pi_{jk} = 0$ if $\{\sigma \in \Sigma : \delta(q_j, \sigma) = q_k\} = \emptyset$.

The state transition matrix, (Π-matrix), is defined as: $\mathbf{\Pi} = \begin{bmatrix} \pi_{11} & \pi_{12} & \cdots & \pi_{1n} \\ \pi_{21} & \pi_{22} & \cdots & \pi_{2n} \\ \vdots & \vdots & \ddots & \vdots \\ \pi_{n1} & \pi_{n2} & \cdots & \pi_{nn} \end{bmatrix}$

2.3   Measure of Attractiveness of the States

A real measure ν_θ^i for state i is defined as $\nu_\theta^i = \sum_{\tau=0}^{\infty} \theta (1 - \theta)^\tau \bar{v}^i \Pi^\tau \bar{\chi}$ (4)

where $\theta \in (0, 1]$ is a user-specified parameter and $\bar{v}^i$ is defined in Eqn. 2.

Remark 1. Physical Significance of Real Measure

Assuming that the current state of the Markov process is i, i.e., the state probability vector is $\bar{v}_i$, at an instant τ time-steps in the future, the state probability vector is given by $\bar{v}^i \Pi^\tau$. Further, the expected value of the characteristic function is given by $\bar{v}^i \Pi^\tau \bar{\chi}$. The measure of state i, described by Eqn. 4, is the weighted expected value of χ over all time-steps in the future for the Markov process that begins in state i. The weights for each time-step $\theta (1 - \theta)^\tau$, form a decreasing geometric series (sum equals 1). The measure in vector form yields

$$\bar{\nu}_\theta = \theta (\mathbb{I} - (1 - \theta) \Pi)^{-1} \bar{\chi} \qquad \text{and} \qquad \bar{\nu}_{norm} = \frac{1}{\sum_k \nu_k} \bar{\nu} \qquad (5)$$

Remark 2. The effect of θ

The parameter θ controls the rate at which the weights decrease with increasing values of τ. Large values of θ forces the weights to decay rapidly, thereby placing more importance to states reachable in the near future from the current state. In fact, $\theta = 1$ implies that $\nu_\theta^i = \chi^i$. On the other hand, small values of θ captures the interaction with a large neighborhood of connected states. As $\theta \to 0$, the dependence on the initial state i is slowly lost (provided Π is irreducible).

The probabilistic transition decisions are dictated by $\bar{\nu}_{norm}$. Higher measure for a state implies that the discounted expected reward from that state is higher; consequently the incentive to transition to that state is proportionately higher.

3   Influence Model

One of the most interesting facets of a networked society is the strong interdependence between rewards and popularity of choices. For example, the reward from joining the revolutionary group ($\chi(R)$) may be small at the initial stages, but as more and more people join, the estimate of $\chi(R)$ as well as the probability of success of the revolution $P(s)$, increases. One of the advantages of the proposed modeling paradigm is that this multi-step reasoning is built into the language theoretic structure. Also, the framework lends itself well to all conventional linear and non-linear influence models such as the Friedkin-Johnsen [8] model or the bounded confidence model developed by Krause [9].

In this paper, it is assumed that the influence is entirely through the characteristic function χ of the states. This assumption is based on the physical insight that the anticipated reward from a state is the most well-discussed and well-broadcast quantity in a social network. In addition, influence of mass media can be accounted for, by assuming that an unbiased reporter reports the mass opinion in the form of a unified reward for the different choices averaged over the entire population. This may be mathematically expressed as

$$\bar{\chi}^i(t+1) = f_i \bar{\chi}^i(t) + g_i \frac{1}{|\mathcal{N}_i|} \sum_{j \in \mathcal{N}_i} \bar{\chi}^j(t) + h_i \frac{1}{N} \sum_{j=1}^{N} \bar{\chi}^j(t) \qquad (6)$$

where, N is the size of the network, $\mathcal{N}_i$ is the set of first order neighbors of node i, f_i is that fraction of the ith individuals opinion about potential value of the available states that is based on his past beliefs, g_i is the fraction derived from the opinion of his acquaintances (network neighbors) and $h_i = 1 - f_i - g_i$ is the fraction formed due to the influence of mass media such as newspapers, television, etc.

4   Progression of the Social Dynamics

The scale-free BA extended model network created with the Pajek [11] software has been used to model the connectivity structure, since many of the real-world networks are conjectured to be scale-free, including the World Wide Web, biological networks, and social networks [10]. The parameters are listed below.

$N \rightarrow$ Number of vertices: 100
$m_0 \rightarrow$ Number of initial, disconnected nodes: 3
$m \rightarrow$ Number of edges to add/rewire at a time ($m \leq m_0$): 2
$p \rightarrow$ Probability to add new lines: 0.3333
$q \rightarrow$ Probability to rewire edges ($0 \leq q \leq 1 - p$): 0.33335

Each individual is issued a random number drawn from a uniform distribution $U(0, 1)$, representing the time remaining before that person makes a decision. This imposes an ordering on the list of people in the network. As soon as someone makes a decision, the time to his next decision, drawn from $U(0, 1)$ is assigned, and the list is updated. Additionally, external events g and $\tilde{g}$ are also associated with a random time drawn from $U(a, b)$. Choosing a and b, the external events can be interspersed more, or less densely. At t_0, all individuals are initialized at state I. Initial values of the true reward vector $\bar{\chi}$ and the true event probabilities are fixed. Individuals receive a noisy estimate of the true probabilities and the rewards. At the time epoch t_k, when it is the ith person's turn to make a decision, he updates his personal estimate of the reward vector according to the influence equation (Eqn. 6). He then calculates the degree of attractiveness of the states based on the normalized measure, using Eqn. 5. The transition probabilities are calculated as $P(q_{t_{k+1}} = q'|q_{t_k} = q, \sigma = \sigma') = \nu_{norm}(q')R(q, \sigma', q')$ where $R(q, \sigma', q') = 1$ if $\sigma' : q \to q'$ exists, otherwise 0. The only difference in the case of an external event such as $g, \tilde{g}, s$ or f is that everyone simultaneously updates their states rather than asynchronously, as in the case of internal events.

5   Simulation Scenarios and Results

5.1   Effect of Global Events

In this experiment, the probabilities of the good and bad external events, $P(g)$ and $P(\tilde{g})$ are varied to observe the effect of long term government policies on a population. Each simulation was run 30 times and the average of all the runs are showed in Fig. 2. When $P(g):P(\tilde{g})$=1:1, the percentage of the population in states R and G are equal, and a large part of the population remains undecided (Fig. 2a). In the absence of any deadline for making a decision this result is only to be expected. As the government starts to push more and more unpopular policies, opinions bifurcate and the revolutionary group starts to become more popular. An interesting aspect of this result is that there is a threshold in the fraction of perceived negative acts by the government, above which the popularity of state R ramps up at a steady rate until everybody converges to a unipolar opinion (Fig. 2c). Below this threshold, order is maintained even when the government is not particularly popular (Fig. 2b).

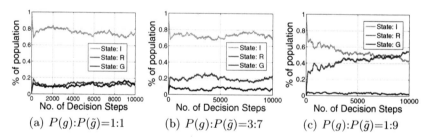

(a) $P(g):P(\tilde{g})$=1:1   (b) $P(g):P(\tilde{g})$=3:7   (c) $P(g):P(\tilde{g})$=1:9

Fig. 2. Effect of external events on population opinion without influence group

5.2   Effect of an Influence Group

An interesting question in these social experiments is whether the mass opinion can be affected by adding a group of agents in the network with a premeditated agenda. These agents passively broadcast their opinion about the values of different decision states. In this paper, the influence has been studied through simulations in two different situations.

In the first case, starting from an even distribution $P(g):P(\tilde{g})$=1:1, the proportion of unpopular acts are increased in two stages, $P(g):P(\tilde{g})$=3:7 and 1:9

respectively. In the second case, the proportion of popular acts exceed unpopular acts and $P(g){:}P(\tilde{g})$ is designed to be equal to 7:3 and 9:1. As a preliminary study, a 3 person group of external agents have been added to a total networked group of 100. This group randomly chooses people to form links, the probability of forming a link being 0.25. At decision step 5000, this influence group is activated, with all external agents initialized to state R. Themselves being in state R and broadcasting a high reward value for R, this group indirectly starts convincing its first order neighbors to increase their personal estimate of $\chi(R)$. Furthermore, in this particular setting, probability of success of the revolution $P(s)$ being linked to the percentage of the population in R, individuals develop a higher estimate of $P(s)$ as a result of being associated with these influence group members. Since the event probabilities affect the estimate of the state measure (μ_{norm}), more people probabilistically convert to state R. In fact, from the simulation a steady trickle of people are observed to convert to state R starting from step $t_k = 5000$. Gradually, the population converges to a state of revolution. As expected, when the government policies are unpopular, the opinion is already in favor of revolution, resulting in a faster transition (Fig. 3a, 3b and 3c) . But interestingly, even when the government decisions were more benevolent than harmful, inclusion of the influence group in the society can have destabilizing effects eventually leading to a social upheaval (Fig. 3d, 3e and 3f).

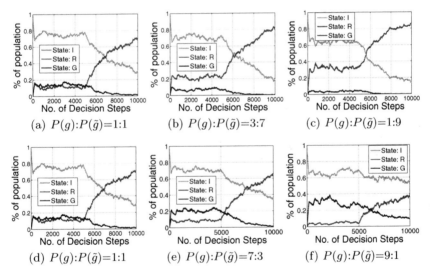

(a) $P(g){:}P(\tilde{g})=1{:}1$ (b) $P(g){:}P(\tilde{g})=3{:}7$ (c) $P(g){:}P(\tilde{g})=1{:}9$

(d) $P(g){:}P(\tilde{g})=1{:}1$ (e) $P(g){:}P(\tilde{g})=7{:}3$ (f) $P(g){:}P(\tilde{g})=9{:}1$

Fig. 3. Effect of external events on population opinion with $|I| = 3$

6   Conclusion and Future Work

This paper introduces a modeling paradigm based on a language-theoretic framework for stochastic simulation of a person's decision-making process in a social setting. It has been conjectured that, even though every individual in a network is influenced to by his/her social interactions, this is not blind following - rather, the influence is through altering the estimate of values or rewards to be obtained through different choices. Once these values are learned, individuals make their own optimal choices based on discounted reward maximization policies. In this paper, a specific problem which deals with the effect of long term government policies on the stability and dynamics of the population opinion has been studied. The effect of including an influencing agent group has also been observed. Specifically one case, where the influence group agenda matches the population sentiment, and another case, where it opposes, have been both reported. The results point to the existence of thresholds in the government policies below which it is possible to maintain an unpopular yet stable government, but above which, transition to an unstable state becomes inevitable. On one side of the threshold, the society ends up divided in opinion and reaches a multi-polar steady state. When the threshold is exceeded, the population converges to one unified opinion.

References

1. Greene, W.: Discrete Choice Modeling. In: Mills, T., Patterson, K. (eds.) The Handbook of Econometrics, Palgrave, London, vol. 2 (March 2009)
2. Bikhchandani, S., Hirshleifer, D., Welch, I.: A Theory of Fads, Fashion, Custom, and Cultural Change as Informational Cascades. The Journal of Political Economy 100(5), 992–1026 (1992)
3. Watts, D.J.: A simple model of global cascades on random networks. PNAS 99(9), 5766–5771 (2002), doi:10.1073/pnas.082090499
4. Dorogovtsev, S.N., Goltsev, A.V., Mendes, J.F.F.: Ising model on networks with an arbitrary distribution of connections. Physical Review E 66(1), 016104 (2002)
5. Klemm, K., Eguíluz, V.M., Toral, R., San Miguel, M.: Nonequilibrium transitions in complex networks: A model of social interaction. Physical Review E 67(2), 026120 (2003)
6. Kempe, D., Kleinberg, J., Tardos, É.: Maximizing the spread of influence through a social network. In: Proceedings of the Ninth ACM SIGKDD International Conference on Knowledge Discovery and Data Mining, pp. 137–146. ACM 1-58113-737-0
7. Ray, A.: Signed real measure of regular languages for discrete event supervisory control. International Journal of Control 78(12), 949–967 (2005)
8. Friedkin, N.E., Johnsen, E.C.: Social Influence Networks and Opinion Change. In: Advances in Group Processes, vol. 16, pp. 1–29. JAI Press Inc. (1999)
9. Hegselmann, R., Krause, U.: Opinion Dynamics and Bounded Confidence Models, Analysis and Simulation. Journal of Artificial Societies and Social Simulation 5 (2002), http://ideas.repec.org/a/jas/jasssj/2002-5-2.html
10. Barabási, A., Albert, R.: Emergence of scaling in random networks. Science 286(5439), 509–512 (1999)
11. Batagelj, V., Mrvar, A.: Pajek - Analysis and Visualization of Large Networks. In: Juenger, M., Mutzel, P. (eds.) Graph Drawing Software. Mathematics and Visualization, pp. 77–103. Springer, Heidelberg (2003) ISBN: 3-540-00881-0

A Comparative Study of Smoking Cessation Intervention Programs on Social Media

Mi Zhang, Christopher C. Yang, and Jiexun Li

College of Information Science and Technology, Drexel University,
19104 Philadelphia PA, United States
{Mi.Zhang,Chris.Yang,Jiexun.Li}@drexel.edu

Abstract. As an alternative to traditional face-to-face counseling and support group, social media has become a new venue for health intervention programs. Different online intervention channels have their own characteristics and advantages. This study focuses on comparing two types of online intervention channels, i.e., forum and Facebook page, for smoking cessation using social network analysis techniques. QuitNet is a popular online intervention program for smoking cessation. In this study, we collected data from QuitNet's forum and its Facebook group page and constructed two social networks, respectively. We analyzed and compared these two social networks with a focus on users' quit statuses and communication patterns in the two different channels.

Keywords: Social Network Analysis (SNA), Social Support, Health Informatics, Smoking, Tobacco, Smoking Cessation, QuitNet.

1   Introduction

It has been demonstrated for a long time that traditional interventions such as face-to-face consulting are effective methods for smoking cessation [1]. However, these traditional methods rely on face-to-face conversation and cannot reach a large amount of people at the same time. With the development of Internet, many online smoking cessation intervention programs have been created. Research shows that these online intervention programs are less effective than face-to-face consulting for individuals but they could reach much more smokers [2].

QuitNet founded in 1995 is one of the most influential online smoking cessation programs. It provides different services to support people to quit smoking and attracts a large number of users. An et al. [3] analyzed the characteristics and behaviors of users on QuitNet, and found it effective to promote smoking cessation. QuitNet owns 10 forums on its website which are professional communities for users to discuss issues on smoking cessation. Recently, QuitNet also built communities on other social media channels like Facebook and Twitter. In this paper, we investigate and compare the user characteristic of quit status, as well as user behaviors on QuitNet Forum (namely The Quitstop)[1] and QuitNet Facebook[2].

---

[1] http://forums.quitnet.com/aspBanjo/Message_List.asp?Conference_ID=10&Forum_ID=8&r=122226
[2] https://www.facebook.com/QuitNet

S.J. Yang, A.M. Greenberg, and M. Endsley (Eds.): SBP 2012, LNCS 7227, pp. 87–96, 2012.

A lot of research has been carried out to study online smoking cessation intervention programs. There are three aspects that most research investigates: user characteristics, program effectiveness, and discussion topics. According to previous research, active users of online intervention programs tend to be female, older, and abstinent of smoking [4], [5]; most online smoking cessation programs are shown effective [2], [3], [6], [7]; and different topics are being discussed online and attract different concentrations [8], [9].

In previous research, data of user behavior are collected through surveys, experiments or Internet contents. Survey [4], [6], [7] and randomized controlled trial [2], [10], [11] are the most common methods to acquire data. Surveys can acquire user information about their characteristics, whereas controlled trials can explore and compare user behaviors under different situations. However, both of the two methods are time-consuming and only refer to a small proportion of users. Another method for data collection is to extract information directly from website contents of online communities, including posts and comments published by users [5], [8], [9]. This method could get raw data of all user behavior during a certain period of time. However, this kind of information is not thoroughly analyzed or fully utilized in most current research. Recently, Coob & Abrams [5] extracted a social network based on data of QuitNet, including information of exchanged messages, buddy list and posts on forums. They studied characteristics of the whole social network as well as five subgroups of it. It was the first study which applied formal social network analysis to online smoking cessation programs.

QuitNet Forum and QuitNet Facebook are smoking cessation intervention communities based on different social media channels. They have their own advantages and shortcomings. As a professional online smoking cessation program, QuitNet Forum is better organized and managed. Some experts are invited to take part in QuitNet Forum regularly so that they could provide professional suggestions for users. In addition, users of QuitNet Forum are much more motivated so they perform more active than users of QuitNet Facebook. Their discussion topics are usually more targeted.

QuitNet Facebook provides a public page on Facebook which is a convenient way for people to communicate on the topic of smoking cessation. Once a Facebook user "likes" the QuitNet Facebook page, she will receive all its discussion threads. Any Facebook user can participate in Quitnet discussion easily while they are interacting with other Facebook friends as their regular online social activities. Facebook has been a popular social media platform for information dissemination and consumer opinion collection. In contrary, QuitNet Forum requires users to register a specific account with a password and go to a specific website to participate in discussion.

There are different self-report measurements for smoking cessation outcomes. Point prevalent abstinence, continuous abstinence and prolonged abstinence are three basic classes of self-report measures [12] [13]. A person is regarded as either "smoking" or "abstinence" according to if he smokes in a certain period of time in these measurements. However, it is difficult to develop a single outcome measure to employ in all cases [13]. In this study, we do not adopt an absolute outcome of either "smoking" or "abstinence" to measure smoking quitters. We investigate quit status - the period length of abstinence as a user characteristic and compare it in different online health intervention programs. Particularly, "quit status" is defined as the number of days that a former smoker has been abstinent from the day he stopped smoking to the day he posted the last message on Quitnet Forum or Quitnet Facebook. We are particularly interested in this characteristic of users as a measure of health outcome in these online intervention programs.

In this research we are aimed at answering the following two research questions:

Question 1. How are users of QuitNet Forum and Facebook different in terms of their quit statuses?

Question 2. How are users interacting with each other on QuitNet Forum and Facebook? And how are these interaction patterns related to their quit statuses?

2   Data Collection and Description

Both the QuitNet Forum and its Facebook page are publically available online, where we collected data for our study. In each QuitNet forum, every registered user can start new threads and comment on others' posts. On QuitNet's Facebook page, only the creator of the public page has the privilege to initiate a new thread on the wall, while other Facebook users who concern this public page can only comment on these posts. Therefore, there are much fewer posts and comments on the Facebook page than on QuitNet Forum. For most users of Quitnet Forum, their quit statuses can be acquired from their profile pages. On its Facebook page, QuitNet launches a post on every Friday, calling users to report their quit statuses. This allows us to collect and analyze most users' valid quit statuses in these two social media channels.

In this study, we collected a 1-month (05/01/2011~05/31/2011) data set of user posts and comments from Quitnet Forum "The Quitstop". In order to construct a social network with sufficient and comparable size (number of users with valid quit status), we collected a 3-month (04/01/2011~06/31/2011) data set from the QuitNet Facebook page. Our study focuses on comparing user quit statuses of two channels, so collecting data of similar user numbers with valid quit statuses is important for two channels. Our study also investigates user interactions within each channel separately, so the different periods of data for each channel would not influence the correctness and efficiency of analysis. Table 1 summarizes some descriptive statistics of the two datasets.

Table 1. Data descriptions

	Time period	# of users	# of users with valid quit status	# of posts	# of comments
QuitNet Forum	05/01/2011-05/31/2011	1,169	534	3,017	24,713
QuitNet Facebook	04/01/2011-06/31/2011	664	394	111	2,574

3   Methods

To answer the first research question, we analyze the quit statuses of users on QuitNet Forum and QuitNet Facebook. We use non-parametric statistical tests to compare them. According to the quit statuses, users are divided into five groups as shown in Table 3. Each group includes users who have been abstinent for a certain period of time. In former research, it was found that smokers move a series of stages to quit smoking, including precontemplation, contemplation, preparation, action and maintenance [12]. Action is the period that people take real actions to quit smoking. It includes an early action period of 0 to 3 month, and a later action period of 3 to 6 month. Our first two groups in Table 3 respectively represent the early and later action periods. Maintenance is the period beginning 6 months after action started. Velicer etc. recommended 6 to 60 months as the duration of the maintenance stage [12]. The third and fourth groups in Table 3 present the maintenance stage. The last group presents stage after maintenance.

To answer the second research question, we construct and analyze two undirected graphs to represent the social networks of QuitNet Forum and QuitNet Facebook, respectively. In these two social networks, each actor (or node) represents a user. Every two actors who have participated in a same post are connected by a tie (or link). Each tie carries a weight representing the frequency that the two actors participate in a same post. The link structure of social networks reflects the interaction and communication patterns among the actors. In particular, we use the following social network analysis metrics or methods to investigate the two social networks [14], [15].

(1) Density:

Density of a network reflects how closely actors in a social network are interacting with each other. This metric is defined by formula (1):

$$\Delta' = \frac{2L}{W_{max}*g(g-1)} \qquad (1)$$

where L represents the sum of weights of all ties, g represents the number of actors in the social network, and W_{max} is the maximum weight of all ties. $0 \le \Delta' \le 1$.

(2) Degree centrality ($C_A(n_i)$):

Node degree centrality of an actor is defined as the sum of weights of ties linked to it. An actor with a high degree has more interactions with others in the network.

(3) Group centralization:

Group centralization is an index that measures the variance of the individual actors' degrees in the social network. It is defined by formula (2).

$$C_A' = \frac{\sum_{i=1}^{g}(C_A(n^*)-C_A(n_i))}{W_{max}*max\sum_{i=1}^{g}(c_A(n^*)-c_A(n_i))} \qquad (2)$$

where $C_A(n^*)$ is the largest degree centrality among all g actors in the network; $\sum_{i=1}^{g}(C_A(n^*) - C_A(n_i))$ is the sum of difference between the largest degree and other actors' degree; $max\sum_{i=1}^{g}(c_A(n^*) - c_A(n_i))$ represents the theoretical maximum sum of differences in actor degree taken over all possible binary graphs with g actors; and W_{max} is the maximum weight of all ties. $0 \le C_A' \le 1$.

Furthermore, we use non-parametric tests and ANOVA with post-hoc tests to compare the degree centrality of individual actors across different quit status groups.

(4) Core/periphery structure:

Core/periphery analysis divides actors in a social network into two discrete groups, a core with a high density of ties and a periphery with a low density of ties [15]. Actors in the core are more actively and closely interacting with each other and tend to be leaders or active participants of online discussions. We also analyze and compare the quit statuses of core actors in both social networks.

In our study, we use SPSS 19 for all statistical analyses and we use a popular tool named UCINET 6 for social network analysis.

4   Results and Discussion

4.1   Comparing Quit Status of Users in the Two Channels

Table 2 shows some descriptive statistics of quit statuses of users on QuitNet Forum and QuitNet Facebook. For the 534 users of QuitNet Forum and 394 users of QuitNet Facebook with valid quit status, the ranges of quit status are [0, 4327] and [0, 4063], respectively. The average and median of quit status of QuitNet Forum are 558.57 and 127; and those of QuitNet Facebook are 567.93 and 234.5. One-sample Kolmogorov-Smirnov tests show that quit status of users in either QuitNet Forum or QuitNet Facebook does not follow a normal distribution. Therefore, we adopt non-parametric methods to test the difference of quit statuses of users between the two channels. Wilcoxon-Mann-Whitney test, Kruskal Wallis test, and Independent Sample Median test all show that there are significant differences on the distribution and median of quit statuses of users from QuitNet Forum and QuitNet Facebook (p-values < 0.001).

Furthermore, we divide the users of the two channels into five groups according to their quit statuses. Table 3 shows the number and percentage of users in each group for the two channels. We observe that the largest group (43.4%) of users of QuitNet Forum includes new quitters who have just been abstinent for no more than 90 days. By contrast, users of QuitNet Facebook are more equally distributed in the five groups.

Table 2. Quit Status of Users

	# of users with valid quit status	Statistics of Quit Status			
		Minimum	Maximum	Mean	Median
QuitNet Forum	534	0	4327	558.57	127
QuitNet Facebook	394	0	4063	567.93	234.5

Table 3. User groups based on quit status for QuitNet Forum and QuitNet Facebook

Quit Status(days)		0-90	91-180	181-720	721-1800	>1800
QuitNet Forum	# of Users	232	69	100	72	61
	Percentage	43.4%	12.9%	18.7%	13.5%	11.4%
QuitNet Facebook	# of Users	98	78	114	70	34
	Percentage	24.9%	19.8%	28.9%	17.8%	8.6%

Both analyses show that users of QuitNet Forum have been abstinent for a shorter period of time than those on QuitNet Facebook. More specifically, almost half of the forum users have only been abstinent for no more than 90 days, whereas users of QuitNet Facebook are distributed more equally at different abstinent stages. In other words, QuitNet Forum tends to attract new quitters, while QuitNet Facebook attracts users with a wider range of smoking statuses. One possible explanation is that new quitters usually need more professional supports and services, which are offered on QuitNet Forum. For people who have been abstinent for a longer time, their enthusiasm and needs for professional supports may decrease, so they are less willing to log in and participate in discussions on QuitNet Forum. By contrast, for users of QuitNet Facebook, Facebook is a platform for them to interact and communicate with their friends and social groups. Even though QuitNet may not be their only interest on Facebook, they receive news feeds from QuitNet (e.g., asking them to report their quit status) regularly and it takes them minimum effort to participate in communications on QuitNet Facebook page. Therefore, smoking quitters at various abstinent stages, who may or may not have strong interest in cessation intervention, are shown more equally on QuitNet Facebook.

4.2   Centrality Analysis of Social Networks

Figure 1 shows the two social networks of QuitNet Forum and QuitNet Facebook. The social network of QuitNet Forum has 1,169 actors and 40,540 ties with a density of 0.15%. The social network of QuitNet Facebook has 664 actors and 19,939 ties in total, and the density is 1.48%. It is difficult to visualize large social networks as shown in Figure 1 although the densities are not high. However, the social network analysis in this study reflects the properties of these two social networks.

Table 4 summarizes the statistics of node-level degree of actors and group centralizations of the two social networks. The social network of QuitNet Forum has a much higher average degree centrality (213.406) as compared to that of QuitNet Facebook (78.610). This reflects that users are interacting with each other more closely on QuitNet Forum than on its Facebook page. However, the group centralization of QuitNet Facebook (10.174%) is more than three times higher than that of QuitNet Forum (2.959%). Group centralization reflects the variance of actors' degree centralities in a social network. The higher group centralization for QuitNet Facebook indicates that this social network is more centralized around a small number of users who tend to be actively interacting with others. In contrast, the social network for QuitNet Forum is more decentralized. This may be due to the fact that users on QuitNet Forum tend to share a common goal of seeking support for smoking cessation and be equally active in online discussion. Facebook users, on the other hand, have a wider variety of interests and habits and therefore exhibit greater variance in participating in online communications about smoking cessation.

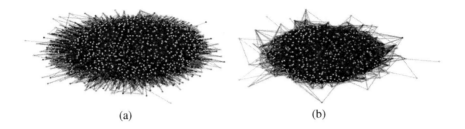

(a)   (b)

Fig. 1. (a) the social network of QuitNet Forum and (b) the social network of QuitNet Facebook

Table 4. Statistics of Degree Centrality

	Mean	Maximum	Minimum	Group centralization
QuitNet Forum	213.406	8735	1	2.959%
QuitNet Facebook	78.610	1153	1	10.174%

Table 5. Statistics of Degree Centrality in Each Group

Quit Status(days)		0-90	91-180	181-720	721-1800	>1800
QuitNet	Mean	406.03	269.68	308.44	294.00	306.21
Forum	Median	67	61	42	32.5	54
Facebook	Mean	88.12	116.87	127.96	117.78	84
QuitNet	Median	67	80	68.5	71.5	37.5

A series of non-parametric tests are implemented to analyze the differences of degree of actors in the five different quit status groups. For both networks, the mean and median of degree centralities of actors in each quit status group are shown in Table 5. Our statistical tests show that, for actors of QuitNet Forum, there is no significant difference on degree across the five groups. For actors of QuitNet Facebook, the p-values of Kruskal-Wallis test and Independent Sample Median test are 0.031 and 0.021, which indicates significant difference on degree centrality of actors across the five groups. Furthermore, we used ANOVA with LSD post-hoc test to analyze the difference on degree means between the five groups of actors in QuitNet Facebook but did not find significant difference (p-value > 0.1). However, the LSD post-hoc tests show significant differences between Group 1 (0~90 days) and Group 3 (181~720 days) with p-value = 0.035, and between Group 3 (181~720 days) and Group 5 (>1800 days) with p-value = 0.1. The p-values between other groups are all above 0.1. This indicates that the actors of Group 3 (181~720 days) are likely to have higher degree centrality than those of other groups.

4.3   Core/Periphery Structure of Social Networks

Using core/periphery structure analysis on the social network of QuitNet Forum, we identify 46 core actors and 42 of them have a valid quit status. For QuitNet Facebook, we identify 67 core actors and 66 of them have a valid quit status. The distributions of the core users in the two social networks according to their quit statuses are shown in Table 6.

Table 6. Core User Groups Based on Quit Status for QuitNet Forum and QuitNet Facebook

Quit Status(days)		0-90	91-180	181-720	721-1800	>1800
QuitNet Forum	User Number	21	4	7	5	5
	Proportion	50.0%	9.5%	16.7%	11.9%	11.9%
QuitNet Facebook	User Number	12	16	23	11	4
	Proportion	18.2%	24.2%	34.8%	16.7%	6.1%

The core/periphery structures of the two social networks concur with our previous findings. On QuitNet Forum, half of the core actors are new quitters who have been abstinent for less than three months. It is about the same proportion of total users on QuitNet Forum. On QuitNet Facebook, users in different quit status groups have significantly different degree centralities. In particular, users in Group 3 (181~720 days) have the highest centrality. Table 6 also shows that most core actors of QuitNet Facebook belong to Group 3. It is an interesting finding that the most interactive users on QuitNet Facebook are those who have been abstinent for about half a year to two years. People of this group are in the early period of maintenance stage of smoking cessation. They have finished action stage and already been abstinent for a relatively long period of time Compared to new smoking quitters at action stage, they are in a stable status, and may not have urgent needs for professional and medical support. Compared to those who have been abstinent for more than two years, people in Group 3 have not completely accomplished the process of smoking cessation. Therefore, they still keep interested in cessation related discussion, including seeking emotional support from long-term quitters and offering help to new quitters. So with less barrier of user participation (logging in to a specific Web site), QuitNet Facebook can better fit their characteristics and attract their participation.

5   Conclusion

In this study we use statistics and social network analysis methods to compare user quit statuses and communication behaviors of two different online channels of a smoke cessation intervention program QuitNet. Our analyses show that QuitNet Forum attracts more new smoking quitters whereas on QuitNet Facebook users are more equally distributed across different stages. While on QuitNet Forum users are interacting with each other more closely, QuitNet Facebook shows a more centralized structure with a small number of users leading communications. In particular, the core users on QuitNet Facebok are mostly those who are at the early period of maintenance stage for smoking cessation.

According to user features of different smoking cessation programs, we could improve current online communities to provide better services to facilitate intervention. QuitNet Forum could focus on smoking quitters at early stages and provide them supports from different perspectives. QuitNet Facebook should take full advantages of "passive" messages because most users receive them on their Facebook walls when they log on Facebook every day. In addition, QuitNet Facebook could take certain measures to promote interactions among users, especially interactions between users at different quit stages.

There are several limitations in this study. First, since the quit status of users is calculated based on information extracted from sources online, some may not be accurate. Second, other than quit status, user characteristics such as gender and age are not considered in this study. In our future research, we plan to adopt survey and other methods for data collection so as to incorporate other user characteristics for more comprehensive analysis. By comparing health intervention via different social media channels, we can better understand their characteristics and advantages for implementing more effective intervention programs.

References

1. Russell, M.A.H., Wilson, C., Taylor, C., Baker, C.: Effect of General Practitioners' Advice Against Smoking. British Medical Journal 2, 231–235 (1979)
2. Shahab, L., McEwen, A.: Online Support for Smoking Cessation: A Systematic Review of the Literature. Addiction 104, 1792–1804 (2009)
3. An, L.C., et al.: Utilization of Smoking Cessation Informational, Interactive, and Online Community researches as Predictors of Abstinence: Cohort Study. J. Med. Internet Res. 10, e55 (2008)
4. Cobb, N.K., Graham, A.L.: Characterizing Internet Searchers of Smoking Cessation Information. J. Med. Internet Res. 8, e17 (2006)
5. Cobb, N.K., Graham, A.L., Abrams, D.B.: Social Network Structure of a Large Online Community for Smoking Cessation. American Journal of Public Health 100, 1282–1289 (2010)
6. Cobb, N.K., Graham, A.L.: Initial Evaluation of A Real-World Internet Smoking Cessation System. Nicotine Tob. Res. 7, 207–216 (2005)
7. Graham, A.L., Cobb, N.K., Raymond, L., Still, S., Young, J.: Effectiveness of An Internet-Based Worksite Smoking Cessation Intervention at 12 Months. Journal of Occupational and Environmental Medicine 49, 821–828 (2007)
8. Burri, M., Baujard, V., Etter, J.F.: A Qualitative Analysis of An Internet Discussion Forum for Recent Ex-Smokers. Nicotine Tob. Res. 8, 13–19 (2006)
9. Selby, P., van Mierlo, T., Voci, S.C., Parent, D., Cunningham, J.A.: Online Social and Professional Support for Smoker Trying to Quit: An Exploration of First Time Posts From 1562 Members. J. Med. Internet Res. 12, e34 (2010)
10. Japuntich, S.J., et al.: Smoking Cessation via the Internet: A Randomized Clinical Trial of an Internet Intervention as Adjuvant Treatment in a Smoking Cessation Intervention. Nicotine Tob. Res. 8, 59–67 (2006)
11. Pike, K.J., Rabius, V., McAlister, A., Geiger, A.: American Caner Society's QuitLink: Randomized Trial of Internet Assistance. Nicotine Tob. Res. 9, 415–420 (2007)

12. Velicer, W.F., Prochaska, J.O., Rossi, J.S., Snow, M.: Assessing Outcome In Smoking Cessation Studies. Psychological Bulletin 111, 23–41 (1992)
13. Velicer, W.F., Prochaska, J.O.: A Comparison of Four Self-report Smoking Cessation Outcome Measures. Addictive Behaviors 29, 51–60 (2004)
14. Wasserman, S., Faust, K.: Social Network Analysis: Methods and Applications. Cambridge University Press, New York (1994)
15. Borgatti, S.P., Everet, M.G.: Models of corerperiphery structures. Social Networks 21, 375–395 (1999)
16. Graham, A.L., Bock, B.C., Cobb, N.K., Niaura, R., Abrams, D.B.: Characteristics of Smokers Reached and Recruited to An Internet Smoking Cessation Trial: A Case of Denominators. Nicotine Tob. Res. 8, 43–48 (2006)

Trends Prediction Using Social Diffusion Models

Yaniv Altshuler, Wei Pan, and Alex (Sandy) Pentland

MIT Media Lab
{yanival,panwei,sandy}@media.mit.edu

Abstract. The importance of the ability to predict trends in social media has been growing rapidly in the past few years with the growing dominance of social media in our everyday's life. Whereas many works focus on the detection of anomalies in networks, there exist little theoretical work on the prediction of the likelihood of anomalous network pattern to globally spread and become "trends". In this work we present an analytic model for the social diffusion dynamics of spreading network patterns. Our proposed method is based on information diffusion models, and is capable of predicting future trends based on the analysis of past social interactions between the community's members. We present an analytic lower bound for the probability that emerging trends would successfully spread through the network. We demonstrate our model using two comprehensive social datasets — the *Friends and Family* experiment that was held in MIT for over a year, where the complete activity of 140 users was analyzed, and a financial dataset containing the complete activities of over 1.5 million members of the *eToro* social trading community.

Keywords: Trends Prediction, Information Diffusion, Social Networks.

1 Introduction

We live in the age of social computing. Social networks are everywhere, exponentially increasing in volume, and changing everything about our lives, the way we do business, and how we understand ourselves and the world around us. The challenges and opportunities residing in the social oriented ecosystem have overtaken the scientific, financial, and popular discourse.

In this paper we study the evolution of trend spreading dynamics in social networks. Where there have been numerous works studying the topic of anomaly detection in networks (social, and others), literature still lacks a theoretic model capable of predicting *how do network anomalies evolve*. When do they spread and develop into global trends, and when are they merely statistical phenomena, local fads that get quickly forgotten? We give an analytically proven lower bound for the spreading probability, capable of detecting "future trends" – spreading behavior patterns that are likely to become prominent trends in the social network.

We demonstrate our model using social networks from two different domains. The first is the *Friends and Family* experiment [1], held in MIT for over a year, where the complete activity of 140 users was analyzed, including data

S.J. Yang, A.M. Greenberg, and M. Endsley (Eds.): SBP 2012, LNCS 7227, pp. 97–104, 2012.

concerning their calls, SMS, MMS, GPS location, accelerometer, web activity, social media activities, and more. The second dataset contains the complete financial transactions of the *eToro community* members – the world's largest "social trading" platform, allowing users to trade in currency, commodities and indices by selectively copying trading activities of prominent traders.

The rest of the paper is organized as follows : Section 2 discusses related works. The information diffusion model is presented in Section 3 and its applicability is demonstrated in Section 4, while concluding remarks are given in Section 5.

2   Related Work

Diffusion Optimization. Analyzing the spreading of information has long been the central focus in the study of social networks for the last decade [2] [3]. Researchers have explored both the offline networks structure by asking and incentivizing users to forward real mails and E-mails [4], and online networks by collecting and analyzing data from various sources such as *Twitter* feeds [5].

The dramatic effect of the network topology on the dynamics of information diffusion in communities was demonstrated in works such as [6] [7]. One of the main challenges associated with modeling of behavioral dynamics in social communities stems from the fact that it often involves stochastic generative processes. While simulations on realizations from these models can help explore the properties of networks [8], a theoretical analysis is much more appealing and robust. The results presented in this work are based on a pure theoretical analysis.

The identity and composition of an initial "seed group" in trends analysis has also been the topic of much research. Kempe et al. applied theoretical analysis on the seeds selection problem [9] based on two simple adoption models: *Linear Threshold Model* and *Independent Cascade Model*. Recently, Zaman et al. developed a method to trace rumors back in the topological spreading path to identify sources in a social network [10], and suggest that methods can be used to locate influencers in a network. Some scholars express their doubts and concerns for the influencer-driven viral marketing approach, suggesting that "everyone is an influencer" [11], and companies "should not rely on it" [12]. They argue that the content of the message is also important in determining its spreads, and likely the adoption model we were using is not a good representation for the reality.

Our work, on the other hand, focuses on predicting emerging trends given a current snapshot of the network and adoption status, rather than finding the most influential nodes. We provide a lower bound for the probability that an emerging trend would spread throughout the network, based on the analysis of the diffusion process outreach, which is largely missing in current literature.

Adoption Model. A fundamental building block in trends prediction that is not yet entirely clear to scholars is the adoption model, modeling individuals' behavior based on the social signals they are exposed to. Centola has shown both theoretical and empirically that a complex contagion model is indeed more precise for diffusion [13, 14]. Different adoption models can dramatically alter the model outcome [15]. In fact, a recent work on studying mobile application diffusions using mobile phones demonstrated that in real world the diffusion process is a far more complicated phenomenon, and a more realistic model was proposed in [16]. Our results also incorporate this realistic diffusion model.

Trends Prediction and Our Proposed Model. In this work we study the following question : Given a snapshot of a social network with some behavior occurrences (i.e. an emerging trend), what is the probability that these occurrences (seeds) will result in a viral diffusion and a wide-spread trend (or alternatively, dissolve into oblivion). Though this is similar to the initial seed selection problem [9], we believe that the key factor to succeed in a viral marketing campaign optimization is a better analytical model for the diffusion process itself.

The main innovation of our model is the fact that it is based on a fully analytical framework with a scale-free network model. Therefore, we manage to overcome the dependence on simulations for diffusion processes that characterizes most of the works in this field [6, 17]. We are able to do so by decomposing the diffusion process to the transitive random walk of "exposure agents" and the local adoption model based on [16]. While there are some works that analyze scale-free network [18] most of them come short to providing accurate results, due to the fact that they calculate the expected values of the global behaviors dynamics. Due to strong "network effect" however, many real world networks display much less coherent patterns, involving local fluctuations and high variance in observed parameters, rendering such methods highly inaccurate and sometimes impractical. Our analysis on the other hand tackles this problem by modeling the diffusion process on scale-free networks in a way which takes into account such interferences, and can bound their overall effect on the network.

3   Trend Prediction in Social Networks

One of the main difficulties of trends-prediction stems from the fact that the first spreading phase of "soon to be global trends" demonstrates significant similarity to other types of anomalous network patterns. In other words, given several observed anomalies in a social network, it is very hard to predict which of them would result in a wide-spread trend and which will quickly dissolve into oblivion.

We model the community, or social network, as a graph G, that is comprised of V (the community's members) and E (social links among them). We use n to denote the size of the network, namely $|V|$. In this network, we are interested in predicting the future behavior of some observed anomalous pattern a. Notice that a can refer to a growing use of some new web service such as *Groupon*, or alternatively a behavior such as associating oneself with the "*99% movement*".

Notice that "exposures" to trends are transitive. Namely, an "exposing" user generates "exposure agents" which can be transmitted on the network's social links to "exposed users", which can in turn transmit them onwards to their friends, and so on. We therefore model trends' exposure interactions as movements of random walking agents in a network, assuming that very user that was exposed to a trend a generates β such agents, on average.

We assume that our network is (or can be approximated by) a scale free network $G(n, c, \gamma)$, namely, a network of n users where the probability that user v has d neighbors follows a power law :

$$P(d) \sim c \cdot d^{-\gamma}$$

We also define the following properties of the network :

Definition 1. *Let $V_a(t)$ denote the group of network members that at time t advocate the behavior associated with the potential trend a.*

Definition 2. *Let us denote by $\beta > 0$ the average "diffusion factor" of a trend a. Namely, the average number of friends a user who have been exposed to the trend will be talking about the trend with (or exposing the trend in other ways).*

Definition 3. *Let P_Δ be defined as the probability that two arbitrary members of the network vertices have degrees ratio of Δ or higher :*

$$P_\Delta \triangleq Prob\,[deg(u) > \Delta \cdot deg(v)]$$

Definition 4. *We denote by σ_- the "low temporal resistance" of the network :*

$$\forall t, \Delta_t \quad , \quad \sigma_- \triangleq \max\left\{ 1 \le \Delta \;\middle|\; 1 - e^{-\Delta \cdot \frac{\beta \Delta_t \cdot |V_a(t)|}{n}} \cdot (1 - P_\Delta) \right\}$$

Definition 5. *Let $P_{Local-Adopt}(a, v, t, \Delta_t)$ denote the probability that at time $t + \Delta_t$ the user v had adopted trend a (for some values of t and Δ_t). This probability may be different for each user, and may depend on properties such as the network's topology, past interactions between members, etc.*

Definition 6. *Let P_{Local} denote that expected value of the local adoption probability throughout the network :*

$$P_{Local} = \underset{u \in V}{E}\,[P_{Local-Adopt}(a, u, t, \Delta_t)]$$

Definition 7. *Let us denote by $P_{Trend}(\Delta_t, \frac{V_a(t)}{n}, \varepsilon)$ the probability that at time $t + \Delta_t$ the group of network members that advocate the trend a has at least $\varepsilon \cdot n$ members (namely, that $|V_a(t + \Delta_t)| \ge \varepsilon \cdot n$).*

We assume that the seed group of members that advocate a trend at time t is randomly placed in the network. Under this assumption we can now present the main result of this work : the lower bound over the prevalence of an emerging trend. Note that we use $P_{Local-Adopt}$ as a modular function in order to allow future validation in other environments. The explicit result is given in Theorem 2.

Theorem 1. *(See [19] for a complete proof) For any values of Δ_t, $|V_a(t)|$, n and ε, the probability that at time $t + \Delta_t$ at least ϵ portion of the network's users would advocate trend a is lower bounded as follows :*

$$P_{Trend}\left(\Delta_t, \frac{|V_a(t)|}{n}, \varepsilon\right) \ge P_{Local}^{\varepsilon \cdot n} \cdot \left(1 - \Phi\left(\frac{\sqrt{n} \cdot (\varepsilon - \tilde{P}_-)}{\sqrt{\tilde{P}_-(1 - \tilde{P}_-)}}\right)\right)$$

where :

$$\tilde{P}_- = e^{-\left(\frac{\Delta_t \cdot \sigma_-}{2} - \rho_{opt_-} + \frac{\rho_{opt_-}^2}{2\Delta_t \cdot \sigma_-}\right)}$$

and where :

$$\rho_{opt_-} \triangleq \underset{\rho}{\mathrm{argmin}} \left(P_{Local}^{\varepsilon \cdot n} \cdot P_{Trend} \left(\Delta_t, \frac{|V_a(t)|}{n}, \varepsilon \right) \right)$$

and provided that :

$$\rho_{opt_-} < \Delta_t \cdot \sigma_-$$

and as σ_- depends on P_Δ, using the following bound :

$$\forall v, u \in V \quad , \quad P_\Delta \le \frac{c^2 \cdot \Delta^{1-\gamma}}{2\gamma^2 - 3\gamma + 1}$$

Recent studies examined the way influence is being conveyed through social links. In [16] the probability of network users to install applications, after being exposed to the applications installed by the friends, was tested. This behavior was shown to be best modeled as follows, for some user v :

$$P_{Local-Adopt}(a, v, t, \Delta_t) = 1 - e^{-(s_v + p_a(v))} \tag{1}$$

Exact definitions and methods of obtaining the values of s_v and $w_{v,u}$ can be found in [16]. The intuition of these network properties is as follows :

For every member $v \in V$, $s_v \ge 0$ captures the individual susceptibility of this member, regardless of the specific behavior (or trend) in question. $p_a(v)$ denotes the *network potential* for the user v with respect to the trend a, and is defined as the sum of network agnostic *"social weights"* of the user v with the friends exposing him with the trend a.

Notice also that both properties are trend-agnostic. However, while s_v is evaluated once for each user and is network agnostic, $p_a(v)$ contributes network specific information and can also be used by us to decide the identity of the network's members that we would target in our initial campaign.

Using Theorem 1 we can now construct a lower bound for the success probability of a campaign, regardless of the specific value of ρ :

Theorem 2. *(See a complete proof in [19]) For any values of Δ_t, $|V_a(t)|$, n and ε, the probability that at time $t + \Delta_t$ at least ϵ portion of the network's users advocate the trend a is :*

$$P_{Trend} \left(\Delta_t, \frac{|V_a(t)|}{n}, \varepsilon \right) \ge e^{-\varepsilon \cdot n \cdot \xi_G \cdot \xi_N^{\rho_{opt_-}}} \cdot \left(1 - \Phi \left(\sqrt{n} \cdot \frac{\varepsilon - \tilde{P}_-}{\sqrt{\tilde{P}_-(1 - \tilde{P}_-)}} \right) \right)$$

where :

$$\tilde{P}_- = e^{-\left(\frac{\Delta_t \cdot \sigma_-}{2} - \rho_{opt_-} + \frac{\rho_{opt_-}^2}{2\Delta_t \cdot \sigma_-}\right)}$$

and where :

$$\rho_{opt_-} \triangleq \underset{\rho}{\text{argmin}} \left(e^{-\varepsilon \cdot n \cdot \xi_G \cdot \xi_N^\rho} \cdot P_{Trend}\left(\Delta_t, \frac{|V_a(t)|}{n}, \varepsilon \right) \right)$$

and provided that :

$$\rho_{opt_-} < \Delta_t \cdot \sigma_-$$

and where ξ_G denotes the network's adoption factor and ξ_N denotes the network's influence factor :

$$\xi_G = e^{-\frac{1}{n} \sum_{v \in V} s_v} \qquad , \qquad \xi_N = e^{-\frac{1}{n} \sum_{e(v,u) \in E} \left(\frac{w_{u,v}}{|\mathcal{N}_v|} + \frac{w_{v,u}}{|\mathcal{N}_u|} \right)}$$

4   Experimental Results

We have validated our model using two comprehensive datasets, the *Friends and Family* dataset that studied the casual and social aspects of a small community of students and their friends in Cambridge, and the *eToro* dataset — the entire financial transactions of over 1.5M users of a "social trading;; community.

The datasets were analyzed using the model given in [16], based on which we have experimentally calculated the values of β, ξ_G, ξ_N and σ_-.

Figures 1 and 2 demonstrate the probabilistic lower bound for trend emergence, as a function of the overall penetration of the trend at the end of the time period, under the assumption that the emerging trend was observed in 5% of the population. In other words, for any given "magnitude" of trends, what is the probability that network phenomena that are being advocated by 5% of the network, would spread to this magnitude. Notice that although a longer spreading time slightly improve the penetration probability, the "maximal outreach" of the trend (the maximal rate of global adoption, with sufficient probability) is dominated by the topology of the network, and the local adoption features.

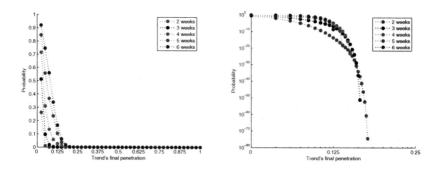

Fig. 1. Trends spreading potential in the *eToro* network, for various penetration rates. Initial seed group is defined as 5% of the population. Each curve represents a different time period, from 2 weeks to 6 weeks.

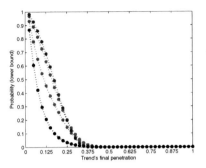
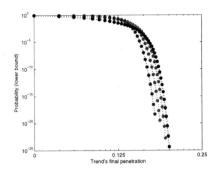

Fig. 2. Trends spreading potential in the *Friends and Family* network, for various penetration rates. Initial seed group is defined as 5% of the population. Each curve represents a different time period, from 2 weeks to 5 weeks.

5   Conclusions and Future Work

In this work we have discussed the problem of trends prediction, that is — observing anomelous network patterns and predicting which of them would become a prominent trend, spreading successfully throughout the network. We have analyzed this problem using information diffusion techniques, and have presented a lower bound for the probability of a pattern to become a global trend in the network, for any desired level of spreading. In order to model the local interaction between members, we have used the results from [16] that studied the local social influence dynamics between members of social networks.

Though our work provides a comprehensive theoretical framework to understand trends diffusion in social networks, there are still a few challenges ahead. For example, we wish to extend our model to other network models such as Erdos-Renyi random networks, as well as Small World networks. This is essential as more evidences are suggesting that some communities involve complex structures that cannot be easily approximated using s simplistic scale-free model [20].

In addition, our results can be used in order to provide answers to other questions, such as what is the optimal group of members that should be used as a "seed group" in order to maximize the effects of marketing campaigns. Another example might be finding changes in the topology of the social network that would influence the information diffusion progress in a desired way (either to encourage or surpass certain emerging trends).

In order to achieve these goals we are planning a large-scale field test with a leading online social platform, that would give us access to collect more empirical supporting evidences, as well as conducting an active experiment in which we would try to predict trends in real time.

Finally, we are interested in comparing the prediction obtained from our model with the actual semantics of the trends, to better understand the connection between the trends semantics and the diffusion process they undergo.

References

1. Aharony, N., Pan, W., Ip, C., Khayal, I., Pentland, A.: Social fmri: Investigating and shaping social mechanisms in the real world. Pervasive and Mobile Computing
2. Huberman, B., Romero, D., Wu, F.: Social networks that matter: Twitter under the microscope. First Monday 14(1), 8 (2009)
3. Leskovec, J., Backstrom, L., Kleinberg, J.: Meme-tracking and the dynamics of the news cycle. In: Proceedings of the 15th ACM SIGKDD International Conference on Knowledge Discovery and Data Mining, pp. 497–506. Citeseer (2009)
4. Dodds, P., Muhamad, R., Watts, D.: An experimental study of search in global social networks. Science 301(5634), 827 (2003)
5. Kwak, H., Lee, C., Park, H., Moon, S.: What is twitter, a social network or a news media? In: Proceedings of the 19th International Conference on World Wide Web, pp. 591–600. ACM (2010)
6. Choi, H., Kim, S., Lee, J.: Role of network structure and network effects in diffusion of innovations. Industrial Marketing Management 39(1), 170–177 (2010)
7. Nicosia, V., Bagnoli, F., Latora, V.: Impact of network structure on a model of diffusion and competitive interaction. EPL (Europhysics Letters) 94, 68009 (2011)
8. Herrero, C.: Ising model in scale-free networks: A monte carlo simulation. Physical Review E 69(6), 67109 (2004)
9. Kempe, D., Kleinberg, J., Tardos, É.: Maximizing the spread of influence through a social network. In: Proceedings of the Ninth ACM SIGKDD International Conference on Knowledge Discovery and Data Mining, pp. 137–146. ACM (2003)
10. Shah, D., Zaman, T.: Rumors in a network: Who's the culprit?, Arxiv preprint arXiv:0909.4370
11. Bakshy, E., Hofman, J., Mason, W., Watts, D.: Everyone's an influencer: quantifying influence on twitter. In: Proceedings of the Fourth ACM International Conference on Web Search and Data Mining, pp. 65–74. ACM (2011)
12. Watts, D., Peretti, J.: Viral marketing for the real world
13. Centola, D., Macy, M.: Complex contagions and the weakness of long ties. American Journal of Sociology 113(3), 702 (2007)
14. Centola, D.: The spread of behavior in an online social network experiment. Science 329(5996), 1194 (2010)
15. Dodds, P., Watts, D.: Universal behavior in a generalized model of contagion. Physical Review Letters 92(21), 218701 (2004)
16. Pan, W., Aharony, N., Pentland, A.: Composite social network for predicting mobile apps installation. In: AAAI (2011)
17. Banerjee, S., Mallik, S., Bose, I.: Reaction diffusion processes on random and scale-free networks, Arxiv preprint cond-mat/0404640
18. Meloni, S., Arenas, A., Moreno, Y.: Traffic-driven epidemic spreading in finite-size scale-free networks. Proceedings of the National Academy of Sciences 106(40), 16897 (2009)
19. Altshuler, Y., Pan, W., Pentland, A.: Trends prediction using social diffusion models, arXiv.org (2011)
20. Leskovec, J., Kleinberg, J., Faloutsos, C.: Graphs over time: densification laws, shrinking diameters and possible explanations. In: Proceedings of the Eleventh ACM SIGKDD International Conference on Knowledge Discovery in Data Mining, pp. 177–187. ACM (2005)

Towards Democratic Group Detection in Complex Networks

Michele Coscia[1], Fosca Giannotti[2], and Dino Pedreschi[1]

[1] Computer Science Dep., University of Pisa, Italy
{coscia,pedre}@di.unipi.it
[2] ISTI - CNR, Area della Ricerca di Pisa, Italy
fosca.giannotti@isti.cnr.it

Abstract. To detect groups in networks is an interesting problem with applications in social and security analysis. Many large networks lack a global community organization. In these cases, traditional partitioning algorithms fail to detect a hidden modular structure, assuming a global modular organization. We define a prototype for a simple local-first approach to community discovery, namely the democratic vote of each node for the communities in its ego neighborhood. We create a preliminary test of this intuition against the state-of-the-art community discovery methods, and find that our new method outperforms them in the quality of the obtained groups, evaluated using metadata of two real world networks. We give also the intuition of the incremental nature and the limited time complexity of the proposed algorithm.

1 Introduction

Complex network analysis has emerged as a popular domain of data analysis over the last decade and especially its community discovery (CD) sub field has been proven useful for many applications, related also to crime prevention. The concept of a "community" in a network is intuitively a set of individuals that are very similar to each other, more than to anybody else outside the community [2]. In network terms, sets of nodes densely connected to each other and sparsely connected with the rest of the network. To efficiently detect these structures is very useful for a number of applications, we recall here information spreading [5] and crime understanding and prevention [11].

The classical problem definition of community discovery finds a very intuitive counterpart for small networks, but for medium and large scale networks the CD problem becomes much harder. At the global level, very little can be said about the modular structure of the network: the organization of the system becomes simply too complex. The graph of Facebook includes more than 800 millions nodes[1], but even in less than 0.002% of the total network no evident organization can be identified easily (Figure 1 on the left), resulting in a structureless hairball. In these cases generic community discovery algorithms tend to fail trying to

---

[1] http://www.facebook.com/press/info.php?statistics

S.J. Yang, A.M. Greenberg, and M. Endsley (Eds.): SBP 2012, LNCS 7227, pp. 105–113, 2012.

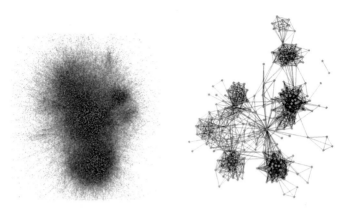

Fig. 1. The real world example of the "local vs global" structure intuition

cluster the whole structure and return some huge communities and a long list of small branches. This approach generally fails for large networks due to the difference in structural organization at global and local scale.

To cope with this difficulty, we propose a change of mentality: since our community definition works perfectly in the small scale, then it should be applied only at this small scale. The structure of cohesive groups of nodes emerges considering a *local* fragment of an otherwise big network. But what does *local* mean? Common sense goes that people are good at identifying the reasons why they know the people they know: each node has presumably an incomplete, yet clear, vision of the social communities that surrounds it. In Figure 1 on the right we chose one node from the example and extracted its ego-network and some groups can be easily spotted. The ego is part of all these communities and knows that particular subsets of its neighborhood are part of these communities too. Different egos have different perspectives over the same neighbors and it is the union of all these perspectives that creates an optimal partition of the network. This is achieved by a *democratic* bottom-up approach: in turn, each node gives the perspective of the communities surrounding it and then all the different perspectives are merged together in an overlapping structure.

We name our algorithm **D**emocratic **E**stimate of the **M**odular **O**rganization of a **N**etwork, or *DEMON*: we extract the ego network of each node and apply a Label Propagation CD algorithm [8] on this structure, ignoring the presence of the ego itself, that will be judged by its peers neighbors. We then combine, with equity, the vote of everyone in the network. The result of this combination is a set of (overlapping) modules, the guess of the real communities in the global system, made not by an external observer, but by the actors of the network itself. *DEMON* is *incremental*, allowing to recompute the communities only for the newly incoming nodes and edges. Nevertheless, *DEMON* has a low theoretical time complexity, and our experiments show that its performance is superior to that of fastest competitor methods, both overlapping and non overlapping. Online social networks have proved that individuals are part of many different communities and groups of interest, therefore the overlapping partition of groups is a crucial feature. The properties of *DEMON* support its use in massive real world scenarios, for example in tasks of group and threat detection.

The paper is organized as follows. In Section 2 we present the related work in the community discovery literature. In Section 3 we have the problem definition. Section 4 describe the algorithm structure. Our experiments are presented in Section 5, and finally Section 6 concludes the paper.

2   Related Work

The problem of finding groups in complex networks has been tackled by an impressive number of valid works. Traditionally, a community is defined as a dense subgraph, in which the number of edges among the members of the community is significantly higher than the outgoing edges. This definition does not cover many real world scenarios, and in the years many different solutions started to explore alternative definitions of communities in complex networks [2].

A variety of CD methods are based on the *modularity* concept, a quality function of a partition proposed by Newman [7]. Modularity scores high values for partitions in which the internal cluster density is higher than the external density. Besides modularity, a particular important field is the application of information theory techniques. In particular, Infomap has been proven to be one amongst the best performing non overlapping algorithms [2]. Modularity approaches are affected by known issues, namely the resolution problem and the degeneracies of good solutions [4]. A very important property for community discovery is the ability to return overlapping partitions [10], i.e., the possibility of a node to be part of more than one community. Specific algorithms developed over this property are Hierarchical Link Clustering [1], HCDF [6] and k-clique percolation [3]. Finally, an important approach is Label Propagation [8]: in this work authors detect communities by spreading labels through the edges of the graph and then labeling nodes according to the majority of the labels attached to their neighbors, until a general consensus is reached. This algorithm is extremely fast and provides a reasonable good quality of the partition.

To find groups in networks and to extract useful knowledge from their modular structure is a prolific track of research with a number of applications. We recall the GuruMine framework, whose aim is to identify leaders in information spread and to detect groups of users that are usually influenced by the same leaders [5]. Many other works investigate the possibility of applying network analysis for studying, for instance, the dynamics of crime related behaviors [11].

3   Networks and Communities

We represent a network as an undirected, unlabeled and unweighted simple graph, denoted by $\mathcal{G} = (V, E)$ where V is a set of nodes and E is a set of edges, i.e., pairs (u, v) representing the fact that there is a link in the network connecting nodes u and v. Our problem definition is to find groups in complex networks. However, this is an ambiguous goal, as the definition itself of "community" in a complex network is not unique [2]. Furthermore, if we want to develop an efficient instrument for criminal profiling and prevention, an analyst may want to cluster many different kinds of groups of suspects for many different reasons. Therefore, we need to narrow down our problem definition as follows.

We define a basic graph operation, namely the $EgoMinusEgo$ operation: it consists in extracting the ego network of a node, i.e. the collection of a node and all its direct neighbors and all the edges between them, eliminating the ego itself. Given a graph $\mathcal{G}$ and a node $v \in V$, the set of *local communities* $\mathcal{C}(v)$ of node v is a set of (possibly overlapping) sets of nodes in $EgoMinusEgo(v, \mathcal{G})$, where each set $c \in \mathcal{C}(v)$ is a community: each node in c is closer to any other node in c than to any other node in $c' \in \mathcal{C}(v)$, with $c \neq c'$. We refer here to the topological distance between nodes in a graph, namely the length of the shortest path connecting any two nodes. Finally, we define the set of *global communities*, or simply communities, of a graph $\mathcal{G}$ as $\mathcal{C} = Max(\bigcup_{v \in V} \mathcal{C}(v))$, where, given a set of sets $\mathcal{S}$, $Max(\mathcal{S})$ denotes the subset of $\mathcal{S}$ formed by its maximal sets only; namely, every set $s \in \mathcal{S}$ such that there is no other set $s' \in \mathcal{S}$ with $s \subset s'$. In other words, we generalize from local to global communities by selecting the maximal local communities that cover the entire collection of local communities, each found in the $EgoMinusEgo$ network of each individual node.

4   The Algorithm

In this section we informally describe our proposed solution to the community discovery problem. *DEMON* cycles over each individual node, it applies the $EgoMinusEgo(v, \mathcal{G})$ operation (we cannot simply extract the ego network, because the ego node is directly linked to all nodes leading to noise). The next step is to compute the communities contained in $EgoMinusEgo(v, \mathcal{G})$. We chose to perform this step by using a community discovery algorithm borrowed from the literature: the Label Propagation (LP) algorithm [8]. This choice has been made for the following reasons: (1) LP shares with this work the definition of what is a community; (2) LP is known as the least complex algorithm in the literature; (3) LP will return results of a quality comparable to more complex algorithms [2]. Reason #2 is particularly important, since this step needs to be performed once for every node of the network and we cannot spend a superlinear time for each node at this stage, if we want to scale up to millions of nodes.

The result of this step of the algorithm is a set of the local communities, from node v point of view. These local communities are the node social identity, and they are then used to get a global perspective of the network, not of the single node per se. These communities are likely to be incomplete and should be used to enrich what *DEMON* already discovered so far. Thus, the next step is to merge each local community of $\mathcal{C}(v)$ into the result set $\mathcal{C}$. The *Merge* operation is here defined:

$$Merge(c, \mathcal{C}) = \begin{cases} \mathcal{C} & \exists c' \in \mathcal{C} : c \subseteq c' \\ \{c\} \cup \{c' \in \mathcal{C} \mid c' \not\subseteq c\} & \text{otherwise} \end{cases}$$

In other words, the community c is added to the collection $\mathcal{C}$ only if it is not covered by any community already in $\mathcal{C}$; in this case, all communities in $\mathcal{C}$ covered by c, if any, are removed. This procedure guarantees that the following property holds.

Property 1. **Incrementality.** Given a graph $\mathcal{G}$, an initial set of communities $\mathcal{C}$ and an incremental update $\Delta\mathcal{G}$ consisting of new nodes and new edges added to $\mathcal{G}$, where $\Delta\mathcal{G}$ contains the entire ego networks of all new nodes and of all the preexisting nodes reached by new links, then

$$DEMON(\mathcal{G} \cup \Delta\mathcal{G}, \mathcal{C}) = DEMON(\Delta\mathcal{G}, DEMON(\mathcal{G}, \mathcal{C})) \qquad (1)$$

This is a consequence of the fact that only the local communities of nodes in $\mathcal{G}$ affected by new links need to be reexamined, so we can run $DEMON$ on $\Delta\mathcal{G}$ only, avoiding to run it from scratch on $\mathcal{G} \cup \Delta\mathcal{G}$.

Intuitively, $DEMON$ algorithm presents also the properties of determinacy, order insensitivity and compositionality, but we leave the proof of there properties for future work. The incrementality property entails that $DEMON$ can efficiently run in a streamed fashion, considering incremental updates of the graph as they arrive in subsequent batches; essentially, incrementality means that it is not necessary to run $DEMON$ from scratch as batches of new nodes and new links arrive: the new communities can be found by considering only the ego networks of the nodes affected by the updates (both new nodes and old nodes reached by new links.)

As for the time complexity of $DEMON$, we note that it is based on the Label Propagation algorithm, whose complexity is $\mathcal{O}(n+m)$ [8], where n is the number of nodes and m is the number of edges. LP is performed once for each node, thus a rough estimate of worst case complexity of $DEMON$ is $\mathcal{O}(n^2 + nm)$. However, LP is not applied to the network as a whole, but only to small ego networks: this bound would be tight if and only if each node has n neighbors, i.e., the entire graph is a clique. A better estimate is $\mathcal{O}(n\bar{k}^2)$, where $\bar{k}$ is the average degree of the network, since it is expected that each ego network will contain $\bar{k} - 1$ nodes and $(\bar{k} - 1)^2$ edges on the average worst case.

5   Experiments

We now present our experimental findings. We considered two networks, a general overview about the statistics of these networks can be found in Table 1. The two networks are:

Table 1. Basic statistics of the studied networks. $|V|$ is the number of nodes, $|E|$ is the number of edges and $\bar{k}$ is the average degree of the network.

Network	$|V|$	$|E|$	$\bar{k}$
Congress	526	14,198	53.98
IMDb	56,542	185,347	6.55

Congress[2], the network of US representatives of the House and the Senate during the 111st US congress (2009-2011): the bills are usually co-sponsored by many politicians and we connected politicians if they have at least 75 co-sponsorships (deleting the connections created only by bills with more than 10 co-sponsors). The set of subjects a politician frequently worked on is the *qualitative attribute* of this network, the *quantitative attributes* are derived from the size of the set of subjects of each politician. **IMDb**[3], the network of actors connected if they appear together in at least two movies from 2001 to 2010. The *qualitative attributes* are the user assigned keywords summarizing the movies each actor has been part of. The total number of movies in which an actor appeared is instead the basis of the *quantitative attributes*. In both networks we expect to find a rich overlap being politicians usually involved in many different topics and actors present in many different crews.

We now evaluate the quality of a set of communities discovered in these datasets, by performing a direct comparison between the discovered communities and the qualitative and quantitative attributes attached to the nodes. We then evaluate our algorithm against two state-of-the-art community discovery methods using the proposed quality measures, namely the Hierarchical Link Clustering [1] as state-of-the-art for overlapping; and Infomap [9], as state-of-the-art of non-overlapping algorithms. Besides output quality, we also compare computational performances. Finally, we present some examples of web related knowledge that we are able to extract with *DEMON* algorithm. The experiments were performed on a Dual Core Intel i7 64 bits @ 2.8 GHz, equipped with 8 GB of RAM and with a kernel Linux 2.6.35-22-generic.

We test the quality of the community sets returned by each algorithm against four main quality functions. The chosen functions do not consider only the network structure, like Modularity [7]. By looking only at the topology of the network we cannot evaluate communities according to their meaning in the real world: using a semantics-based set of quality measures is particularly critical in the case of social and web networks. The evaluation measures used in this paper are directly inherited from [1]. We present their formulation and we point to that reference for further details. A key concept of these evaluation measures is the definition of a non-trivial community, i.e. a community with at least three nodes.

Community Quality. The networks studied here possess *qualitative attributes* that attaches a small set of annotations or tags to each node. We state that "similar" nodes share more *qualitative attributes* than dissimilar nodes, quantifying this intuition by evaluating how much higher are on average the Jaccard coefficient of the set of *qualitative attributes* for pair of nodes inside the communities over the average of the entire network.

Overlap Quality. We can use the *quantitative attributes* to quantify how many different communities a node should be part of. We define the Overlap Quality as follows: $OQ(X; MD) = \sum_{y \in MD} \sum_{x \in X} p(x, y) \log \frac{p(x, y)}{p(x) \, p(y)}$, where

---

X is the vector that assigns to each node the number of nontrivial communities extracted by the algorithm, MD is the vector of the quantitative attributes, and $p(x, y)$ is the probability of each variable value co-occurrence.

Community Coverage. To measure community coverage, we simply count the fraction of nodes that belong to nontrivial community.

Overlap Coverage. We count the average number of memberships in nontrivial communities for each node. This measure shows how much information is extracted from that portion of the network that the particular algorithm was able to analyze.

5.1 Evaluation

In Figure 2a we report the runtimes of the tested algorithms on the networks we analyzed. One important caveat regards the Infomap, that is very dependent on random walks and greedy heuristics, therefore Infomap needs to be performed several times to get reasonable results. In Figure 2 we report the cost of one single iteration, but for Infomap the experiments need to run at least 50 or 100 iterations, making the total running time of Infomap in the same order of magnitude of *DEMON*. Also, Infomap is not incremental as *DEMON*. On the other hand, in general HLC is 2x to 100x slower than *DEMON*.

In Figure 2b we report the scores of the combination algorithm-network using the evaluation measures we introduced in this section. We stacked the score of each measure in one single bar and we normalized the values of each measure (since some of them do not take values from 0 to 1): from bottom to top they are Community Quality (white), Overlap Quality (light gray), Community Coverage (dark gray) and Overlap Coverage (black). The normalization was done by dividing each score of the algorithm by the maximum score registered among *DEMON*, HLC and Infomap. *DEMON* algorithm performs better according to the combined score over both datasets. In general, Community Quality is always very good, while a complete constant is better scores in the Overlap Coverage. Also the Overlap Quality is generally better except in IMDb. Overlap is a crucial aspect of real world data. From this analysis we can conclude that not only

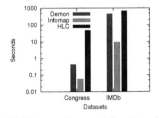
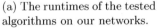
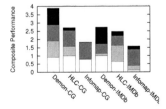

(a) The runtimes of the tested algorithms on our networks.

(b) The composite performances of the tested algorithms on our networks.

Fig. 2. Comparison between *DEMON*, HLC and Infomap

DEMON is in general a better algorithm, but it is especially the optimal choice if we are particular interested in a better description of the overlapping and complex reality.

5.2   Case Study

In this Section we present a brief case study using the communities extracted from the IMDb network. In Figure 3 left we represent one chosen community, IMDb #276. All actors are related with the Star Wars saga. Using the power of overlapping we can jump from this community to the related communities, i.e. the other communities of its members.

Fig. 3. Representations of some communities extracted from the IMDb network

In Figure 3 right we chose to visualize some of the surrounding communities of the British actor Christopher Lee: he is part also of a wider and more complete community regarding the whole new Star Wars trilogy, besides the communities regarding the trilogy of Lord of the Rings. Another interesting galaxy of communities is composed by different groups of actors that usually work together with director Tim Burton. The picture we get from Figure 3 right is a decent summary of the movie acting career of sir Christopher Lee for the past decade. If we have reliable data about criminal activities, collaborations and affiliations, a similar set of results can be obtained to profile criminals and to obtain a similarity measure for criminal groups, therefore providing a fundamental tool to law enforcement and crime prevention.

6   Conclusion and Future Works

In this paper we proposed a new method for solving the problem of finding significant groups in complex networks, aiming at real world applications such as crime prevention. We propose a democratic approach, where the peer nodes judge where their neighbors should be clustered together. This approach is fast and incremental. We have shown in the experimental section that this method allows a discovery of communities in different real world networks: the quality of the overlapping partition is improved w.r.t state-of-the-art algorithms.

Many lines of research remain open for future work. We want to prove other fundamental properties of our algorithm, namely determinacy, order insensitivity and compositionality: these properties can be exploited to apply *DEMON* on huge networks. A more comprehensive experimental section, with more and larger datasets, is also in the roadmap to test *DEMON* in a wider set of domains.

Acknowledgments. Michele Coscia is a recipient of the Google Europe Fellowship in Social Computing, and this research is supported in part by this Google Fellowship.

References

1. Ahn, Y.-Y., Bagrow, J.P., Lehmann, S.: Link communities reveal multiscale complexity in networks. Nature 466(7307), 761–764 (2010)
2. Coscia, M., Giannotti, F., Pedreschi, D.: A classification for community discovery methods in complex networks. SAM 4(5), 512–546 (2011)
3. Derényi, I., Palla, G., Vicsek, T.: Clique Percolation in Random Networks. Physical Review Letters 94(16), 160202 (2005)
4. Fortunato, S., Barthélemy, M.: Resolution limit in community detection. Proceedings of the National Academy of Sciences 104(1), 36–41 (2007)
5. Goyal, A., On, B.-W., Bonchi, F., Lakshmanan, L.V.S.: Gurumine: A pattern mining system for discovering leaders and tribes. In: International Conference on Data Engineering, pp. 1471–1474 (2009)
6. Henderson, K., Eliassi-Rad, T., Papadimitriou, S., Faloutsos, C.: Hcdf: A hybrid community discovery framework. In: SDM, pp. 754–765 (2010)
7. Newman, M.E.J.: Modularity and community structure in networks. Proceedings of the National Academy of Sciences 103(23), 8577–8582 (2006)
8. Raghavan, U.N., Albert, R., Kumara, S.: Near linear time algorithm to detect community structures in large-scale networks. Physical Review E (2007)
9. Rosvall, M., Bergstrom, C.T.: Maps of random walks on complex networks reveal community structure. PNAS 105(4), 1118–1123 (2008)
10. Shen, H.-W., Cheng, X.-Q., Guo, J.-F.: Quantifying and identifying the overlapping community structure in networks. J. Stat. Mech. (2009)
11. Yonas, M.A., Borrebach, J.D., Burke, J.G., Brown, S.T., Philp, K.D., Burke, D.S., Grefenstette, J.J.: Dynamic Simulation of Community Crime and Crime-Reporting Behavior. In: Salerno, J., Yang, S.J., Nau, D., Chai, S.-K. (eds.) SBP 2011. LNCS, vol. 6589, pp. 97–104. Springer, Heidelberg (2011)

Addiction Dynamics May Explain the Slow Decline of Smoking Prevalence

Gaurav Tuli[1], Madhav Marathe[1], S.S. Ravi[2], and Samarth Swarup[1]

[1] Virginia Bioinformatics Institute, Virginia Tech, Blacksburg, VA 24061, USA
{gtuli,mmarathe,swarup}@vbi.vt.edu
[2] Computer Science Department, University at Albany – SUNY,
Albany, NY 12222, USA
ravi@cs.albany.edu

Abstract. The prevalence of cigarette smoking in the United States has declined very slowly over the last four decades, despite much effort by multiple governmental and non-governmental institutions. Peer influence has been shown to be the largest contributing factor to the spread of smoking behavior, which suggests the use of epidemic models for understanding this phenomenon. Here we develop a *structured resistance model*, which is an *SIS* model extended to include multiple *S* and *I* states corresponding to different levels of addiction. This model exhibits a backward bifurcation, which means that once the behavior is endemic, it can be very difficult to remove entirely from the population. We do numerical experiments with the Framingham Heart Study social network to show that the resulting epicurve closely matches empirical data on the overall decline in smoking behavior.

1 Introduction

Nicotine, in the form of cigarette smoking or chewing tobacco, is one of the most heavily used addictive drugs, and the leading preventable cause of disease, disability, and death in the U.S. [15]. It imposes a significant health-care burden on the population. The economic costs of smoking in the United States are estimated at $193 billion annually ($97 billion in productivity losses from premature death and $96 billion in health-care expenditures).

Like any other addictive drug, cigarette smoking behavior becomes compulsive and difficult to cease even after knowing the substantial health benefits of quitting [13]. Recent studies show that 35 million smokers express a desire for quitting smoking each year, but more than 85 percent of those who try to quit on their own relapse within a week [16]. Nevertheless, despite sustained and significant efforts by governmental and non-governmental institutions, smoking prevalence among youth and adult smokers has only declined slowly from 45% to 21% in the past 45 years [14,6,7] (see figure 1).

It has been repeatedly shown that smoking behavior is contagious, i.e., that peer influence (including family members) is the strongest factor in both initiation and cessation of smoking [11,4,8,9]. From an epidemiological viewpoint, the

S.J. Yang, A.M. Greenberg, and M. Endsley (Eds.): SBP 2012, LNCS 7227, pp. 114–122, 2012.

SIS model seems appropriate for modeling the contagion of smoking behavior, since smokers who quit can relapse. Here, the S state (which stands for "Susceptible") corresponds to non-smokers and the I state ("Infected") stands for smokers. However, the slow decline of smoking prevalence is puzzling from this perspective, as we shall discuss in the next section.

Our main contribution in this work is to introduce an extension to the SIS model, which we call the *structured resistance model*, to account for the addictive nature of smoking behavior. In this model we have multiple S and I states corresponding to increasing levels of addiction. We present this model in section 3 and we present simulations with this model on the Framingham Heart Study social network [5] in section 4. We end with a discussion of the model and possible directions for future work.

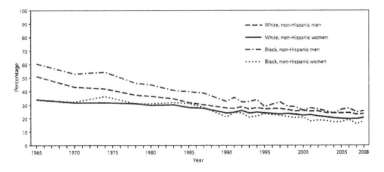

Fig. 1. Smoking prevalence has declined slowly over the course of four decades. Source: CDC (*http://www.cdc.gov/mmwr/pdf/other/su6001.pdf*, p. 109).

2   Modeling the Smoking Epidemic

For the standard SIS model of epidemics, it is well known that there is an epidemic threshold, β_c, such that if the probability of infection is below β_c, the only steady state is the disease free state, while for values of the probability of infection above β_c, the only steady state corresponds to a finite fraction of the population being in the I state [2]. Figure 2 illustrates this scenario numerically. It can also be calculated analytically for some classes of networks.

If we assume that the effect of anti-smoking efforts is to reduce the infectiousness of smoking, then we would expect the prevalence of smoking to reduce fairly sharply, and to disappear once the infectiousness becomes low enough. This expectation stands in contrast to the reality depicted in figure 1, which shows that after much effort, the prevalence of smoking has declined steadily, but very slowly over more than four decades.

This puzzling contrast suggests that the SIS model is not quite right for understanding the contagion of smoking behavior. We argue that it misses the defining feature of smoking behavior, which is that smoking is addictive. Therefore we need a new model.

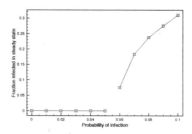

Fig. 2. Bifurcation diagram for the SIS model

3   The Structured Resistance Model

To model the dynamics of addictive behavior, we adapt a model developed by Reluga et al. [17]. Their study investigates the dynamics of disease immunity. We invert its semantics to model resistance to addictive behavior. This *structured resistance model* is shown in figure 3a. The multiple S states correspond to increasing levels of susceptibility to the behavior, and the multiple I states correspond to increasing levels of addiction. Initially, an individual starts out in state S_1, and moves to state I_1 upon adopting the behavior. The probability of this transition is given by $\beta \sigma_1$, where β is a multiplier on all $S \to I$ transitions and will be taken to correspond to R_0. The rate at which individuals quit, i.e., transition from an I state to an S state, is given by the corresponding γ_i parameter. Crucially, since the behavior is addictive, transitions from I_i only go to $S_{j>i}$. This means that when an individual quits, his susceptibility level is at least as high (and possibly higher) than it was before, but never lower. The probability of making the transition $I_i \to S_{j>i}$ is given by f_{ij}. The only way to recover to a lower level of susceptibility is via the $S_j \to S_{j-1}$ transitions, the probability of which is given by g_j. This is meant to model the fact that if an individual stays free of the addictive behavior for a long time, his level of susceptibility can decrease.

For a fully mixed population, the state update equations can be written as follows [17].

$$\frac{dS_j}{dt} = g_{j+1}S_{j+1} - g_j S_j - \beta \sigma_j I_{Total} S_j + \sum_{i=1}^{j} \gamma_i I_i f_{ij}, \tag{1}$$

$$\frac{dI_j}{dt} = \beta \sigma_j I_{Total} S_j - \gamma_j I_j, \tag{2}$$

$$I_{Total} = \sum_{j=1}^{n} I_j, \tag{3}$$

for $j = 1, ..., n$. We assume that $g_1 = g_{n+1} = 0$, since these transitions correspond to states that do not exist in the model.

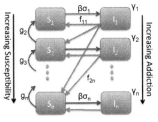

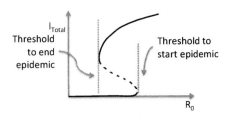

(a) The Structured Resistance Model

(b) Schematic of a backward bifurcation

Fig. 3. The structured resistance model is shown on the left. Some parameters are not marked for clarity, but these correspond to the ones that are shown. See text for details. A schematic of a backward bifurcation is shown on the right. The solid lines indicate stable steady states and the dashed line indicates an unstable steady state.

Note that an individual makes an $S \to I$ transition only if one or more of his neighbors are in an I state. However, an individual makes the $S_j \to I_j$ transition with probability $\beta\sigma_j$, no matter *which* I state its neighbor is in. In other words, the contagion spreads from individuals in I states to individuals in S states, but does not depend on the details of which I states the spreaders are in. This is why we have an I_{Total} term in the equations instead of terms for each of the I states.

From these equations, it can be shown that when the quantity,

$$Q = -\sigma_1 - \sum_{j=2}^{n} \frac{\beta\sigma_1}{g_j} \left(\frac{\sigma_1}{\gamma_1} - \frac{\sigma_j}{\gamma_j} \right) \left(1 - \sum_{k=1}^{j-1} f_{1k} \right) \qquad (4)$$

is positive and increasing in β, then the epidemic bifurcation is a backward bifurcation. This means that the bifurcation looks like in figure 3b (the "threshold to start epidemic"). There are actually multiple bifurcations and multiple steady states in this scenario. The solid lines in figure 3b indicate the stable steady states and the dashed portion indicates an unstable steady state.

The bifurcation diagram can be effectively divided into three regions. From $R_0 = 0$ to the "threshold to end epidemic", there is only one stable steady state, which corresponds to the entire population being in the susceptible state. In this range, the contagion does not take off, no matter what the initial state may be. Similarly, from the "threshold to start epidemic", for higher values of R_0, there is only one stable steady state, which corresponds to the contagion becoming endemic, i.e. if even only very few individuals are initially in an I state, a finite fraction will be in the I states in the long run. In between these two regions, we have a region where there are two stable steady states and one unstable steady state. In this region, if the initial state starts above the dashed curve, the population will move to the upper steady state, while if it starts below the dashed curve, the population will move to the lower steady state.

Practically, this means that for a new addictive behavior to become endemic in the population, its "infectiousness" must be higher than the threshold to start the epidemic (which is higher), but once it becomes endemic, for efforts to counter it to be successful, they must succeed in reducing the infectiousness of the behavior to below the threshold to end the epidemic (which is lower). Intuitively, this means that significantly more effort might be required to end the epidemic than expected.

The equations above describe the behavior of the model in a fully-mixed population, or equivalently on a fully-connected network. In the next section, we do simulations to investigate the model on a more realistic network. Since we are interested in understanding the decline in smoking prevalence over a period of decades, we use a dynamic network that has been constructed from the Framingham Heart Study [5], which is a longitudinal study spanning precisely that period.

4   Simulations

Since the increasing levels in the structured resistance model represent increasing levels of susceptibility and addiction, we choose the probability of $S_i \rightarrow I_i$ to be increasing with i, and the recovery rates γ_i to be decreasing with i. The f_{ij} values, which control the $I_i \rightarrow S_j$ transitions, are chosen to be zero when $j < i$, as mentioned earlier. The entire set of parameters is shown in table 1. It turns

Table 1. Parameters for simulations with the structured resistance model

Level, i	Infection probability, σ_i	Recovery rate, γ_i	f_{i1}, f_{i2}, f_{i3}	Resistance waxing rate, g_i
1	0.05	0.7	0.4, 0.4, 0.2	0.0
2	0.5	0.5	0.0, 0.7, 0.3	0.2
3	0.7	0.3	0.0, 0.0, 1.0	0.1

out that any set of parameters chosen according to these conditions will result in the model exhibiting a backward bifurcation. We can verify this is the case by substituting these parameter values into equation 4.

We conduct simulations using the Framingham Heart Study (FHS) social network. The FHS is a longitudinal study that gathered data on many health characteristics and health behaviors. The social network is a time-varying network spanning the years 1971-2008 (i.e., starting with the "offspring cohort" of the FHS). The offspring cohort were recruited into the study at an early age (the lowest age in the data is 5 years), thus giving us a network with children and adolescents as well as adults. Edges in the network correspond to various social and familial relationships. For the present work, we assume each edge to be an undirected edge along which the contagion can spread in either direction. For each edge present in the network, the data provides a start month and an end month. Thus edges are present at different times and for different durations.

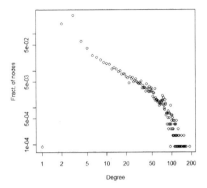

Fig. 4. The degree distribution of the Framingham Heart Study union graph. It is not scale-free.

The degree distribution for the union graph (which assumes all the edges are present for all time) is shown in figure 4. The Framingham Heart Study data has been used in other instances of the study of the spread of smoking [4].

The bifurcation diagram for the structured resistance model on the Framingham Heart Study social network is shown in figure 5a, using the parameters in table 1. The diagram is obtained by doing simulations with two different initial conditions. In the first condition, the number of initial infections is very small, and in the other, the number of initial infections is large. In the region where there are two stable steady states, the first initial condition converges to the lower steady state and the second to the upper one. The lower threshold, shown in blue, corresponds to the upper steady state, while the upper threshold, shown in red, corresponds to the lower stationary state (the backward bifurcation). We see that there is a large gap between the two thresholds, which suggests that once the behavior is endemic, a large amount of effort is required (β has to be brought down a lot) to entirely eliminate the behavior from the population. Note that the unstable equilibrium is not shown because it is hard to determine numerically.

Figure 5b shows a sample epicurve obtained as follows. We initialize the population by randomly setting 5% of the nodes to be in state I_1 while the rest are in state S_1. The value of β is chosen to be 1.3, which is well above the upper threshold. We run the model until it reaches a stationary state, which corresponds to about 42% of nodes in I states. Then β is decreased slowly to simulate increasing awareness of the dangers of smoking and increased resistance to initiation. This causes the proportion of nodes in I states to decrease along the blue curve in the bifurcation diagram, which results in a slow decline in smoking prevalence. The epicurve is plotted from this point on in figure 5b. This qualitatively matches the empirical data reported by the CDC (figure 1).

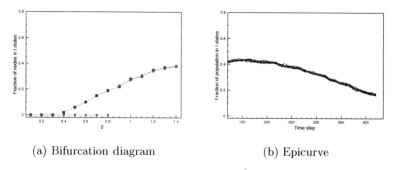

(a) Bifurcation diagram   (b) Epicurve

Fig. 5. The bifurcation diagram for the structured resistance model on the Framingham Heart Study social network is shown on the left. A sample epicurve is shown on the right, which exhibits the slow decline in smoking prevalence as β is decreased after time step 120. The green curve shows the average of the black curves (which show individual simulation results).

5   Conclusion

We have presented an extended SIS model, called the structured resistance model, to capture the dynamics of addictive behavior. Levels in the model correspond to increasing susceptibility and addiction to a behavior. This model exhibits a backward bifurcation, which suggests a possible reason for the slow decline of smoking prevalence in the United States.

This basic model can be extended in various ways. There are factors other than peer influence that affect smoking behavior, such as socioeconomic status and marital status [3], access to cigarettes and exposure to advertising [10], and prices and policies [12]. Data about all these factors can be included into an agent-based model driven by the basic structured resistance model. Detailed synthetic information environments can be constructed by fusing data about these behaviors with other data sets on demographics, locations, and activities to build a complete picture of the ecology of a smoker [1].

Smoking is a complex, "policy-resistant" problem. We believe that mathematical modeling and simulation-based approaches are essential to understanding such systems and to achieving lasting social benefits.

Acknowledgments. This work has been partially supported by NSF PetaApps Grant OCI-0904844, NSF Netse Grant CNS-1011769, NSF SDCI Grant OCI-1032677, DTRA R&D Grant HDTRA1-0901-0017, DTRA R&D Grant HDTRA1-11-1-0016, DTRA CNIMS Grant HDTRA1-07-C-0113, DTRA CNIMS Grant HDTRA1-11-D-0016-0001, and NIH MIDAS project 2U01GM070694-7.

References

1. Barrett, C., Eubank, S., Marathe, A., Marathe, M., Pan, Z., Swarup, S.: Information integration to support policy informatics. The Innovation Journal 16(1), article 2 (2011)
2. Boguñá, M., Pastor-Satorras, R., Vespignani, A.: Epidemic spreading in complex networks with degree correlations. In: Pastor-Satorras, R., Rubi, J.M., Diaz-Guilera, A. (eds.) Statistical Mechanics of Complex Networks. Lecture Notes in Physics, vol. 625, pp. 127–147 (2003)
3. Broms, U., Silventoinen, K., Lahelma, E., Koskenvou, M., Kaprio, J.: Smoking cessation by socioeconomic status and marital status: The contributions of smoking behavior and family background. Nicotine and Tobacco Research 6(3), 447–455 (2004)
4. Christakis, N.A., Fowler, J.H.: The collective dynamics of smoking in a large social network. N. Engl. J. Med. 358(21), 2249–2258 (2008)
5. Feinleib, M., Kannel, W.B., Garrison, R.J., McNamara, P.M., Castelli, W.P.: The Framingham Offspring Study: Design and preliminary data. Prev. Med. 4, 518–525 (1975)
6. Centers for Disease Control and Prevention. Cigarette use among high school students – United States, 1991-2009. Technical Report MMWR 2010 59,797–801, Centers for Disease Control and Prevention (2010)
7. Centers for Disease Control and Prevention. Early release of selected estimates based on data from the 2010 National Health Interview Survey: current smoking (2010)
8. Gilman, S.E., Rende, R., Boergers, J., Abrams, D.B., Buka, S.L., Clark, M.A., Colby, S.M., Hitsman, B., Kazura, A.N., Lipsitt, L.P., Lloyd-Richardson, E.E., Rogers, M.L., Stanton, C.A., Stroud, L.R., Niaura, R.S.: Parental smoking and adolescent smoking initiation: An intergenerational perspective on tobacco control. Pediatrics 123, e274–e281 (2009)
9. Go, M.-H., Green Jr., H.D., Kennedy, D.P., Pollard, M., STucker, J.: Peer influence and selection effects on adolescent smoking. Drug and Alcohol Dependence 109, 239–242 (2010)
10. Henriksen, L., Feighery, E.C., Schleicher, N.C., Cowling, D.W., Kline, R.S., Fortmann, S.P.: Is adolescent smoking related to the density and proximity of tobacco outlets and retail cigarette advertising near schools? Prev. Med. 47(2), 210–214 (2008)
11. Hoffman, B.R., Sussman, S., Unger, J.B., Valente, T.W.: Peer influences on adolescent cigarette smoking: A theoretical review of the literature. Substance Use and Misuse 41, 103–155 (2006)
12. Liang, L., Chaloupka, F., Nichter, M., Clayton, R.: Prices, policies and youth smoking, May 2001. Addiction 98(suppl.1), 105–122 (2003)
13. U.S. Department of Health and Human Services. The health benefits of smoking cessation: A report of the Surgeon General. Atlanta: Centers for Disease Control and Prevention, National Center for Chronic Disease Prevention and Health Promotion, Office on Smoking and Health

14. U.S. Department of Health and Human Services. Reducing the health consequences of smoking: 25 years of progress. a report of the Surgeon General. U.S. Department of Health and Human Services, Public Health Service, Centers for Disease Control, Center for Chronic Disease Prevention and Health Promotion, Office on Smoking and Health. Technical Report DHHS Publication No. (CDC) 89-8411, U.S. Department of Health and Human Services, Public Health Service, Centers for Disease Control, Center for Chronic Disease Prevention and Health Promotion, Office on Smoking and Health (1989)
15. National Institute on Drug Abuse. Infofacts: Cigarettes and other tobacco products (2011)
16. National Institute on Drug Abuse. Tobacco addiction - research report series (2011)
17. Reluga, T.C., Medlock, J., Perelson, A.S.: Backward bifurcations and multiple equilibria in epidemic models with structured immunity. Journal of Theoretical Biology 252, 155–165 (2008)

Love All, Trust a Few: Link Prediction
for Trust and Psycho-social Factors in MMOs

Muhammad Aurangzeb Ahmad[1], Zoheb Borbora[1], Jaideep Srivastava[1],
and Noshir Contractor[2]

[1] Department of Computer Science, University of Minnesota, Minneapolis, MN, USA
{mahmad,zborbora,srivastav}@cs.umn.edu
[2] School of Communication, Northwestern University, Evanston, IL, USA
nosh@northwestern.edu

Abstract. Massively Multiplayer Online Games (MMOGs) where millions of people can interact with one another have been described as mirrors of human societies and offer excellent venues to analyze human behavior at both the psychological as well as the social level. Within the context of predictive analysis (link prediction as a classification task) in MMOGs, the connection between psycho-sociological theories of communication networks. A mapping of how various elements of trust and other social interactions (mentoring, adversarial relationship, trade) relate to prediction tasks is also established. Results from classification experiments indicate that social environments affect prediction tasks in cooperative vs. adversarial environments in MMOGs and the implications of these results for generalizability of link prediction algorithms is also analyzed.

Keywords: Trust in social networks, Prediction and Psycho-Social Theories, Adversarial environments, MMOs.

1 Introduction

One of the seminal events of the last decade has been the explosion of myriad arrays of various form of social media which generate gigabytes of data every hour and thus provide an unprecedented opportunity to analyze human behavior on a massive scale. Mainly because of this data revolution it is now possible to not just build better theories regarding human behavior but also move from a descriptive analysis of social data to a predictive analysis. One issue which is usually coterminous with predictive modeling is that it is often the case that the models do not explain the psychological and social reasons behind *why* the model is successful in predictive analysis and thus essentially a black box. We consider these issues in the context of the link prediction problem.

While the problem of link prediction has been studied before in a number of contexts in social networks, we note that this problem has not been addressed with respect to the role of social science theories to explain the efficacy of featuresets in prediction tasks. One step in that direction is work by Ahmad et al [4]

S.J. Yang, A.M. Greenberg, and M. Endsley (Eds.): SBP 2012, LNCS 7227, pp. 123–130, 2012.

who try to incorporate Monge and Contractor's Multi-Theoretical Multi-Level framework [10] in the link prediction tasks. We take their work one step further by linking the feature space to theory space and additionally describe how the results of prediction tasks can be interpreted in terms of social science theories.

2 Background

The link prediction problem consists of a family of prediction problems which may range from predicting the formation [9], breakage [11], change of links to recurrence in the edge formation [12]. The link prediction problem was first described by Liben-Nowell and Kleinberg [9] and the Inter-Network Link Prediction Problem was first described by Ahmad et al [4] who also proposed a social science theory based approach to address that problems. In a follow up work Borbora et al [5] explored the problem of efficacy of feature space associated with link prediction to determine a robust set of features for link prediction.

Model based explanations for predictive modeling can be divided into three main types: (i) Explanations regarding how the algorithm works (ii) Explanations regarding how the model explains the phenomenon, such explanations are usually absent from black box models e.g, Neural Nets. (iii) In social, psychological and cognitive domains explanations that link the model to motivations that can be ascribed to intentional agents (people) or groups of such agents (society). In recent years there has been a move towards linking prediction algorithms, models and feature spaces to explanations in terms of social and psychological theories when these involve social phenomenon. That is mainly because an explanation agnostic model would not gain much currency in the social science domain where the primary goal is to not just study these phenomenon but also provide explanations with respect to why things happen. Borbora et al [6] thus note the distinction between theory driven and data driven models and how one can inform the other in creating better predictive models.

3 A Psycho-Social Framework for Link Prediction

The MTML framework [10] describes the creation, maintenance and development of linkages in social networks in organizational and inter-organizational contexts and links together various theories in the sociology literature which also harkens to psychological motivations regarding why people form relationships with one another. We refer the reader to the text by Monge and Contractor [10] for a detailed descritionb of the theories of communication networks. We take these theories as starting point in feature set construction and also the partition of the feature space based on the appropriate features. We use the feature-set scheme used by Hasan et al [7] and modified by Ahmad et al [4] as our starting point but we expand it to include additional features which are more appropriate for a larger social space. They divide their feature space into three sets of features as follows: (i) Proximity Features (ii) Aggregated Features (iii) Topological Features. We note that this classification scheme is based on how the featuresets are

Table 1. Mapping Between Feature-sets and Theories in the MTML Framework (i)

	Self-Intr.	Cognition	Evolution	Exchange	Contag.	Homo.	Prox.
Ascribed							
Human Gender						X	
Avatar Gender						X	
Avatar Race						X	
Country						X	X
Σ Human Age						X	
Σ Avatar Age						X	
Human Age Diff.						X	
Avatar Age Diff.						X	
Σ Joining Age	X						X
Joining Age Diff.	X						X
Acquired							
Char Class Ind.	X						
Char Level Sum		X					
Char Level Diff.		X					
Guild Indicator		X			X		X
Guild Rank Sum	X	X					X
Guild Rank Diff.	X	X					

constructed with minimal or no regard to their relationship to motivations with respect to why people form links. We expand their scheme and extend the set of features and first divide them based on how they are described in the sociology literature. Thus the Proximity features can be mapped to Ascribed (attributes based on some intrinsic node characteristics) and Acquired characteristics (node characteristics which can change in time) [2]. The topological features mostly map onto the social neighborhood based characteristics. Additionally we introduce a new class of characteristics i.e., trans-social characteristics which span multiple social networks and are defined as indicator functions i.e., if the node n belongs to the network A then the value of the function is 1. The mappings between the theories and the featuresets is given in Tables 1 and 2.

4   Experiments

We use a dataset from a massively multiplayer online game called EverQuest II (EQ2) where players can interact with one another in multiple ways and there are many avenues of socialization so that it is possible to construct multiple coextensive social networks between them. To check how well the classification tasks do in different social environments, we use data from two different servers or social environments in EQ2. One of the servers (called guk) represents a cooperative or neutral environment, called Player vs. Environment (PvE). The other server (called Nagafen) represents an adversarial environment, Player vs. Player (PvP). Our main motivation behind using different social environments

Table 2. Mapping Between Feature-sets and Theories in the MTML Framework (ii)

	Self-Intr.	Cognition	Evolution	Exchange	Contag.	Homo.	Prox.
Social Neighborhood							
Degree Cent. Diff	X		X				
Betwn. Cent. Diff	X		X				
Σ Degree			X			X	X
Degree Diff.			X			X	X
Shortest distance					X		
Σ Clustering Ind.	X		X	X			
Common Neighbors				X			
Salton Index	X	X	X				
Jaccard Index	X	X	X				
Sorensen Index	X	X	X				
Adar-Adamic Index	X	X	X				
Resource Alloc. Index	X	X	X				
Trans-Social							
Trust Link		X	X	X			
Mentor Link		X	X	X			
Trade Link		X	X	X			
Group Link		X	X	X			
Combat Link		X	X				X

was that the social relationships would form differently in the two networks and thus that would be reflected in the efficacy of the prediction algorithms even though the same feature sets are used in the feature space. The network characteristics of these networks are given in Table 3 where NCC refers to the number of connected components. We use a binary classification approach for link prediction as proposed by Hasan et al [7] for link prediction within and across social networks [4]. The dataset is divided into training period and test period. For each of the tasks 60,000 instances are prepared for prediction. A positive example is when the edge does not exist in the training period but exists in the test period. In the case of the negative example the edge does not exist in either periods. We used a standard set of classifiers (Naive Bayes, Bayes Net, KNN, SVM, JRip, J48, Adaboost) for our experiments and report for best results for each classification task.

The results of the experiments for the two networks are given in Table 4. The source network refers to the network which is used to construct the training examples and the destination network is the network for whom the prediction has to be made and is from the test period. The main thing to note here is that while the results for many of the prediction tasks remain more or less the same, in a subset of the cases there is a marked difference between the results that we get for the adversarial environment vs. the cooperative environment. The cases which are markedly different for the two environments are highlighted in a different color in Table 4. Thus consider the prediction results for the mentoring network, as noted in previous work [4] and [5] the prediction performance for

Table 3. Network Statistics for all the networks used

Type	Network	Nodes	Edges	Diameter	NCC
PvE	Trust	15,465	23,145	37	1,488
PvP	Trust	13,184	15,945	27	2,237
PvE	Mentor	23,207	93,079	39	316
PvP	Mentor	36,973	187,452	≤ 27	97
PvE	Trade	31,900	1,796,438	≤ 24	11
PvP	Trade	49,132	2,142,832	≤ 24	20
PvP	Combat	59,468	3,767,395	≤ 24	32

the mentoring network is relatively low as compared to the other networks. However the results for the same prediction task but in the adversarial network are much better. This difference can be attributed to the fact that just as the adversarial environment results in greater competition between players who are in opposing teams and thus adversaries, the opposite is also true for people in the same teams i.e., one would expect greater loyalty for players in the same teams in adversarial environments as compared to people who are in cooperative environments. This results in overall better prediction results for the mentoring network prediction task. A similar difference is noted for the prediction tasks from mentoring to trade as well as from the trade to the mentoring network. Again, in both the cases the results for the adversarial environment are better as compared to the cooperative environment. The main take away from these observations is that the mentoring network is a better predictor for links in the trade network and vice versa in the adversarial environment as compared to the cooperative environment and for the same reasons. While the mentoring and trade results are commutative in this case, this is not true for the prediction tasks for the trade and trust networks i.e., there is a marked difference in the prediction results for trade to trust and not vice versa for the two environments. The main reason for this result is that a trade edge has a low transaction cost associated with it as compared to a trust edge which has a high cost associated with it. Thus a trust relationship is likely to have a corresponding trade relationship associated with it but not vice versa. Theories of co-evolution [10] would imply that in cooperative environments neutral and positive interactions (trade and trust respectively) are likely to percolate from one dimension to another but this is less likely in adversarial environments which explains the results.

We note that given the nature of the two environments the Combat network is not present in the PvE server. Additionally we have access to another network in the PvE environment, called the grouping network, which was not extracted for the PvP environment at the time of these experiments. The group network refers to an ingame network formed by players who group together to complete quests. These are analogous to military missions or other logistical missions in the offline world. The results for the Combat network in the adversarial environment and the results for the Grouping network in the cooperative environment are given in Table5. Overall the results for the combat network as well as the grouping

Table 4. Results for Link Prediction in Adversarial vs. Cooperative Environments

Networks		Cooperative			Adversarial		
Source	Destination	Precision	Recall	F-Score	Precision	Recall	F-Score
Trust	Trust	0.79	0.69	0.74	0.79	0.66	0.72
Mentor	Mentor	0.63	0.48	0.54	0.77	0.71	0.74
Trade	Trade	0.80	0.78	0.79	0.86	0.85	0.86
Trust	Mentor	0.67	0.43	0.52	0.64	0.47	0.54
Trust	Trade	0.75	0.73	0.74	0.78	0.79	0.78
Mentor	Trust	0.88	0.76	0.82	0.85	0.67	0.75
Mentor	Trade	0.74	0.74	0.74	0.84	0.85	0.84
Trade	Trust	0.89	0.83	0.86	0.88	0.75	0.81
Trade	Mentor	0.67	0.55	0.60	0.81	0.75	0.78

network are quite good even when compared against other prediction tasks. The main exception in this case is again the mentoring network where the prediction results for grouping to mentoring network are not as good as the other prediction results. The main reason for this observation is that while a large number of mentoring instances co-occur with the grouping instances i.e., mentoring occurs in the context of grouping in such cases but the opposite is not necessarily true i.e., grouping usually does not co-occur with mentoring [3].

5   Interpretation and Methodological Issues

We have considered the problem of link prediction in the context of two different social environment and a feature space mapped onto different social science theories. Our main motivation for using two different social environments is to highlight the hazards of generalization without considering the social environments associated with the prediction task. Thus consider previous results reported by Hasan et al [7], Ahmad et al [4] and Borbora [5] using similar techniques and link prediction tasks in general, the generalizability of the feature space is assumed without the social context. Theories in the social sciences e.g., the MTML framework [10] imply that social networks in different social environments evolve differently which is in turn reflected in their network structure. The differences in the network structures are also likely to affect prediction and this is in line with some of the observations that we made in the results.

There are additional methodological issues with respect to generalizing across MMOG environments. Thus consider the case of modeling of team formation dynamics in the online world by Johnson et al [8] who show that the same generative models can be used to explain guild formation in World of Warcraft and street gangs in Los Angeles. Based on their observations they generalize that there must be some common generative mechanism for team formation in online guilds and offline street gangs. Ahmad et al [1] replicated their results in EQ2 and discovered that the generalization does not carry over to EQ2. More research is required to settle this issue conclusively but both of these cases highlight the fact

Table 5. Results for Link Prediction for the Group and Combat Networks

Networks		Adversarial			Networks		Cooperative		
Source	Destination	Precision	Recall	F-Score	Source	Destination	Precision	Recall	F-Score
Group	Group	0.88	0.90	0.89	Trust	Combat	0.88	0.91	0.89
Trust	Group	0.88	0.90	0.89	Mentor	Combat	0.84	0.85	0.84
Mentor	Group	0.85	0.83	0.84	Trade	Combat	0.88	0.90	0.89
Trade	Group	0.86	0.86	0.86	Combat	Combat	0.89	0.91	0.90
Group	Trust	0.87	0.75	0.80	Combat	Trust	0.88	0.74	0.80
Group	Mentor	0.61	0.47	0.53	Combat	Mentor	0.83	0.78	0.81
Group	Trade	0.81	0.83	0.82	Combat	Trade	0.86	0.88	0.87

that generalizations are unwarranted especially in contexts where social contexts are not taken into account.

6 Conclusion

Predictive analysis, especially classification, is an important aspect of machine learning and while the internal mechanism of most classification algorithms are well understood, a mapping of feature spaces to social and psychological theories is not well understood. In this paper we considered such a mapping and used two datasets representing two social environments in an MMOG. The results showed that for a subset of the prediction tasks the prediction models perform differently using the same feature set. This implies that the network structures associated with the adversarial versus the cooperative environments are different and should inform the selection of features for future work.

Acknowledgment. The research reported herein was supported by the AFRL via Contract No. FA8650-10-C-7010, and the ARL Network Science CTA via BBN TECH/W911NF-09-2-0053. The data used for this research was provided by the SONY corporation. We gratefully acknowledge all our sponsors. The findings presented do not in any way represent, either directly or through implication, the policies of these organizations.

References

1. Ahmad, M.A., Borbora, Z., Shen, C., Srivastava, J., Wiliams, D.: Guilds Play in MMOs: Rethinking Common Group Dynamics Models SofInfo 2011, October 6-8 (2011)
2. Ahmad, M.A., Ahmad, I., Srivastava, J.: Marshall Poole Trust me, I m an Expert: Trust, Homophily and Expertise. In: MMOs IEEE SocialCom 2011, Boston, MA, October 9-11 (2011)
3. Ahmad, M.A., Huffaker, D., Wang, J., Treem, J., Kumar, D., Poole, M.S., Srivastava, J.: The Many Faces of Mentoring in an MMORPGs. In: IEEE Social Computing Workshop on Social Intelligence in Applied Gaming, SocialCom 2010, Minneapolis, MN, USA, August 20-22 (2010)

4. Ahmad, M.A., Borbora, Z., Srivastava, J.: Noshir Contractor Link Prediction Across Multiple. In: Social Networks Domain Driven Data Mining Workshop (DDDM 2010), ICDM 2010 Sydney, Australia (2010)
5. Borbora, Z.H., Ahmad, M.A., Haigh, K.Z., Srivastava, J., Wen, Z.: Exploration of robust features of trust across multiple social networks. In: Fifth IEEE Conference on Self-Adaptive and Self-Organizing Systems Workshops (SASOW 2011), October 3-7. Ann Arbor, Michigan (2011)
6. Borbora, Z., Hsu, K.-W., Srivastava, J., Williams, D.: Churn Prediction in MMORPGs using player motivation theories and ensemble approach. In: Third IEEE International Conference on Social Computing. MIT, Boston (2011)
7. Al Hasan, M., Chaoji, V., Salem, S., Zaki, M.: Link prediction using supervised learning. In: Workshop on Link Counter-terrorism and Security. SIAM (2006)
8. Johnson, N.F., Xu, C., Zhao, Z., Ducheneaut, N., Yee, N., Tita, G., et al.: Human group formation in online guilds and offline gangs driven by a common team dynamic. Physical Review E 79(6), 066117 (2009)
9. Liben-Nowell, D., Kleinberg, J.M.: The link prediction problem for social networks. In: CIKM 2003 (2003)
10. Monge, P., Contractor, N.: Theories of Communication Networks. Oxford University Press, Cambridge (2003)
11. Sharan, U., Neville, J.: Exploiting Time-Varying Relationships in Statistical Relational Models. In: SNA-KDD 2007 (2007)
12. Tylenda, T., Angelova, R., Bedathur, S.: Towards Time-aware Link Prediction in Evolving Social Networks. In: KDD-SNA 2009 (2009)

Effect of In/Out-Degree Correlation on Influence Degree of Two Contrasting Information Diffusion Models

Kouzou Ohara[1], Kazumi Saito[2], Masahiro Kimura[3], and Hiroshi Motoda[4]

[1] Department of Integrated Information Technology, Aoyama Gakuin University
ohara@it.aoyama.ac.jp
[2] School of Administration and Informatics, University of Shizuoka
k-saito@u-shizuoka-ken.ac.jp
[3] Department of Electronics and Informatics, Ryukoku University
kimura@rins.ryukoku.ac.jp
[4] Institute of Scientific and Industrial Research, Osaka University
motoda@ar.sanken.osaka-u.ac.jp

Abstract. How the information diffuses over a large social network depends on both the model employed to simulate the diffusion and the network structure over which the information diffuses. We analyzed both theoretically and empirically how the two contrasting most fundamental diffusion models, Independent Cascade (IC) and Linear Threshold (LT) behave differently or similarly over different network structures. We devised two rewiring structures, one preserving in/out-degree correlation and the other changing in/out-degree correlation while both preserving their in/out-degree distributions, and analyzed how co-link rate and in/out-degree correlation affect the influence degree of each diffusion model using two real world networks, each as the base network on which rewiring is imposed. The results of the theoretical analysis qualitatively explain the empirical results, and the findings help deepen the understanding of complex diffusion phenomena.

Keywords: Information diffusion, network structure, influence degree, node degree distribution.

1 Introduction

The emergence of Social Media such as Facebook, Digg and Twitter has provided us with the opportunity to create large social networks, which are becoming an important medium for spreading information. Recently, substantial attention has been devoted to analyzing and mining social networks from the point of information diffusion [14,15,11,19,2,1,16]. One of the most well studied problems is the influence maximization problem, *i.e.*, the problem of finding a limited number of influential nodes that are effective for the spread of information. Many algorithms have been proposed to solve the problem using probabilistic information diffusion models on a network [8,12,5,9,6,4]. In order to investigate diffusion phenomena using probabilistic models, it is indispensable to understand the behavioral differences among models, and provide an effective method for selecting the most appropriate model for a particular task we want to analyze.

S.J. Yang, A.M. Greenberg, and M. Endsley (Eds.): SBP 2012, LNCS 7227, pp. 131–138, 2012.
© Springer-Verlag Berlin Heidelberg 2012

There are two contrasting fundamental probabilistic models that have been widely used by many researchers. One is the *independent cascade (IC)* model [7,8] and the other is the *linear threshold (LT)* model [18,8]. The IC model takes a sender-centered approach such that each information sender independently influences its neighbors with some probability (*information push style model*). The LT model is a receiver-centered approach such that each information receiver adopts the information if and only if the number of its neighbors that have adopted the information exceeds some threshold, where the threshold is treated as a random variable (*information pull style model*). We analyze how the IC and the LT models differ from or similar to each other in terms of information diffusion for a wide range of social networks with different structures.

In this paper, we compare *influence degree* obtained by the IC and the LT models from the network structure perspective. Here, the influence degree of a node v under a probabilistic diffusion model in a network is defined to be the expected number of *active* nodes at the end of the information diffusion process that starts from the initial active node v, where nodes that have been influenced with the information are referred to as being active. First, we theoretically analyze the properties of the IC and the LT models on scale-free networks, and derive the following two properties: 1) as the in/out-degree correlation decreases, the influence degree decreases for the IC model but it does not change for the LT model and 2) as the co-link (bidirectional link) rate decreases, the influence degree increases for both the IC and the LT models, but the IC model is much less sensitive than the LT model. To verify these properties, we systematically generated a series of scale-free networks with varying in/out-degree correlation and co-link rate, applying two rewiring strategies, one preserving in/out-degree correlation and the other changing in/out-degree correlation while both preserving their in/out-degree distributions. We used two real world scale free networks as the bases to apply these strategies, and experimentally confirmed that the above two properties indeed hold.

2   Diffusion Models

Let $G = (V, E)$ be a directed network, where V and E ($\subset V \times V$) are the sets of all the nodes and links, respectively, and $|V| \le |E|$ can be naturally assumed for commonly-seen social networks. We recall the definition of the IC and the LT models according to the literatures [8,9]. In these models, the diffusion process proceeds from an initial active node in discrete time-step $t \ge 0$, and it is assumed that nodes can switch their states only from inactive to active (*i.e.*, the SIR setting).

The IC model has a *diffusion probability* $p_{u,v}$ with $0 < p_{u,v} < 1$ for each link (u, v) as a parameter. Suppose that a node u first becomes active at time-step t, it is given a single chance to activate each currently inactive child node v, and succeeds with probability $p_{u,v}$. If u succeeds, then v will become active at time-step $t + 1$. If multiple parent nodes of v first become active at time-step t, then their activation trials are sequenced in an arbitrary order, but all performed at time-step t. Whether u succeeds or not, it cannot make any further trials to activate v in subsequent rounds. The process terminates if no more activations are possible.

The LT model has a *weight* $q_{u,v}$ (> 0) with $\sum_{u \in B(v)} q_{u,v} \le 1$ for each link (u, v) as a parameter, where $B(v) = \{u \in V; (u, v) \in E\}$ is the set of parent nodes of node v. First, for any node $v \in V$, a *threshold* θ_v is chosen uniformly at random from the interval $[0, 1]$. An inactive node v is influenced by its active parent nodes. If the total weight from v's active parent nodes at time-step t is no less than θ_v, i.e., $\sum_{u \in B_t(v)} q_{u,v} \geq \theta_v$, then v will get activated at time-step $t + 1$. Here, $B_t(v)$ is the set of all the parent nodes of v that are active at time-step t. The process terminates if no more activations are possible.

3 Analysis of Local Influence Degree

We first define local influence degree of node u, denoted by $\sigma_L(u)$, as the expected number of u's child nodes directly activated by u. For the IC model, $\sigma_L^{IC}(u)$ is given by $\sigma_L^{IC}(u) = \sum_{v \in F(u)} p_{u,v}$, where $F(u)$ stands for the set of u's child nodes defined by $F(u) = \{v \in V; (u, v) \in E\}$. For the LT model $\sigma_L^{LT}(u)$ is given by $\sigma_L^{LT}(u) = \sum_{v \in F(u)} q_{u,v}$ because each weight $q_{u,v}$ is regarded to be the probability that the threshold θ_v is chosen from the interval $[0, q_{u,v}]$. Then, we can calculate the average local influence degree over all nodes, denoted by $\bar{\sigma}_L(G)$. For the LT model, if we impose the condition $\sum_{u \in B(v)} q_{u,v} = 1$ for any node $v \in V$, we can prove $\bar{\sigma}_L^{LT}(G) = 1$ from the following relations.

$$\bar{\sigma}_L^{LT}(G) = \frac{1}{|V|} \sum_{u \in V} \sigma_L^{LT}(u) = \frac{1}{|V|} \sum_{u \in V} \sum_{v \in F(u)} q_{u,v} = \frac{1}{|V|} \sum_{(u,v) \in E} q_{u,v} = \frac{1}{|V|} \sum_{v \in V} \sum_{u \in B(v)} q_{u,v} = 1.$$

For the IC model, if we impose the uniform diffusion probability setting, i.e., $p_{u,v} = p$ for any link $(u, v) \in E$, which has been employed in many previous studies (*e.g.*, [8]), we can calculate $\bar{\sigma}_L^{IC}(G)$ as follows:

$$\bar{\sigma}_L^{IC}(G) = \frac{1}{|V|} \sum_{u \in V} \sigma_L^{IC}(u) = \frac{1}{|V|} \sum_{u \in V} \sum_{v \in F(u)} p_{u,v} = \frac{1}{|V|} \sum_{(u,v) \in E} p = \frac{|E|}{|V|} p,$$

where $\frac{|E|}{|V|}$ is equal to the average degree $d = \frac{1}{|V|} \sum_{u \in V} |B(u)| = \frac{1}{|V|} \sum_{u \in V} |F(u)| = \frac{|E|}{|V|}$, and is no less than 1 as we assume $|V| \leq |E|$. Thus, by setting the uniform diffusion probability to the inverse of average degree, i.e., $p = \frac{1}{d} = \frac{|V|}{|E|}$, we obtain $\bar{\sigma}_L^{IC}(G) = 1$. This makes the IC and LT models equivalent in terms of the average local influence degree. Hereafter, we impose these settings to evaluate the influence degree. Note that local influence degree of node u for the IC model becomes $\sigma_L^{IC}(u) = \sum_{v \in F(u)} p_{u,v} = p|F(u)|$.

So far we focused on local influence degree of node $u \in V$ under the condition that the node u has become active. However, when considering the cascade of information diffusion, we need to consider the probability $r(u)$ that the node u is activated by its parent nodes. Namely, we consider cascading local influence degree defined by $\sigma_{CL}(u) = r(u)\sigma_L(u)$. As the simplest case, we employ the probability $r(u)$ that the node u is activated at the next time step by some active node selected uniformly at random from the node set V. For the IC model, $r^{IC}(u)$ is given by $r^{IC}(u) = \frac{1}{|V|} \sum_{s \in B(u)} p_{s,u} = \frac{p|B(u)|}{|V|}$, and for the LT model, $r^{LT}(u)$ is given by $r^{LT}(u) = \frac{1}{|V|} \sum_{s \in B(u)} q_{s,u} = \frac{1}{|V|}$. Thus we obtain the average cascading local influence degree $\bar{\sigma}_{CL}$ for the IC and LT models as follows:

$$\bar{\sigma}_{CL}^{IC}(G) = \frac{1}{|V|} \sum_{u \in V} r^{IC}(u)\sigma_L^{IC}(u) = \frac{p^2}{|V|^2} \sum_{u \in V} |B(u)||F(u)|, \tag{1}$$

$$\bar{\sigma}_{CL}^{LT}(G) = \frac{1}{|V|} \sum_{u \in V} r^{LT}(u)\sigma_L^{LT}(u) = \frac{1}{|V|^2} \sum_{u \in V} \sigma_L^{LT}(u) = \frac{1}{|V|}. \tag{2}$$

Therefore, by noting that the in/out-degree correlation $dc_{I/O}(G)$ is quantified by

$$dc_{I/O}(G) = \frac{\frac{1}{|V|} \sum_{u \in V} |B(u)||F(u)| - d^2}{\sqrt{\frac{1}{|V|} \sum_{u \in V} |B(u)|^2 - d^2} \sqrt{\frac{1}{|V|} \sum_{u \in V} |F(u)|^2 - d^2}},$$

and the denominator of $dc_{I/O}(G)$ is determined by the standard deviations of in/out-degree distributions, we can see that the average cascading local influence degree of the IC model is affected by the in/out-degree correlation $dc_{I/O}(G)$ when the standard deviations are fixed, as shown in Eq. (1), while that of the LT model is not affected, as shown in Eq. (2). Namely, we can conjecture that influence degree of the IC model also decreases when the in/out-degree correlation decreases.

Another important factor affecting influence degree is the co-link rate $cr(G)$ which is defined by $cr(G) = \frac{1}{|E|} \sum_{u \in V} |B(u) \cap F(u)|$. Evidently, for a bidirectional network G, we obtain $cr(G) = 1$ because $B(u) = F(u)$ for any $u \in V$. Assume a node $v \in B(u) \cap F(u)$; if v succeeds activating u, then the reverse link (u, v) never contributes to increasing an active node, conversely, if u succeeds activating v, then the reverse link (v, u) never does so. Thus, we conjecture that influence degree of the IC and LT model increases when the co-link rate $cr(G)$ decreases. However, there is a subtle difference between the IC and the LT models. Think of the network with co-link rate close to 1. Evidently the in/out-degree correlation is also close to 1. Assume that k parents of a node v which has a large degree $D = |F(v)| = |B(v)|$ get activated. The expected probability that the node v becomes activated is $1 - (1 - 1/d)^k$ for the IC model and k/D for the LT model where d is the average node degree. For the IC model the probability is large for a small number of k and insensitive to $|D|$. Thus, once it gets activated, the reverse k links which do not contribute further activation is small. On the other hand, for the LT model the node v is not activated unless k is large. Thus, once it gets activated, the reverse k links do not contribute further activation is also large. This implies that the IC model is less sensitive to the change of co-link rate than the LT model.

4 Experiments

To confirm our conjectures in Section 3, we conducted extensive experiments using both synthetic and real world large networks, rewiring their links according to the two strategies presented in this section. However, due to the page limitation, we show only the results for the two real world networks: one bidirectional and the other directional[1].

4.1 Rewiring Strategies

We devised two rewiring strategies. Both preserve the in/out-degree distribution. The first one rewires links of a given network G preserving the in/out-degrees of each node, which is equivalent to the method of generating randomized networks presented in [13]. We implemented this strategy by swapping the two destination nodes v and v'

---

[1] The networks we omitted here include synthetic networks generated by the BA model [3] and the CNN model [17], and four other networks derived from the real world data.

of links $e = (u, v)$ and $e' = (u', v')$ from two starting nodes u and u'. The links are chosen uniformly at random. Obviously, this never changes $dc_{I/O}(G)$, but does change $cr(G)$. We refer to this rewiring strategy as the DCP (in/out-Degree Correlation Preserved) method, and denote the network G rewired by this method by $dcp_\alpha(G)$, where α is the link rewiring probability, $i.e.$, v of e and v' of e' are swapped with the probability α. The larger α is, the smaller $cr(G)$ is. Thus, the DCP method allows us to investigate how the co-link rate affects the influence degree of the IC and the LT models. The second one rewires links changing the in/out-degree correlation. This is to confirm our conjecture that the in/out-degree correlation affects the influence degrees of the IC model. We implemented this by swapping $E_I(v)$, all the incoming links to a node v, and $E_I(v')$, all the incoming links to a node v' with a probability α. Nodes v and v' are randomly chosen. Namely, $E_I(v)$ becomes $\{(u, v); u \in B(v')\}$, and $E_I(v')$ becomes $\{(s, v'); s \in B(v)\}$ after swapping. This method changes the in-degree of chosen nodes without changing their out-degree while preserving the in/out-degree distributions of the network G. We refer to this method as the DCU (in/out-Degree Correlation Unpreserved) method, and denote the network G rewired by the DCU method with a link rewiring probability α by $dcu_\alpha(G)$. The larger α is, the smaller the in/out-degree correlation is.

4.2 Datasets and Network Structure

In this section, we explain the two real world networks for which we present the experimental results. The first one is a bidirectional network derived from the Enron Email Dataset [10]. We regarded each email address as a node, and constructed a bidirectional link between two email addresses u and v only if u sent an email to v and received an email from v. After that, we extracted the maximal strongly connected component. We refer to this bidirectional network as the Enron network, which has $4,254$ nodes and $44,314$ directed links. The second one is a directional network derived from a Japanese word-of-mouth communication site for cosmetics, "@cosme"[2], where each user page is associated with *fan links*. A fan link from user u to user v is generated if user v registers user u as his/her favorite user. We extracted a fan network from @cosme by tracing up to ten steps in the fan links starting from a randomly chosen user in December 2009. The resulting network has $45,024$ nodes and $351,299$ directed links. We refer to this network as the Cosme network.

For these networks, we investigated the influence degree $\sigma(v)$ of each node v of the networks $dcp_\alpha(G)$ and $dcu_\alpha(G)$ under the IC and the LT models, varying α from 0.0 to 1.0 by 0.1. Note that $dcp_{0.0}(G) = dcu_{0.0}(G) = G$. The influence degree $\sigma(v)$ was estimated by the empirical mean of the number of active nodes obtained from 10,000 independent runs of information diffusion based on the bond percolation technique [9]. According to the discussion in Section 3, we set a unique value $p = 1/d$ to every $p_{u,v}$ for the IC model. Namely, p was set to 0.10 for the Enron network, and 0.13 for the Cosme network.

---

[2] http://www.cosme.net/

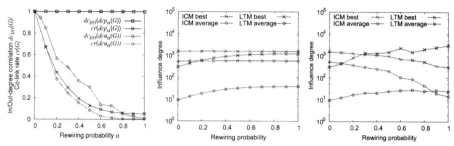

(a) The change of In/Out-degree correlation and Co-link rate.

(b) The best and the average influence degrees with the DCP method.

(c) The best and the average influence degrees with the DCU method.

Fig. 1. Experimental results for the Enron network

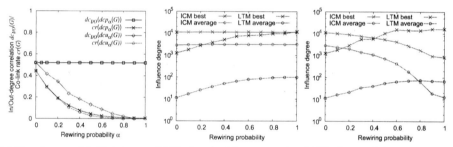

(a) The change of In/Out-degree correlation and Co-link rate.

(b) The best and the average influence degrees with the DCP method.

(c) The best and the average influence degrees with the DCU method.

Fig. 2. Experimental results for the Cosme network

4.3   Experimental Results

Figures 1a and 2a show how the in/out-degree correlation $dc_{I/O}(G)$ and the co-link rate $cr(G)$ of a given network G change with the two rewiring methods, DCP and DCU, for the Enron and the Cosme networks, respectively. We see that both methods work just as we intended: $cr(G)$ decreases in a similar fashion for both the DCP and the DCU methods, as the rewiring probability α becomes larger, while $dc_{I/O}(G)$ does not change with the DCP method, but it does decrease similarly to $cr(G)$ with the DCU method. Note that both $dc_{I/O}(G)$ and $cr(G)$ of the Enron network are 1.0 for $\alpha = 0.0$ because it is bidirectional.

Figure 1b illustrates how the DCP method affects the best and the average influence degrees over all the nodes of the Enron network. As we expected, both influence degrees of the LT model become larger as the rewiring probability becomes larger, and the co-link rate becomes smaller. The influence degrees of the IC model does not seem to increase, but indeed they slightly increase within the range of $\alpha = 0.0$ to 0.6 where the co-link rate drastically decreased. This qualitatively supports the analysis in Section 3. The same tendencies can be found in the result for the Cosme network as shown in Fig. 2b. We also observed the same tendencies for the other networks we omitted here.

Figures. 1c and 2c show how the DCU method affects the best and the average influence degrees of the IC and the LT models. Both $dc_{I/O}(G)$ and $cr(G)$ decrease with α. This imposes two conflicting factors for the IC model, but the effect of $dc_{I/O}(G)$ surpasses and the influence degrees of the IC model decrease for both the Enron and the Cosme networks. On the other hand, the influence degrees of the LT model are affected by only $cr(G)$. Thus, they increase in the same way as in Figs. 1b and 2b. The same observation is obtained for the other networks. This also qualitatively supports the analysis in Section 3.

5   Conclusion

Understanding how information diffuses over a large social network is important to do any kind of social network analysis, but it is difficult because actual diffusion depends on both the diffusion model employed and the properties of the network structure over which the information diffuses. Independent Cascade (IC) and Linear Threshold (LT) models have been used widely by many researchers. Both are probabilistic models but have contrasting properties, *i.e.*, information push (IC) and information pull (LT). Social networks have common characteristics. The most important one would be the scale free property. There can be many structures that hold this property. We devised two rewiring strategies that can systematically transform one network structure to another structure preserving the scale free property, one preserving in/out-degree correlation (DCP method) and the other changing in/out-degree correlation (DCU method). Each strategy was successively applied with different probabilities to two real world social networks, generating a series of networks, each with a gradually changing structure. We chose co-link rate and in/out-degree correlation as the two parameters that characterize the network structure, and investigated how these parameters affects the influence degree of the two models (IC and LT). The major new findings are 1) the IC model is sensitive to in/out-degree correlation and the influence degree is positively correlated to it, whereas the LT model is insensitive to it and 2) Both the IC and the LT models are negatively correlated to co-link rate, but its dependency is much less sensitive in the IC model. These properties can be qualitatively derived by the theoretical analysis and verified by the extensive experiments using the above networks as well as others not reported in this paper. These findings are useful in deepening our understanding of the complex information diffusion phenomena over a social network.

Acknowledgments. This work was partly supported by Asian Office of Aerospace Research and Development, Air Force Office of Scientific Research under Grant No. AOARD-11-4111, and JSPS Grant-in-Aid for Scientific Research (No. 23700181).

References

1. Bakshy, E., Hofman, J., Mason, W., Watts, D.: Everyone's an influencer: Quantifying influences on twitter. In: Proceedings of the 4th International Conference on Web Search and Data Mining (WSDM 2011), pp. 65–74 (2011)
2. Bakshy, E., Karrer, B., Adamic, L.A.: Social influence and the diffusion of user-created content. In: Proceedings of the 10th ACM conference on Electronic Commerce, pp. 325–334 (2009)
3. Barabási, A.L., Albert, R.: Emergence of scaling in random networks. Science 286, 509–512 (1999)
4. Chen, W., Wang, C., Wang, Y.: Scalable influence maximization for prevalent viral marketing in large-scale social networks. In: Proceedings of the 16th ACM SIGKDD International Conference on Knowledge Discovery and Data Mining (KDD 2010), pp. 1029–1038 (2010)
5. Chen, W., Wang, Y., Yang, S.: Efficient influence maximization in social networks. In: Proceedings of the 15th ACM SIGKDD International Conference on Knowledge Discovery and Data Mining (KDD 2009), pp. 199–208 (2009)
6. Chen, W., Yuan, Y., Zhang, L.: Scalable influence maximization in social networks under the linear threshold model. In: Proceedings of the 10th IEEE International Conference on Data Mining (ICDM 2010), pp. 88–97 (2010)
7. Goldenberg, J., Libai, B., Muller, E.: Talk of the network: A complex systems look at the underlying process of word-of-mouth. Marketing Letters 12, 211–223 (2001)
8. Kempe, D., Kleinberg, J., Tardos, E.: Maximizing the spread of influence through a social network. In: Proceedings of the 9th ACM SIGKDD International Conference on Knowledge Discovery and Data Mining (KDD 2003), pp. 137–146 (2003)
9. Kimura, M., Saito, K., Nakano, R., Motoda, H.: Extracting influential nodes on a social network for information diffusion. Data Mining and Knowledge Discovery 20, 70–97 (2010)
10. Klimt, B., Yang, Y.: The Enron Corpus: A New Dataset for Email Classification Research. In: Boulicaut, J.-F., Esposito, F., Giannotti, F., Pedreschi, D. (eds.) ECML 2004. LNCS (LNAI), vol. 3201, pp. 217–226. Springer, Heidelberg (2004)
11. Leskovec, J., Adamic, L.A., Huberman, B.A.: The dynamics of viral marketing. In: Proceedings of the 7th ACM Conference on Electronic Commerce (EC 2006), pp. 228–237 (2006)
12. Leskovec, J., Krause, A., Guestrin, C., Faloutsos, C., VanBriesen, J., Glance, N.: Cost-effective outbreak detection in networks. In: Proceedings of the 13th ACM SIGKDD International Conference on Knowledge Discovery and Data Mining (KDD 2007), pp. 420–429 (2007)
13. Melo, R., Shen-Orr, S., Itzkovitz, S., Kashtan, N., Chklovskii, D., Alon, U.: Network motifs: Simple building blocks of complex networks. Science 298, 824–827 (2002)
14. Newman, M.E.J., Forrest, S., Balthrop, J.: Email networks and the spread of computer viruses. Physical Review E 66, 035101 (2002)
15. Richardson, M., Domingos, P.: Mining knowledge-sharing sites for viral marketing. In: Proceedings of the 8th ACM SIGKDD International Conference on Knowledge Discovery and Data Mining (KDD 2002), pp. 61–70 (2002)
16. Romero, D., Meeder, B., Kleinberg, J.: Differences in the mechanics of information diffusion across topics: Idioms, political hashtags, and complex contagion on twitter. In: Proceedings of the 20th International World Wide Web Conference (WWW 2011), pp. 695–704 (2011)
17. Vázquez, A.: Growing network with local rules: Preferential attachment, clustering hierarchy, and degree correlations. Physical Review 67(5), 056104 (2003)
18. Watts, D.J.: A simple model of global cascades on random networks. Proceedings of National Academy of Science, USA 99, 5766–5771 (2002)
19. Watts, D.J., Dodds, P.S.: Influence, networks, and public opinion formation. Journal of Consumer Research 34, 441–458 (2007)

A Computer-in-the-Loop Approach
for Detecting Bullies in the Classroom

Juan F. Mancilla-Caceres, Wen Pu, Eyal Amir, and Dorothy Espelage

University of Illinois at Urbana-Champaign, Urbana IL 61801, USA
{mancill1,wenpu1,eyal,espelage}@illinois.edu

Abstract. Bullying is a social phenomenon that is highly prevalent within the school population. To study this phenomenon, social scientists traditionally use questionnaires that are costly to administer and that cannot provide detailed information about children's interactions without causing a large amount of fatigue to the participants. An on-line computer game has been developed to aid social scientists in observing, in a non-intrusive way, children's behaviors and roles within their peer group. Participants solve a collaborative and an adversarial task, and are allowed to communicate only through a chat system. Observable data from the game, such as the amount of messages sent and received and points transactions, correlates well with questionnaire data while providing more detailed information about participants' interactions. The online game is a new tool that alleviates the cost of obtaining data and considerably reduces the fatigue of the participants while providing sound results.

Keywords: Group interaction and collaboration, Influence process and recognition, Methodological innovation.

1   Introduction

Computer involvement in adolescent networks has been growing in recent years, giving rise to new group dynamics and new opportunities to study group interactions and individual preferences. Reseach in computer science already takes advantage of the availability of this information, but is biased towards visible computer-mediated social networks when friendship nominations are assumed equal and meaningful. Social science research, on the other hand, applies well founded psychological techniques, but finds difficulty using the data available to computer scientists because of limited computational tools and the partial nature of the data.

In this paper, we present a new method for gathering indirect information about friendships in classrooms and for observing real interactions between adolescents using an on-line social game. The game is used as a tool for observing the behavior of the participants in a non-intrusive way and for gathering information about friendships and roles that each participant plays within their social groups. Current research suggests that peer socialization is one of the key

S.J. Yang, A.M. Greenberg, and M. Endsley (Eds.): SBP 2012, LNCS 7227, pp. 139–146, 2012.
© Springer-Verlag Berlin Heidelberg 2012

factors that determines risky behavior in adolescent social networks [6]. However, in order to understand the roles and behaviors of adolescents, traditional methods involve surveys and questionnaires that can yield limited information due to their cost and potential participant fatigue. These surveys are usually time-consuming and in some cases, they are simply not feasible to use with large school populations. In addition, real social-interaction data (e.g., adolescent manipulations) are difficult to collect, evaluate, and validate, and it is difficult to use available data (e.g., Facebook friendships) to draw sound conclusions.

Instead, we propose that the roles of the participants in the classroom (*bully*, *victim*, or *bystander*) are observable through social interactions within a game designed to simulate common real-world situations, such as collaboration and competition within groups. The data obtained using the game consists on teammate nominations of participants, text messages amongst the players (which can be either public or private), and point transactions. We show that game outcomes are correlated with survey data gathered previously from the same participants, and that they can be eventually used to predict the values of the surveys, and to provide updates to the roles found using survey data (as will be shown in a case analysis).

2   Related Work

Game-based methods for data collection have been previously used as an alternative method to crowdsourcing. One of the first successful examples of what has been called *Games with a Purpose* (GWAPs) was the ESP game [8] used for labeling images by volunteer contributors. In the context of social networks, GWAPs have been created to take advantage of people's engagement on such networks. For example, Collabio [1] is a social tagging game within the online social network Facebook, which encourages friends to tag one another. Collabio's encourages members of the social network to generate information about each other by producing tags that describe another individual. Also, the Turing Game [5] is a Facebook multiplayer game that encourages players to verify common-sense knowledge by carefully designing the rules and stages of the game so that the players have an incentive to generate the appropriate data.

The game described in this paper differs from those previously described in the sense that the desired outcome of the game is not the specific action the players take on the game (i.e., the classification of the image or text by the player), but the interactions amongst the players themselves. This is a novel approach to gather data and information because, although the design of the game is important (in order to encourage participation, engagement, and messages), the actual game task is not crucial (in this case, answering trivia questions). Most of the other games for data collection can be played by a single person, but our game is played in teams, which is necessary to gather information about social group dynamics. Our data directly reflects the actual interactions of the players, thus reducing the noise in the data. This kind of game is especially useful for non-intrusive data-collection in the social sciences because the games provide a natural data-intensive interaction with participants. The game is applied here to learn about social communications, manipulation methods, and reactions to natural occurrences of friendliness or aggression.

With respect to using automatic methods to detect bullying, previous research has focused on cyber-bullying and not on physical bullying as is our case. Some examples are [4] and [7], which focus exclusively on observable behavior (e.g., insulting or racist messages), whereas our approach is aimed at finding the latent behavior associated to children's roles within the classroom, which may or may not be explicit in the game.

3   Game Design

The proposed data collection platform is a multiplayer game played in teams of 3 or 4 members. The game is designed to answer the following questions: 1. What are the friendship relationships amongst the participants, and what kind of interaction do they have (cordial, aggressive, polite, etc.)? 2. How are the loyalties and trust placed amongst the participants, and how do the rules of the game encourage leadership, competitiveness, etc.?

The game is played via a computer network and follows the steps described below:

1. Each user has the opportunity to nominate other participants whom they would or would not like to have on their team. In this stage of the game, we obtain information about task-directed peer nomination. Each team is currently created using a priori information gathered through surveys, ensuring that on each team there is at least one bully and one victim on each team.
2. The second stage of the game consists on collaborating to answer a set of trivia questions, ensuring that all members of the team submit the same answer in order to obtain a reward (in this case, points in the form of coins).
3. The third stage is competitive or adversarial. During this task, each member of the team must provide a different answer to the question while one team member must choose a clearly wrong answer, effectively losing points while the rest of the team wins points. In contrast to the collaborative task, in which the entire team must work together to maximize their individual reward, in the competitive task players are encouraged to directly oppose other members by convincing (or coercing) them to pick the wrong answer.
4. There are two winners in each game: a winning team (summing up all the individual rewards) and a winning player (the one with the larger amount of coins).

During both tasks, participants are encouraged to use the chat system to coordinate their answers and to trade coins (points) amongst themselves. Figure 1 shows a screen shot of the nomination screen (left) and of the current interface of the game (right).

During the collaborative stage of the game, team members work together to answer a set of 5 trivia questions (about topics such as history, geography, pop

Fig. 1. On the left: Nomination screen shot. Each player has the opportunity of stating with whom they want or don't want to play; or alternatively, of stating that they have no preference. On the right: Screen shot of one question during the collaborative stage.

culture, etc.). Each question has four possible answers, and only one of them is correct. For each question, all the team members must agree on the right answer or otherwise no points are awarded to anyone in the team. These rules ensure that the players in the team must communicate and collaborate to agree on an answer, or everybody loses. The fact that the team shares the payoffs and outcomes guarantees that everyone shares the same interest. For each question, team members have the option to peek at the correct answer in exchange for points from one of the players. In order to maximize the utility of the game, players must balance the peek penalty by sharing points amongst themselves. It is in the best interest of each player to share their knowledge about the question and not to let other players peek at the answers (in order to retain points).

During the competitive stage, each team receives a set of 5 trivia questions with four possible answers. Three of the four answers are correct and one of them is incorrect. In this task, each player on a team must choose a different answer for each question, with the added constraint that at least one member of the team must pick the wrong answer. Only the players that choose a correct answer get points. The wrong answer is marked explicitly (written in bold letters) in order to make it obvious and to encourage players to discuss who will pick such answer. It is in the best interest of each player not to pick the wrong answer, but also to ensure that someone else in their team picks it. This can only be accomplished by negotiating (either aggressively or non-aggressively) through the chat channel.

The game is intended to emulate the circumstances of natural interactions amongst participants. Its features include the presence of limited resources (i.e., points and coins), a collaborative task, and a competitive task. By restricting the channel of communication to text messages, we have a non-intrusive way to monitor and analyze the interactions of participants with their peers (a method previously approved by an IRB).

Each game provides the following output:

- Users' team preferences: friends/rivals nominations and the order in which they are selected.
- Raw messages from users: all chat messages both in public and private channels along with the time at which the message was sent.
- Points transactions: transfer and forfeiture of points (e.g. in exchange for information).

4   Correlating Game Behaviors with Questionnaire Data

4.1   Data Collected through Surveys

Three different surveys were administered to ninety-six students from six different 5th grade classrooms in two Midwestern middle schools. These surveys are aimed at measuring aggression and delinquency, as described in [2] and [3]. The surveys include the following scales:

- The *Bully Scale*, which measures the frequency of teasing, name-calling, social exclusion, and rumor spreading.
- The *Fight Scale*, which measures the frequency of physical fighting.
- The *Victimization Scale*, which assesses verbal and physical peer victimization.
- The *Positive Attitude Toward Bullying and Willingness to Intervene*, which evaluates participants attitudes toward bullying, and the extent to which they are willing to assist a victim.
- The *Need for Control and Dominance*, which assesses self-perceptions of dominance and control within one's peer group.

Using the values of these scales, an expert labeled each participant as either a *bully* or a *non-bully*. We hypothesized that there is a subset of students who are bullies that have a need for control and dominance, will engage in coercive tactics directed toward non-friends, and will solicit support for these tactics from friends within and outside their group. The survey data was used to evaluate the psychometric properties of variables yielded by the computer game with the ultimate goal of being able to develop measures of interaction patterns solely through the computer game.

4.2   Data from the Game

Using a 2(bully/non-bully) x 2(collaborative/competitive) ANOVA we studied the interactions and differences in the behaviors of Bullies and Non-Bullies (classified as such according to the data obtained through the surveys and analyzed by an expert) during both Competitive and Collaborative tasks. Results show that those two kinds of players behave differently during both tasks.

The features used for this analysis were (abbreviated name of the features is shown in parenthesis):

- The amount of private messages sent during the collaborative and the competitive stage (*prsent*).
- The number of private messages received (*prrec*).
- The number of public messages sent and received (*pusent* and *purec*).
- The number of times a player peeked at the answer (*peeked*).
- The number of points sent and received (*credsent* and *credrec*).
- Number of positive nominations sent and received, i.e., stating with how many people they want to play and how many want to play with them (*pnsent* and *pnrec*).
- Number of negative nominations sent and received, i.e., stating with how many people they don't want to play and how many do not want to play with them (*nnsent* and *nnrec*).
- Reciprocated nominations, i.e., number of people that nominated each other positively or negatively (*bpn* and *bnn*).
- Unreciprocated nominations, i.e., number of positive nominations to people that nominated the player negatively (*un*).

Table 1 shows the average of the variables per type of player (i.e., Bully or Non-Bully), and the average per stage. There was a significant main effect of bully/non-bully on the amount of private messages sent (*prsent*), and the amounts of times peeked at the answer (*peeked*). Participants labeled as bullies sent more private messages and peeked at the answer a greater number of times than non-bullies. There was also a significant effect of bully/non-bully on the amount of negative nominations sent (*nnsent*); bullies sent more than non-bullies.

There was a significant main effect of collaborative/competitive on all variables. All players sent and received more messages (both public and private), sent and received more coins and peeked at the answer more during the collaborative stage. There were no significant interactions between bully/non-bully and collaborative/competitive. Taken together, the results suggest that bullies tend to send more private messages, to peek at answers more, and to send more negative nominations than non-bullies. Future analysis of the contents of those private messages will provide more insight into the reasons for these behavioral differences.

The previous analysis shows that the game can effectively be used to detect bullies in the classroom by observing those behaviors in which they differ from other players that have different social roles.

5   Observations of Informative Behavior

Several interesting interaction patterns can be observed during the game. We present one example, which we consider specially interesting because it clearly shows a type of bullying behavior that cannot be detected using traditional research methods. Figure 2 shows a subgraph of the nomination network of one classroom. Solid arrows are positive nominations, doted arrows are negative nominations. According to the survey, the individuals labeled as *211* and *214*

Table 1. Results of 2x2 ANOVAs of Bully/Non-Bully, Collaborative/Competitive and the interaction variables. *p<0.1, ** p<0.05.

	prsent	pusent	prrec	purec	credsent	credrec	peeked
Bullies	23.12	20.69	19.79	50.86	2.57	3.5	1.55
Non-Bullies	15.77	17.93	16.71	50.82	3.39	3.25	1.06
p-value	0.037**	0.221	0.326	0.994	0.302	0.735	0.023**

	prsent	pusent	prrec	purec	credsent	credrec	peeked
Collaborative	20.18	22.35	20.22	61.62	3.98	4.01	2.01
Competitive	14.61	14.72	14.56	40.04	2.43	2.6	0.32
p-value	0.057*	<0.001**	0.03**	<0.001**	0.019**	0.022**	<0.001**

	pnsent	pnrec	nnsent	nnrec	bpn	bnn	un
Bullies	6.19	6.05	5.05	3.71	3.05	1.33	1.33
Non-Bullies	6.02	5.97	3.31	3.79	3.26	1.08	1.18
p-value	0.791	0.925	0.09*	0.914	0.671	0.566	0.685

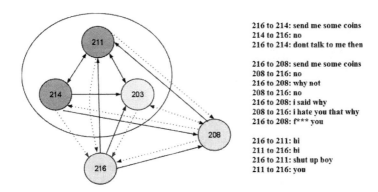

Fig. 2. Example of victimization observed during game, but not captured by survey. Solid arrows show positive nominations, doted arrows show negative nominations. Chat messages show aggressiveness towards *216*.

are bullies, *203* is a victim, and *208* and *216* are bystanders. We can observe that participant *216* positively nominated *211*, *203* and *208*, but was negatively nominated by all of them. It is important to notice that that *211*, *203*, *208*, and *214* almost form a clique. By observing their chat messages, it is clear that *216* is experiencing some kind of victimization, but the survey cannot detect this type of pattern.

6   Conclusions and Future Work

The ultimate goal of this research is to be able to create a tool that can be used broadly to help social scientists and educators better understand and prevent bullying. Currently, our system has shown that bullies, bystanders and victims behave differently while playing the game, and therefore it can be used to identify bullies in classrooms that have not been surveyed yet. For this, a reliable model to identify individual bullies using only data from the game is being developed.

The data collected in the game includes a large amount of text messages sent amongst the participants. So far, the messages have been classified by a team of 20 raters (each message by at least two raters), but these features have not yet been included in the model. In the future, NLP techniques for discourse analysis will be used to increase the efficiency of bully/victim behavior analysis.

The game itself might still be improved. The rules of the games determine the kind of information that is obtained. New game tasks will be developed in order to increase participants' engagement in the game and the amount of information gathered. By changing the rules, modifying the tasks, and changing the way teams are created, it is possible to observe the key interactions that provoke or invoke bullying. Possible variations include: (1) changing the order of the tasks and determining how the change affects the way participants interact with one another; (2) having a finite number of points for an entire team and only changing the distribution of the points according to the performance in the game; and (3) forcing roles onto the members of the team to determine how quickly participants lead or follow others.

Acknowledgement. The authors would like to thank John Elliott and Tony Winkler for their help during various stages of this study.

References

1. Bernstein, M., Tan, D., Smith, G., Czerwinski, M.: Collabio: A Game for Annotating People within Social Networks. In: Proceedings of ACM UIST (2009)
2. Espelage, D., Holt, M.: Bullying and victimization during early adolescence: Peer influences and psychosocial correlates. Journal of Emotional Abuse 2, 123–142 (2001)
3. Espelage, D., Mebane, S., Adams, R.: Empathy, caring, and bullying: Toward an understanding of complex associations. In: Espelage, D., Swearer, S. (eds.) Bullying in American Schools: A Social-Ecological Perspective on Prevention and Intervention, pp. 37–61 (2004)
4. Lieberman, H., Dinakar, K., Jones, B.: Let's Gang Up on Cyberbullying. Computer 44(9), 93–96 (2011)
5. Mancilla-Caceres, J.F., Amir, E.: Evaluating commonsense knowledge with a computer game. In: Campos, P., Graham, N., Jorge, J., Nunes, N., Palanque, P., Winckler, M. (eds.) INTERACT 2011, Part I. LNCS, vol. 6946, pp. 348–355. Springer, Heidelberg (2011)
6. Mayberry, M.L., Espelage, D.L., Koenig, B.: Multilevel modeling of direct effects and interactions of peers, parents, school, and community influences on adolescent substance use. Journal of Youth Adolescence 38, 1038–1049 (2009)
7. Ptaszynski, M., Dybala, P., Matsuba, T., Masui, F., Rzepka, R., Araki, K., Momouchi, Y.: In the Service of Online Order: Tackling Cyber-Bullying with Machine Learning and Affect Analysis. International Journal of Computational Linguistics Research 1(3), 135–154 (2010)
8. von Ahn, L., Dabbish, L.: Labeling Images with a Computer Game. In: CHI 2004: Proceedings of the SIGCHI Conference on Human Factors in Computing Systems, pp. 319–326 (2004)

Beyond Validation: Alternative Uses and Associated Assessments of Goodness for Computational Social Models

Jessica Glicken Turnley[*], Peter A. Chew, and Aaron S. Perls

Galisteo Consulting Group, Inc., 2403 San Mateo Blvd NE, Suite W-12,
Albuquerque, NM 87110, USA
{JGTurnley,PeterAChew,ASPerls}@aol.com

Abstract. This discussion challenges classic notions of validation, suggesting that 'validity' is not just an attribute of a model. It is a function of the relationship of a particular characteristic of the model (the probability that the model will produce a data set that will match a set it has not yet seen) and a user. If we shift our focus from the model to the user, we can identify other ways in which models can be of use as well as start clarifying how we can identify their goodness in multiple use scenarios. In this discussion, we distinguish validation from calibration, and research or generalizable models from site-specific models. We use literature from fields as varied as physics, systems ecology and computational linguistics to characterize the process of validation. Finally, by extending the use space for models beyond prediction, we introduce the possibility of assessments of goodness other than validation.

Keywords: Social modeling, validation, computational models.

1   Introduction

Computational representations of social phenomena (computational social models) have become increasingly visible in a variety of planning and analysis environments, including corporate, environmental and military/national security arenas. The increased visibility and the increasing sophistication of the models themselves have raised a variety of methodological questions. Questions of validation or, more broadly put, questions around how to assess their goodness are among these. This is particularly important in the national security arena where the models may be asked to contribute to solutions that have high consequence outcomes, such as actions which might jeopardize human lives.

We start with a brief disclaimer. We are not addressing verification at all, although verification and validation are often treated without distinction in the literature. Verification compares the model to the developer's requirements [1]. Validation is a

---

[*] Corresponding author.

S.J. Yang, A.M. Greenberg, and M. Endsley (Eds.): SBP 2012, LNCS 7227, pp. 147–155, 2012.
© Springer-Verlag Berlin Heidelberg 2012

significantly different process as it compares the model to the real world. It is these comparisons to the real world that are of interest to us here.

Since our focus is on the national security application of computational social models, we considered definitions of validation from both the U.S. Department of Defense (DoD) and Department of Energy (DOE). Their definitions are very similar. DoD defines validation as "the process of determining the degree to which a model or simulation and its associated data are an accurate representation of the real world *from the perspective of the intended uses of the model.*" (ibid., emphasis added). DOE's definition of validation (which it borrowed from the American Institute of Aeronautics and Astronautics - AIAA) is the same, with the exception of the italicized words [2]. Note particularly that validation is a process, not a determination of a quality or characteristic of a model. Note also that the definition of validation is conditional on its use. These points will be important later in the discussion.

2   Validation versus Calibration

Validation differs from calibration. According to [3], calibration is an operation that establishes a relationship between a measuring system (such as a computational social model) applied under certain conditions (some definition of time and space) and some set of standards (the 'real world' at that time and in that space). A very simple example of calibration would be to match a measuring instrument against the standards for that instrument maintained by an organization such as the National Institute of Standards. Any subsequent measurement by that instrument would be considered 'true'. If we matched a computational social model that was applied to a problem specific to a place and time (insurgency in Afghanistan's eastern provinces in 2011, for example) to data from that place and time, we could say that subsequent results of that model were 'true' for the time and place described by our standard. However, this process does *not* allow us to say that the model results will be 'true' for Libya or Somalia in 2012. As we will see later, calibration (tests against real world data sets) is necessary to validate principles that can be applied to multiple data sets. But this does not mean that calibration and validation are the same thing. We elaborate on this further later in this paper. We thus will argue that the DoD addition of the 'intended uses of the model' to its definition of validation changes the definition to one of calibration not validation.

2.1   Validation and Calibration in Environmental Science

The environmental arena has a great deal of experience in dealing with these two types of assessment of goodness (validation and calibration). Ecological models used for regulatory or monitoring purposes (as distinct from those developed for research purposes) are usually developed in reference to some particular locality. The data set is local, the system definition is local, and the resulting model is checked against a local data set. Although the model includes representation of fundamental processes, the processes are parameterized by local constraints which reduce the model's universality. This 'uniqueness of place,' as [4] phrased it, is found in social systems also. While patron-client networks in northern New Mexico and Afghanistan share certain characteristics, the social and historical exigencies of the local context may make them look and function quite differently at times. A model of patron-client networks in northern New Mexico cannot be matched against a data set from Afghanistan. In these cases, what is called 'validity' relies heavily on an assessment of goodness of fit of a model to an existing data set [5], [6], rather than the goodness of fit to multi-site and multi-variable data. Again writing of ecological models, [7] introduces the notion of engineering or site-specific models where one can talk of the validity "within certain domains of applicability and associated with specified accuracy limits." Domain of applicability and accuracy limits are developed in consultation with the model user. We would go further and argue that what they define as engineering models calls for calibration, not validation. In sharp distinction to a research model, these engineering models are not meant to be generalizable.

2.2   Validation in Physics

The physics domain has had decades of experience with computational models of universal principles and their applications to domain-specific problems. The ban on underground testing of nuclear weapons instituted in 1996, and the resulting increased reliance on computational simulations of various aspects of nuclear explosions, stimulated an interest in the physics community in the validation process itself. This led to publications that became seminal in other fields (such as systems ecology) as they addressed validation.

Extensive reviews and critiques of the physics literature on validation have been prepared [8], [9], [10], and [11]; these are critiques which are referenced in the systems ecology literature as applicable to models in that domain. In those reviews and critiques, we find the following: experimental data, they state, are accessible data that best represent reality. The data generated by computational models is thus compared against that developed by experiments. Some quantifiable measure of (lack of) isomorphism is developed to represent the 'accuracy' of the model under a specified set of conditions. The accuracy measurement must account for uncertainty in both the experimental and model data. Ideally, data from a particular model would be compared against an infinite number of experimental data sets representing all possible conditions to determine the universality of the principles the model embodies (see, for example, [11]). (Note that this indicates that we are talking of a research model, not an engineering model.)   As comparison against infinite datasets is obviously impossible, the model will be compared against some subset of the data. It will meet the accuracy threshold in each subsequent test against a data set in that subset of all data – either until the model breaks down or until someone decides that it has met the threshold in enough tests to be considered 'accurate enough.' At this point, some determination (measure) is made of the likelihood or probability that the model will meet the accuracy threshold in yet one more set of conditions for which it has not yet been tested, which includes the use environment. The result of the validation process is a measure of probability that the model-generated data will agree with the next set of 'real-world' data it is given. However, just as you never really know if there's water in a swimming pool until you jump, so one never really knows if a model will work in practice until it is used. Additional tests can lead us to increase our confidence in the likelihood that the model's output will conform to experimental results, but they can never really tell us that it will – in other words, models can only be disproven, not proven [8]. This describes the 'process' part of both the DoD and DOE definitions of validation. Using computational models is a risky business. The calibration and validation processes are designed to reduce that risk, but never eliminate it.

2.3   The Relationship of Validity to the User

Validity in the context of computational models is not just an attribute of a model. It is a function of the relationship of a particular characteristic of the model (the probability we spoke of earlier) and a user. If we shift our focus then from the model to the user, we can identify other ways in which models can be of use as well as move toward clarifying how we can identify their goodness in multiple use scenarios.

A great deal of attention has been paid to the model in the validation process, but not much to the user. It is argued by [12] that "the main benefit of V&V is not (perhaps counter-intuitively) increased focus on the model but the contextual issue of how the model will be used and, therefore, how the organization and its members identify what decisions they are responsible for making and how they negotiate acceptable levels of risk." (ibid, p.10).

Suppose, then, we focus on the user. Let's start with how people learn – since we are interested in how people learn something from models to help in decision-making. To speak very broadly, there are two general approaches to learning. In one, generally labeled cognitivism (see, for example, [13], [14], [15]), the learner is a passive agent who receives context-free information from his environment. His brain processes, which can be explained through measurement and experimentation, incorporate this information into existing or use it to create new schemata, which in turn drive his behavior. The information provider – the instructor, a computer program – is actively providing information to a relatively passive learner. The other, constructivism [15], [16], [17], [18], argues that we generate or create knowledge and meaning from the interaction of current knowledge with new knowledge or experience. Context is critical, and the history of the individual learner matters. The active agent is the learner, not the information provider. Clearly, most learning situations involve some of both approaches.

2.4   Modeling 'Wicked' Problems

Cognitivist approaches work well on relatively stable problems. But suppose that the problem changes as the learner works it. This is (and has been) common in the world of socio-cultural analysis and is characteristic of a complex adaptive system. Such problems have recently been described in some circles as 'wicked problems' [19]. Suppose, for example, I am considering a sociocultural intervention (such as building a schoolhouse) in an occupied area. I begin to do an assessment of the target population so I know what kind of school I need to build – how big, and so on. However, just the fact that I am considering this action (which the target population knows because I must come into contact with it to gather data for the assessment) can change relationships, power structures, local economics and the like. The problem changes as I work it. This is a characteristic of complex adaptive systems which, as [20] pointed out, "never get there. They continue to evolve, and they steadily exhibit new forms of emergent behavior" (ibid, p. 20). These systems never can achieve an optimal state. There is no 'answer' to wicked problems, just ways to manage the system.

In these cases, we think of the model as an advisor rather than either decision-maker or a provider of a piece of context-free information. As [21] noted, "Advisory systems do not make decisions but rather help guide the decision maker in the decision-making process, while leaving the final decision-making authority up to the human user" (ibid, p. 361; see also [22]). As [12] pointed out, "models don't forecast because people forecast" (ibid, p. 4). Models provide information to people who then make the forecast.

3   Implications for Models

From Prediction to Providing Clarity. The use space for a computation model thus can broaden from providing a description of the future (prediction) with more or less certainty into other roles. Certainly, one important role a model can fulfill is to provide information about a possible future, where its predictions are associated with some level of certainty. A model also can provide the decision-maker clarity about the structure or logic of the target complex system. The model thus makes visible the unseen, showing us connections and relationships of which we might otherwise not have been aware. Literature from computational linguistics (a field we have compared to computational social modeling as it shares a number of common characteristics) describes this use as a straitjacket, forcing rigorous consistency and explicitness on the narrative world of (social) theory [23]. (We note here that the multivocality provided by the lack of clarity of narrative structures also has its place). This allows the decision-maker to develop 'what if' scenarios by identifying key input variables in the system to change, and by illustrating the flow of those changes through the system. In these cases, the model does not predict a future with some level of probability, but produces a set of futures, all of which are possible – a possibility space. The value of the model in these cases would be determined by the change in the knowledge of the user, pre- and post-engagement with the model.

The Process of Modeling as Learning. Identifying and describing entities in a system, the relationships among them, and the processes that can affect system functioning can itself be a learning process. This dimension of learning is taken even further in participatory modeling (also known as companion modeling). This is an approach in which the stakeholders in the decision (who themselves may also be decision-makers) take an active part in the construction of the model [24]. As [25] pointed out in his discussion of a workshop focusing on the relationships between computational modelers and policymakers, "[p]articipatory modeling can benefit those involved in building the model ... The very act of constructing a model requires learning, in a structured way, how the modeled system works. Thus the mental model that the participant leaves with may be more sophisticated and more reflective of the best knowledge on the subject—the analyst or policymaker becomes a more proficient expert himself" (ibid, p. 12). The modeler himself learns about the target system through the act of modeling. And when the modeler also is the decision-maker, the decision can change profoundly.

The nature of the model itself can also lead to learning as the user engages with it. It is argued quite strongly by [26] that a model itself is an analogy or metaphor for the real world, and that theory is the medium by which the metaphor is implemented. Any model, including a computational model, is a selection of some of the things and some of the relationships in the real world, presenting not an isomorphism, but a story (a logic of selection of data) about the real world. If a target problem (e.g., 'why do people join extremist groups?') is framed in economic terms (e.g., 'because individuals have limited economic options'), application of economic theory forces collection of economic data – an entirely different data set than would be demanded if the same target problem were framed in religious terms (e.g., 'because individuals are motivated by religious ideology').

Metaphors are a very powerful way to learn. They also require active engagement of the learner in the learning process and make explicit which elements of the target system are of importance to the creator of the metaphor. Metaphors that speak to me may be quite different from those that speak to you. I may see the United States as a melting pot, taking heterogeneous inputs and creating something quite new with them. Others may see the United States as a salad bowl, a tasty dish in which each element retains its own integrity and flavor. These metaphors drive feelings about national language, the importance of participation in ethnic festivals and rituals, the wearing of particular clothing, and the like, in all of which we are highly – but differently – invested. It is the discussions around these different metaphors that force us to question our assumptions and the constructions of the social realities on which they are based. The models /metaphors need not be accurate. The United States is neither a melting pot nor a salad bowl. It is enough that the metaphors be true – that they are both explanations of how this country is struggling with ways to accommodate difference.

4   Conclusions

This discussion addresses several dimensions of the validation problem for computational social modeling. We begin by distinguishing between research models, designed to present the workings of general social principles, and engineering models, designed to present the workings of specific social constructions at specific times and places. Research models are *validated* by developing a set of engineering models which are *calibrated* against real-world data sets. A user thus can develop some measure of *confidence* that the research model (the general principles) will hold the next time it is instantiated with real-world data sets. The *user* can then make a decision to use or not use the results of the model, based on the nature of the use case. The focus of validation thus shifts from a characteristic of a model to a risk calculation made by a user in a given use environment. Thus the determination of validation of a computational model is highly user-centric [27].

We also suggested an additional use for a computational social model that engages us differently with the validation debate. If we are shifting the focus from the model-as-artifact to the model-in-use and including the user, we introduce the possibility of an assessment of goodness which includes the changes in the user as he engages with the model. Research clearly shows that learning takes place as models are constructed. An active view of the modeling process highlights the role the modeler has in selecting what goes into the modeling, making certain simplifying assumptions, and the like [29]. Constructivist approaches to learning and an a characterization of most social problems as 'wicked problems' assign a very active and powerful role to the model user in the management of these problems, and orient us towards ways in which use of a model can change a user's schema and problem-solving techniques.

The ability to predict is critical as governments exercise their national security regimes. That said, we also note that [28] stated "We believe the 9/11 attacks revealed four kinds of failures: in imagination, policy, capabilities, and management." The report asserts that we had most of the data necessary to anticipate such an attack. In fact, earlier administrations had discussed that eventuality. The failure was not a failure to predict. It was a failure to imagine.

We raise the failure of imagination to counterbalance the importance of prediction in an environment as high-consequence as national intelligence and to emphasize the inclusion of the user in the model environment. Our aim here is not to discredit or discount the predictive capability of computational social models. Rather our purpose is to extend the utility of these models by suggesting alternate ways in which they can also be useful.

Acknowledgments. This paper was prepared partially with funds from the Defense Threat Reduction Agency, U.S. Department of Defense. The views expressed herein are those of the authors and do not necessarily reflect the official policy or position of the Defense Threat Reduction Agency, the Department of Defense, or the United States Government.

References

1. U.S. Department of Defense: Instruction No. 5000.61 (December 9, 2009)
2. American Institute of Aeronautics and Astronautics, AIAA guide for the verification and validation of computational fluid dynamics simulations. G-077-1998e (1998)
3. U.S. Department of Commerce, National Institute of Standards and Technology: State Weights and Measures Laboratories Program Handbook. NIST Handbook 143 (2006), http://www.ncbi.nlm.nih.gov (accessed)

4. Beven, K.: Towards a coherent philosophy for modeling the environment. Proceedings of the Royal Society of London 458, 2465–2484 (2002)

5. McDonnell, J.J., Sivapalan, M., Vaché, K., Dunn, S., Grant, G., Haggerty, R., Hinz, C., Hooper, R., Kirchner, J., Roderick, M.L., Selker, J., Weiler, M.: Moving beyond heterogeneity and process complexity: A new vision for watershed hydrology. Water Resources Research 43, W07301 (2007), doi:10.1029/2006WR005467

6. Vogel, R.M., Sankarasubramanian, A.: Validation of a watershed model without calibration. Water Resources Research 39(10), 1292–1300 (2003)

7. Refsgaard, J.C., Henriksen, H.J.: Modeling guidelines – terminology and guiding principles. Advances in Water Resources 27, 71–82 (2004)

8. Oberkampf, W.L., Trucano, T.G.: Verification and validation in computational fluid dynamics. Progress in Aerospace Sciences 38(3), 209–272 (2002)

9. Oberkampf, W. L., Trucano, T. G.: Verification and validation benchmarks. Sandia report, SAND 2007-0853, U.S. Department of Energy, Sandia National Laboratories, Albuquerque, NM (2007)

10. Oberkampf, W. L., Barone, M. F.: Measures of agreement between computation and experiment: validation measures. Sandia report, SAND 2005-4302, U.S. Department of Energy, Sandia National Laboratories, Albuquerque, NM (2005)

11. Oberkampf, W. L., Trucano, T. G., Hirsch, C.: Verification, validation, and predictive capability in computational engineering and physics. Sandia report, SAND 2003-3769, U.S. Department of Energy, Sandia National Laboratories, Albuquerque, NM (2003)

12. McNamara, L.: Why models don't forecast. Draft Paper Prepared for the Workshop Unifying Social Frameworks, August 16-17. National Research Council, Washington, DC (2010), http://www7.nationalacademies.org/bbcss/Why%20Models%20Dont%20Forecast-McNamara.pdf

13. Atkinson, R., Shiffrin, R.M.: Human memory: A proposed system and its control processes. In: Spence, K.W. (ed.) The Psychology of Learning and Motivation: Advances in Research and Theory, vol. 2, pp. 89–195. Academic Press, New York (1968)

14. Dewey, J.: How we think. FQ Books, New York (1910, 2010)

15. Ormrod, J.E.: Educational psychology: developing learners, 4th edn. Merrill Prentice Hall, Upper Saddle River (2003)

16. Bruner, J., Goodnow, J., Austin, A.: A study of thinking. Wiley, New York (1956)

17. Piaget, J.: The Psychology of Intelligence. Routledge, New York (1950)

18. Richardson, V.: Constructivist pedagogy. Teachers College Record 105(9), 1623–1640 (2003)

19. Horst, W., Rittel, J., Webber, M.M.: Dilemmas in a general theory of planning. Policy Sciences 4, 155–169 (1973)

20. Holland, J.: Complex adaptive systems. Daedalus 121(1), 17–30 (1992)

21. Beemer, B., Gregg, D.G.: Advisory systems to support decision making. In: Handbook on Decision Support Systems. Springer, London (2008)

22. Sniezek, J.A.: Judge Advisor Systems Theory and Research and Applications to Collaborative Systems and Technology. In: Proceedings of the 32nd Hawaii International Conference on System Sciences, pp. 1057–1066. IEEE (1999)

23. Shieber, S.: Criteria for designing computer facilities for linguistic analysis. Linguistics 23, 189–211 (1985)

24. Janssen, M.A., Ostrom, E.: Empirically based, agent-based models. Ecology and Society 11(2) (2006), http://www.ecologyandsociety.org/vol11/iss2/art37/ (accessed)

25. Karas, T.: Modelers and policymakers: improving the relationships. Sandia report, SAND 2004-2888, U.S. Department of Energy, Sandia National Laboratories, Albuquerque, NM (2004)
26. Turnley, J.G., Perls, A.S.: What is a computational social model anyway? A discussion of definitions, a consideration of challenges, and an explication of process. Report No. ASCO 2008-009, Defense Threat Reduction Agency, Advanced Systems and Concepts Office, US Department of Defense, Washington DC (2008)
27. Konikow, L.F., Bredehoeft, J.D.: Ground-water models cannot be validated. Advances in Water Resources: Validation of Geo-hydrological Models Part 1 15(1), 75–83 (1992)
28. Zelikow, P., Jenkins, B.D., May, E.R.: National Commission on Terrorist Attacks upon the United States. The 9/11 Commission Report. W.W. Norton & Company, New York (2004)
29. Morgan, M.S., Morrison, M. (eds.): Models as Mediators: Perspectives on Natural and Social Science. Cambridge University Press, Cambridge (1999)

Predicting Recent Links in FOAF Networks

Hung-Hsuan Chen[1], Liang Gou[2], Xiaolong (Luke) Zhang[2], and C. Lee Giles[1,2]

[1] Computer Science and Engineering
[2] Information Sciences and Technology
Pennsylvania State University
hhchen@psu.edu, {lug129,lzhang,giles}@ist.psu.edu

Abstract. For social networks, prediction of new links or edges can be important for many reasons, in particular for understanding future network growth. Recent work has shown that graph vertex similarity measures are good at predicting graph link formation for the near future, but are less effective in predicting further out. This could imply that recent links can be more important than older links in link prediction. To see if this is indeed the case, we apply a new relation strength similarity (RSS) measure on a coauthorship network constructed from a subset of the CiteSeer^X dataset to study the power of recency. We choose RSS because it is one of the few similarity measures designed for weighted networks and easily models FOAF networks. By assigning different weights to the links according to authors coauthoring history, we show that recency is helpful in predicting the formation of new links.

Keywords: Social Network Analysis, Graph Analysis, Vertex Similarity, Coauthor Network Analysis, Relation Strength Similarity, RSS.

1 Introduction

A network is a set of vertices connected by links that formally modeled the relationship between objects. Using the terminology of graph theory, the objects are usually called nodes or vertices, and the links are usually called edges or links (we will use these terms interchangeably in the paper). It has been shown that several vertex properties and the relationship between vertices can be inferred by statistical measures, such as degree, average path length, and clustering coefficient of the vertices.

One measure that has recently attracted much attention is vertex similarity, which measures how similar two vertices are. Vertex similarity measures is usually used to predict the missing or future links of the networks based on the idea that two vertices tend to have a link connection if they are more similar. Recent studies have shown that vertex similarity measures are good at predicting the links of the near future, but they are less effective for the further future link prediction [3,5]. This implies that recent links could be more important indicators than the old links in the link formation process. However, there has been no systematic study of the effect of the recency factor for missing or future link prediction. We address this question with an empirical study and use a subset

S.J. Yang, A.M. Greenberg, and M. Endsley (Eds.): SBP 2012, LNCS 7227, pp. 156–163, 2012.
© Springer-Verlag Berlin Heidelberg 2012

of the CiteSeer$^{X 1}$ dataset to build a coauthorship network. Applying the relation strength similarity as the vertex similarity measure to explore the FOAF network, we show that the performance of link prediction can be improved by assigning more weights to the new links than the old links.

The rest of the paper is organized as follows. In Section 2, we introduce previous work about link formation, link prediction, and vertex similarity measures. Section 3 introduces the new relation strength similarity measure, and the integration of a recency factor. The experiments described in Section 4 demonstrate the power of recency in terms of its ability to predict future collaboration behavior. Discussions and future work appear in Section 5.

2   Related Work

Sociologists have long studied the question - what influences people to make friends? Studies have shown that people sharing mutual friends will be more likely to have these mutual friends become friends in the future [14]. This phenomena, called "triadic closure" [12], has been observed in several types of networks such as coauthorship networks [3,4], social networks [10], and information networks [15]. Based on this observation, local structure based vertex similarity measures such as Jaccard similarity [18], cosine similarity [17], and Adamic-Adar's measure [1] have been suggested as an important measure. The principle behind these methods is that two non-adjacent nodes are more likely to connect if they share more common neighbors. Furthermore, these types of measure are usually computationally efficient.

Instead of using the local structure information such as the number of mutual friends, a global structure has been suggested as influential. Zhou introduced such a global structure by suggesting that two vertices n_i and n_j are similar if the average distance from n_i to any other node n_k is closer to that from n_j to n_k for $i \neq j \neq k$ [21]. Others defined the similarity index recursively: two vertices are similar if their corresponding neighbors are similar [7,9]. Katz proposed an index based on the number and the lengths of the simple paths between two vertices [8]. Although global structure based methods consider the complete graph, empirical studies showed that the global structure based methods are worse than the local structure ones in predicting the missing links [13,3,5].

Actual applications have been developed using vertex similarity measures. CollabSeer[2], an academical collaborator recommendation system, analyzed both the researchers' research interests and the structure FOAF coauthorship network [4]. The "Don't forget Bob!" and "Got the wrong Bob?"[3] are two Gmail Lab features helping users to identify the right mail receivers by analyzing the email network [16].

---

[1] http://citeseerx.ist.psu.edu/

[2] http://collabseer.ist.psu.edu/

[3] http://gmailblog.blogspot.com/2011/04/dont-forget-bob-and-got-wrong-bob.html

Another research topic related to this work is network evolution. Erdős-Rényi graph (ER graph) [6] is probably the simplest random graph generating model. It has been studied for decades thus several properties were carefully analyzed [11]. However, observations showed that several characteristics of real networks don't fit ER model. Other network generating models, such as the Watts-Strogatz graph (WS graph) and Barabási-Albert model (BA model) [20,2], were proposed to fit the the real network better.

3   RSS and Recency Factor

We propose to study the effect of recency on the formation of new edges. To do this, we assign different weights to different edges based partially on their ages. The weighted network is used to compute the similarity score between the vertices. Previous studies showed that the local structure based vertex similarity measures are more computationally efficient and better than global structure based measures in terms of link prediction [3]. However, most of the local structure based similarity measures, such as Jaccard similarity, cosine similarity, and Adamic Adar similarity, consider only the number of mutual friends between two vertices; thus, the edge weights cannot be integrated into these models. Therefore, we use relation strength similarity (RSS), a similarity measure that is designed for weighted networks. The discovery range parameter of RSS can be adjusted so it becomes a local structure based similarity measure, which is computationally efficient. In this section, we first briefly introduce RSS and then integrate into RSS a recency factor.

Relation strength similarity was proposed and analyzed in Chen [3,4,5]. Given a network, RSS is calculated based on the following intuitions: two non-neighboring vertices v_i and v_j are more similar if 1) the path length (number of hops) between v_i and v_j is shorter; 2) the number of distinct paths between v_i and v_j is larger; and 3) the relation strength of the neighbor vertices along the paths from v_i to v_j is larger.

We construct the coauthorship network as a weighted network as follows. Each identical author is regarded as a vertex in the graph. Two vertices v_i and v_j are connected if the two authors have previously coauthored. The edge weight is assigned as the number of coauthored papers. The relation strength is defined as the normalized weight [3] as follows.

$$R(v_i, v_j) := \frac{n_{ij}}{n_i}, \qquad (1)$$

where n_{ij} is the number of v_i and v_j's coauthored papers, and n_i is the number of v_i's publications.

A researcher's research interests may vary over time. Most likely, a recent publication is more representative of a researcher's latest interests. Thus, new collaborators should be better for inferring a researcher's future collaboration preferences. Previous work showed that local structure based similarity measures are better in predicting coauthoring behavior in the near future [3,5]. This could imply that the new collaborators are more important than the old collaborators. To introduce a recency factor to the graph, we model the edge weights as an exponential decay function over time with a half life time of T_h. Let $n_{i,j}(t)$ denote the number of coauthored papers between v_i and v_j at year t. The decay rate λ of the exponential decay function can be derived by Equation 2.

$$
\begin{aligned}
n_{i,j}(t + T_h) &= \tfrac{1}{2} n_{i,j}(t) \\
\Rightarrow n_{i,j}(t) \exp(-\lambda T_h) &= \tfrac{1}{2} n_{i,j}(t) \\
\Rightarrow \lambda &= \tfrac{\ln 2}{T_h}.
\end{aligned}
\tag{2}
$$

Let's assume author v_i and v_j coauthored $n_{i,j}^{(1)}, n_{i,j}^{(2)}, \ldots, n_{i,j}^{(K)}$ papers in year $y_1, y_2, \ldots, y_K$ respectively. The edge weight at time t_{now} is defined as

$$
\begin{aligned}
n_{i,j}(t_{now}) &= \sum_{k=1}^{K} n_{i,j}^{(k)} \exp(-\lambda(t_{now} - t_k)) \\
&= \sum_{k=1}^{K} n_{i,j}^{(k)} \exp\left(\tfrac{-\ln 2}{T_h}(t_{now} - t_k)\right).
\end{aligned}
\tag{3}
$$

Instead of equation 1, the new relation strength considering both the number of coauthored papers and the recency factor between v_i and v_j is defined as Equation 4.

$$
R(v_i, v_j) := \frac{n_{i,j}(t_{now})}{\sum_{\forall k \in N(v_i)} n_{i,k}(t_{now})},
\tag{4}
$$

where $N(v_i)$ returns all the neighbors of v_i.

RSS uses the relation strength between neighbor vertices as the foundation to calculate the similarity score between non-neighboring vertices. Assume v_i can arrive v_j through path p_m, which is formed by vertices $v_i(= u_1)$, u_2, u_3, ..., u_{K-1}, $v_j(= u_K)$. The general relation strength from v_i to v_j through p_m is defined in Equation 5.

$$
R_{p_m}^*(v_i, v_j) := \begin{cases} \prod_{k=1}^{K-1} R(u_k, u_{k+1}) & \text{if } K \le r \\ 0 & \text{otherwise,} \end{cases}
\tag{5}
$$

where r is the discovery range parameter controlling the maximum degree of separation for collaborator recommendation. The parameter plays a tradeoff between the computation efficiency and relation discovery range.

Assuming there are M distinct paths from v_i to v_j, the relation strength similarity is calculated by Equation 6.

$$
S(v_i, v_j) := \sum_{1}^{M} R_{p_m}^*(v_i, v_j).
\tag{6}
$$

Since RSS guarantees the vertex similarity measures within 0 and 1 as long as the relation strength is normalized [4], it is easy to integrate RSS with other scoring.

Table 1. Information and statistical measures of training networks G_1, G_2, and G_3

	G_1	G_2	G_3
Year	$1995 - 1997$	$1999 - 2001$	$2003 - 2005$
Number of Vertices	1,019	2,556	2,198
Number of Edges	2,286	5,308	4,303
Average Degree	4.49	4.15	3.92
Average Clustering Coefficient	0.55	0.55	0.54
Average Shortest Path Length	13.14	14.44	14.19
Diameter	37	45	40

4   Experiments

To show the power of recency, we use a subset of the CiteSeer^X dataset to build coauthorship networks and study the performance before and after introducing the recency factor in terms of their ability to predict future collaboration behavior. To eliminate the author ambiguity problem, random forest learning methods [19] are used to disambiguate different authors with similar names and authors whose names have several variations.

4.1   Experiment Setup

We retrieve the authors who published at least 5 papers between 1995 and 1997 from CiteSeer^X dataset and build a coauthorship network among the authors. The giant component of the network is called network G_1. The same process is performed from 1999 to 2001 and from 2003 to 2005 to generate two more networks G_2 and G_3. The networks G_1, G_2, and G_3 are the training networks because they are used to calculate the the similarity scores between non-neighboring vertices. The information and the statistical measures of the training networks are shown in Table 1.

We create a testing network H_1 from the coauthorship network of the authors who have publications in 1998. The authors who have publications in 1998 but not in interval $[1995, 1997]$ are disregarded since they are not presented in the training network. The edges that already appeared in $[1995, 1997]$ are also disregarded because we are only interested in predicting new collaboration behavior. By similar manner, we created two more testing networks H_2 of year 2002 and H_3 of year 2006. The information and statistical measures of the testing networks are shown in Table 2. Note that the average shortest path length and the diameter are not shown because each of H_1, H_2, and H_3 is not a connected component.

To test the power of recency, we assign different values to the half life time parameter T_h in the calculation. Specifically, we assign T_h to be 0.5, 1.0, 1.5, 2.0, and ∞ (years). When $T_h = \infty$, the model considers only the number of coauthored papers between two authors. We use RSS with discovery parameter

Table 2. Information and statistical measures of testing networks H_1, H_2, and H_3

	H_1	H_2	H_3
Year	1998	2002	2006
Number of Vertices	656	1,613	1,205
Number of Edges	1,255	2,991	2,034
Average Degree	3.83	3.71	3.38
Average Clustering Coefficient	0.54	0.52	0.52

$r = 2$ to calculate the similarity scores between vertices, and claim the top-n similar non-neighboring vertices will connect. The vertex similarity scores calculated from G_1, G_2, and G_3 are used to predict the links in H_1, H_2, and H_3 respectively.

4.2 Experimental Results

Previous studies showed that the precision of link prediction is usually very low [3,5,13]. This is because the sparsity of the links makes a naïve random guess very unlikely to be correct.

As mentioned in last section, we claim the top-n similar non-neighboring vertices to be connected. Different n will cause different precision. To be fair, we show the precisions of different n (from 1 to 100) in Figure 1.

The five different lines in each sub-graph represent $T_h = \infty$ (years), $T_h = 2.0$ (years), $T_h = 1.5$ (years), $T_h = 1.0$ (years), and $T_h = 0.5$ (years) respectively. The lower the value of the half life parameter T_h, the more important the recent edges are. In general, a smaller half life parameter yields better precision in all three experiments. This means the recent edges do play a more important role in future link formation.

5 Discussion and Future Work

Although the evolution of networks has been well studied, most work only considers how the network grows. One interesting topic rarely discussed is whether the nodes or edges degenerate and therefore gradually lose their influence over time? Probably because most of the available network datasets don't contain such information, little has been done to explore this question.

In this paper, we try approach the recency problem by introducing a recency factor to the edges. We construct the coauthorship network and assign the initial weight of an edge to be proportional to the number of coauthored papers between two authors. The weight decays exponentially as time unfolds, and the weight can be strengthen again if the two authors recently have coauthored new papers. By integrating the recency factor, we show that future links can be better predicted. This demonstrates that recent links should be more representative than the older

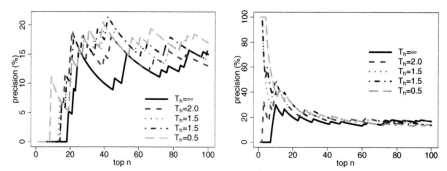

(a) Using G_1 to calculate the similarity scores and predict links in H_1

(b) Using G_2 to calculate the similarity scores and predict links in H_2

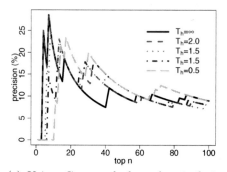

(c) Using G_3 to calculate the similarity scores and predict links in H_3

Fig. 1. The accuracy of different half time values

links for the formation of future links, and implies that the links and nodes may gradually lose their influence and predictive power over time, i.e. they age.

For future work, the effect of recency and aging factor can be further investigated by various machine learning methods. A user survey can also show the effectiveness of link prediction. Since networks do age, it would be interesting to investigate not only the growth but also the degeneration of networks.

Acknowledgements. We gratefully acknowledge partial support from Alcatel-Lucent and NSF.

References

1. Adamic, L., Adar, E.: Friends and neighbors on the web. Social Networks 25(3), 211–230 (2003)
2. Barabási, A., Albert, R.: Emergence of scaling in random networks. Science 286(5439), 509 (1999)

3. Chen, H.-H., Gou, L., Zhang, X., Giles, C.L.: Capturing missing edges in social networks using vertex similarity. In: The 6th International Conference on Knowledge Capture. ACM (2011)
4. Chen, H.-H., Gou, L., Zhang, X., Giles, C.L.: Collabseer: A search engine for collaboration discovery. In: Proceedings of the 11th ACM/IEEE-CS Joint Conference on Digital Libraries. ACM (2011)
5. Chen, H.-H., Gou, L., Zhang, X., Giles, C.L.: Discovering missing links in networks using vertex similarity measures. In: The 27th ACM Symposium on Applied Computing. ACM (2012)
6. Erdös, P., Rényi, A.: On random graphs, i. Publicationes Mathematicae (Debrecen) 6, 290–297 (1959)
7. Jeh, G., Widom, J.: Simrank: a measure of structural-context similarity. In: Proceedings of the Eighth ACM SIGKDD International Conference on Knowledge Discovery and Data Mining, pp. 538–543. ACM (2002)
8. Katz, L.: A new status index derived from sociometric analysis. Psychometrika 18(1), 39–43 (1953)
9. Leicht, E., Holme, P., Newman, M.: Vertex similarity in networks. Physical Review E 73(2), 026120 (2006)
10. Leskovec, J., Backstrom, L., Kumar, R., Tomkins, A.: Microscopic evolution of social networks. In: Proceeding of the 14th ACM SIGKDD International Conference on Knowledge Discovery and Data Mining, pp. 462–470. ACM (2008)
11. Newman, M.E.J.: Random graphs as models of networks. In: Handbook of Graphs and Networks, pp. 35–68 (2003)
12. Nguyen, V.-A., Leung, C.W.-K., Lim, E.-P.: Modeling Link Formation Behaviors in Dynamic Social Networks. In: Salerno, J., Yang, S.J., Nau, D., Chai, S.-K. (eds.) SBP 2011. LNCS, vol. 6589, pp. 349–357. Springer, Heidelberg (2011)
13. Nowell, D., Kleinberg, J.: The link prediction problem for social networks. In: CIKM 2003: Proceedings of the Twelfth International Conference on Information and Knowledge Management, pp. 556–559 (2003)
14. Rapoport, A.: Spread of information through a population with socio-structural bias: I. assumption of transitivity. Bulletin of Mathematical Biology 15(4), 523–533 (1953)
15. Romero, D., Kleinberg, J.: The directed closure process in hybrid social-information networks, with an analysis of link formation on twitter. In: Proceedings of the 4th International AAAI Conference on Weblogs and Social Media, pp. 138–145 (2010)
16. Roth, M., Ben-David, A., Deutscher, D., Flysher, G., Horn, I., Leichtberg, A., Leiser, N., Matias, Y., Merom, R.: Suggesting friends using the implicit social graph. In: Proceedings of the 16th ACM SIGKDD International Conference on Knowledge Discovery and Data Mining, pp. 233–242. ACM (2010)
17. Salton, G.: Automatic text processing: the transformation, analysis, and retrieval of information by computer (1989)
18. Tan, P., Steinbach, M., Kumar, V., et al.: Introduction to data mining. Pearson Addison Wesley, Boston (2006)
19. Treeratpituk, P., Giles, C.: Disambiguating authors in academic publications using random forests. In: Proceedings of the 9th ACM/IEEE-CS Joint Conference on Digital Libraries, pp. 39–48. ACM (2009)
20. Watts, D., Strogatz, S.: Collective dynamics of small-world networks. Nature 393(6684), 440–442 (1998)
21. Zhou, H.: Distance, dissimilarity index, and network community structure. Physical Review E 67(6), 061901 (2003)

Mapping the Twitterverse in the Developing World: An Analysis of Social Media Use in Nigeria

Clayton Fink, Jonathon Kopecky, Nathan Bos, and Max Thomas

Johns Hopkins University Applied Physics Lab
{clayton.fink,jonathon.kopecky,nathan.bos,
max.thomas}@jhuapl.edu

Abstract. There is growing interest in the use of social media in the developing world. For example, the Arab Spring was widely considered to have been heavily influenced by information that propagated via social media such as Twitter and Facebook. Researchers are understandably eager to begin utilizing this growing collection of data to help inform some of their research questions. In this paper we discuss some of the methodological issues researchers need to consider when analyzing social media. We will focus on techniques to determine demographic characteristics of Twitter users, such as ethnicity and location, and why these are necessary. We will discuss some of the results from applying our methods to social media. Looking at social media in Nigeria, we present results showing the geographic distribution of Twitter users, the contribution of mobile users to the Twitter stream, and the estimated ethnic makeup of Twitter users.

Keywords: Methodology, Twitter, Social Media, Ethnicity, Geolocation, Nigeria.

1   Introduction

Social media has become an important communication medium in the developing world, affecting political, economic and social processes, and providing a window into societies where traditional data collection such as polling can be difficult. In the Arab Spring revolutions, social media is believed to have played a role in many aspects of the revolutions: in the beginning information was propagated about the self-immolation of the Tunisian Mohammed Bouazizi via Facebook and Twitter and videos of people protesting in his name were viewed in other Arab countries; later, social media provided a gathering place for dissidents and ultimately a means of mobilization for protest organizers. Social media in the developing world has also had significant influence in less radical ways, too, from helping to root out corruption and incompetence in the high-speed rail network in China, resulting in the resignation of the train minister, and to monitoring elections in notoriously corrupt Nigeria and contributing to the most transparent election in Nigeria's history.

These examples of the influence of social media illustrate the importance of social media for researchers and analysts who focus on a particular region for tracking

S.J. Yang, A.M. Greenberg, and M. Endsley (Eds.): SBP 2012, LNCS 7227, pp. 164–171, 2012.
© Springer-Verlag Berlin Heidelberg 2012

critical issues, gauging the mood of their populace, and monitoring sentiment toward public figures and events. These data are public and readily available, but it is not a safe assumption that social media users are going to be representative of the whole populace. People too poor to afford electricity, people who are illiterate, people too scared to express their beliefs online are just some of the people who will be missed by any analysis of social media. An understanding of who is represented in this online population is required to form valid conclusions about the people in the region of interest and techniques are thus needed to characterize the online population in these regions in terms of geographic distribution and demographics.

We will discuss the methods we used to analyze social media use in Nigeria. Nigeria is the dominant country in West Africa, the most populated country on the continent, has numerous ethnic groups, and has a distinct religious divide between the Christian South and Muslim North. Nigeria is a relatively new democracy, transitioning from military dictatorship to a multi-party system in 1999, but it also has enormous challenges relating to endemic corruption, an inefficient and undependable power grid, inter-religious violence, and ongoing insurgencies in the Muslim North and the petroleum-rich Niger Delta. The communications infrastructure in Nigeria has seen high levels of investment [1] and made wireless technology available across many regions of the country. This technology has given many Nigerians access to the Internet and social media sites such as Twitter and Facebook.

We focus on Twitter data over Facebook data mainly because of difficulties in accessing Facebook data. Although Facebook is the most popular social media platform in Nigeria (with over 4 million users as of October 2011, a 300% increase since early 2010), a large portion of Facebook content is private and the available metadata for users is limited. Instead, we focus on Twitter which is generally public and has richer metadata.

In this paper, we discuss our approach for collecting and analyzing Twitter data from Nigeria, determining the geolocation of users, assessing mobile Twitter usage in the country, and our machine-learning-based approach for tagging Twitter users by ethnic group.

2   Data Collection

Twitter provides a set of public application program interfaces (APIs) for accessing their data. These allow queries against recent public tweets using keywords, hash tags, location and other criteria. They also allow for access to the social graph and the raw Twitter stream. For all of the examples referenced in this paper, these APIs were used to collect data.

To gather tweets from Nigeria, we used the Twitter Search API[1] and its geocode method that accepts a position in latitude and longitude and a radius in miles. For collecting data from Nigeria, we ran searches for 45 Nigerian cities with populations

---

[1] https://dev.twitter.com/docs/using-search

over 100,000 and a radius of 40 miles. For users whose tweets were captured using this method, we used the search API search method to obtain their other tweets.

3   Identifying User Locations

The metadata returned for each tweet via the geocode method contains metadata about user location, language and the client application used when posting the tweet. Location information is available from two sources: a field containing the location from the user's profile and a field containing a latitude/longitude pair populated using the optional geotagging feature supported by mobile Twitter clients (users must opt-in to use this feature). We use these fields to calculate a user's location at the time of the tweet. If the profile location is a latitude/longitude pair (either from a mobile client updating the profile location or from the geotagging feature) we use the geonames.org web service[2] to find all locations within 25 miles of the coordinates and choose the closest populated place. If the location is given as a string, geonames.org is queried with the string for a match. Resolving location names gives us 86% accuracy, based on whether the returned location matched the location intended in the profile. We base the accuracy on the performance against a list of 200 randomly selected, hand-checked profile locations and their resolved locations. The geographic information returned by geonames.org includes a place name, second administrative level (state or province), country code, latitude and longitude. For our work, we used the geotagging location, if present. Otherwise, user location was based on the profile location.

We determined the location for 398,534 unique users as being in Nigeria. For each user, we determined which of the seven geopolitical zones the user was in, based on the Nigerian state containing their resolved location, and retained this value for our analysis.

Although the Twitter API allows location-based searches, it is error prone and frequently returns tweets that obviously do not come from the queried location. We found error rates of around 30% (i.e. 30% of the tweets that were supposed to come from a given region actually came from a different region entirely). A separate issue is that it is possible for people to deliberately misrepresent where they are located, such as a person claiming to be in Egypt's Tahrir Square even though they are actually located in the USA. Thus, when examining social media after a significant news event one should be somewhat wary of relying solely on the locations given in Twitter user profiles.

4   Spatial Distribution of Users

We looked at the distribution of users across the seven geopolitical zones of Nigeria. For all users, we used as their location the resolved location (as mapped to geopolitical zone) of their most recent tweet as of October 6, 2011. Figure 1 shows a

---

[2] http://www.geonames.org/

map of Nigeria and the distribution of the 347,441 Twitter users that we were able to geolocate to the zone level. The South West geopolitical zone, which includes Lagos, is dominant. Significant numbers of users are represented in the South East zone and the Federal Capitol Territory (Abuja). The South South zone (encompassing the Niger Delta) and the North West zone (which includes the major cities of Kaduna and Kano) show much less representation. The North Central and North East zones show the least representation, although we still found at least 4000 unique Twitter users even in these zones.

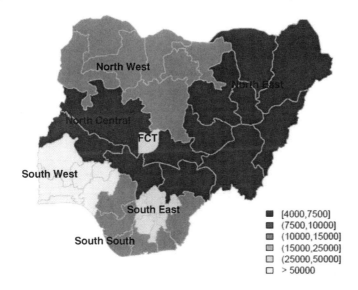

Fig. 1. Distribution of Nigerian Twitter users by geopolitical zone (based on 347,000 users)

These results indicate that we are under-representing one of the major ethnic groups: the Hausa ethnic group whose people tend to live in the northern part of Nigeria, tend to be poorer, and have higher illiteracy rates than other major ethnic groups. Hausa people often have strikingly different views than other Nigerians (for instance Hausa tend to be proponents of traditional Islamic law, Shariah) so when analyzing Twitter content we need to realize we are under-representing certain views.

5   Mobile Usage

Mobile technology has had a significant impact in the developing world [2] and the number of people who use a mobile device to update social media is also growing at a high rate worldwide [3]. We were interested in the pattern of mobile usage in Nigeria. The metadata returned for each tweet contains a field giving the client that was used to update a user's status, capturing whether the user was using a mobile client or not. This information let us determine the extent to which mobile users contributed to the Twitter stream. We looked at 365,000 users with locations resolved to Nigeria, including those users who only gave their location as "Nigeria" in their profile. The results are shown in Figure 2. We found that 79% of users use a mobile device at least part of the time and 66% use a mobile device exclusively. We also found that 17% of users used the geotagging feature, giving some information about the level of usage of smartphones. These results suggest that, indeed, mobile devices are a significant factor in Twitter usage. This finding has implications for surveying online populations since that portion of the data stream associated with mobile usage can be geolocated with high accuracy, given that coordinates are often available in the tweet's metadata.

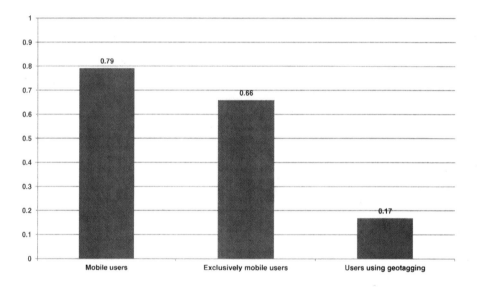

Fig. 2. Mobile usage of Twitter in Nigeria (based on 365,000 users)

6   Ethnicity of Twitter Users

Ethnicity as a demographic category has historically been of great importance in Nigeria. There are approximately 300 ethnic groups in the country, but three ethnicities are dominant in terms of representation in the population: Hausa-Fulani, Igbo, and Yoruba. The Hausa group is predominately Islamic and is dominant in the North. The Igbo, predominately Christian, is dominant in the South East. The Yoruba are split equally between Islam and Christianity and dominate the South West. Each ethnicity also represents a linguistic group and, in addition to English, Hausa-Fulani, Igbo and Yoruba are all recognized as the official languages of Nigeria. Because a person's name can be an indicator of their ethnicity, we used machine learning to classify user names as belonging to one of these three ethnicities, or into a fourth "other" category comprising names from other Latin-script-based languages.

We used Support Vector Machines (SVM) [4] to train two classifiers: a binary classifier that discriminates between Hausa/Igbo/Yoruba or "other" (NG-OTHER), and a multi-class classifier that discriminates between Hausa, Igbo, and Yoruba (IHY). Training data for Igbo and Hausa names was taken from online baby name lists[3], Hausa names were taken from the follower list of the BBC Hausa service's Twitter account[4], and "Other" names were taken from the follower list of an EMACS user group's Twitter account[5]. As features, we used letter n-grams of length one through five and affixes of length one through four.

Using the LIBSVM package we performed 10-fold cross validation runs on balanced training data and obtained 86% accuracy on the NG-OTHER classifier and 90% on the IHY classifier. Previous work using a hierarchal Bayesian approach to classifying Igbo and Hausa names from Facebook profiles [5] gave an accuracy of 81%.

We used these classifiers to label names from the profiles of Twitter users whose locations were resolved to a geopolitical zone. Names representing organizations were dropped and all honorifics and suffixes were removed. This resulted in names for 83,579 users. Only the leftmost (first) and rightmost (surname) names were considered. If the name was a single word, it was used for first name and surname. The NG-OTHER classifier was run on both names for each user. If the user's surname was classified as Nigerian, the IHY classifier was run on both names. A user was assigned the ethnicity result for the surname, except if the surname was classified as Hausa and the first name was classified as Yoruba. In this case, the user was assigned Yoruba as their ethnicity, reflecting the fact that a significant number of Yoruba are Muslim.

The results of this classification are shown in Figure 3 for users classified as Hausa, Igbo, or Yoruba by geopolitical zone. Little ground truth exists for the population distribution of ethnicities in Nigeria at a micro-level, but at a coarse level we know that Hausa-Fulani tend to live in the North, Igbo in the South East region, and Yoruba in the South West region. As a surrogate for ground truth, we used aggregated results from the Afrobarometer survey [6] from the years 1999 through 2008 based on a survey question concerning which languages were spoken in the home. Comparing these numbers by geopolitical zone, we obtained a correlation of 0.65. For most regions, our classifiers correctly identify the dominant ethnic group. As we can see, we tend to predict too many Hausa and too few Yoruba. This is likely because some of the Muslim Yoruba are being misclassified as Hausa.

---

[3] http://www.onlinenigeria.com/nigeriannames/
[4] http://twitter.com/#!/bbchausa
[5] https://twitter.com/#!/emacs

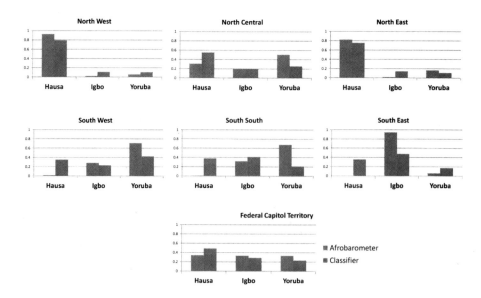

Fig. 3. Twitter users by ethnicity across geopolitical zones vs. Afrobarometer results

7   Conclusion

Social media is a very promising source of social and cultural data for studying the developing world. Twitter appears particularly useful for analysts because of its open nature and useful metadata that provides the ability to determine important demographic characteristics of the data. This paper provides researchers improved methods to utilize Twitter data for their own research questions. Once demographic and geographic information is determined, then those tweets can be examined for content, which could then be used to, say, determine sentiment toward particular entities or trust in the government. Of course, it is imperative to determine if the views in social media are those of the educated elite or the mainstream populace. The methods described in this paper address this issue by showing how certain aspects of an online population can be characterized.

Acknowledgements. This work was supported by Office of Naval Research grant N00014-10-1-0523.

References

1. Nigeria is increasingly connected (n.d.),
   http://www.itu.int/net/itunews/issues/2010/01/31.aspx
   (retrieved November 11, 2011)
2. Donner, J.: Research Approaches to Mobile Use in the Developing World: A Review of the Literature. The Information Society 24(3), 140–159 (2008), doi:10.1080/01972240802019970

3. Top mobile internet trends (n.d.),
   `http://www.firsteditiondesign.com/kpcbtop10mobiletrends02101`
   `1finalpdf-110210002130-phpapp02.pdf` (retrieved November 11, 2011)
4. Cortes, C., Vapnik, V.: Support-vector networks. Machine Learning 20(3), 273–297 (1995),
   doi:10.1007/BF00994018
5. Rao, D., Fink, C., Oates, T.: Hierarchical Bayesian Models for Latent Attribute Detection in
   Social Media. Artificial Intelligence (2011)
6. Afrobarometer (2008), `http://www.afrobarometer.org`

Modeling Infection with Multi-agent Dynamics

Wen Dong[1], Katherine Heller[2], and Alex (Sandy) Pentland[1]

[1] MIT Media Laboratory
{wdong,sandy}@media.mit.edu
[2] Department of Brain and Cognitive Sciences, MIT
kheller@gmail.com

Abstract. Developing the ability to comprehensively study infections in small populations enables us to improve epidemic models and better advise individuals about potential risks to their health. We currently have a limited understanding of how infections spread within a small population because it has been difficult to closely track and infection within a complete community. This paper presents data closely tracking the spread of an infection centered on a student dormitory, collected by leveraging the residents' use of cellular phones. This data is based on daily symptom surveys taken over a period of four months and proximity tracking through cellular phones. We demonstrate that using a Bayesian, discrete-time multi-agent model of infection to model the real-world symptom report and proximity tracking records can give us important insights about infections in small populations.

Keywords: human dynamics, living lab, stochastic process, multi-agent modeling.

1 Introduction

Modeling contagions in social networks can help us facilitate the spread of valuable ideas and prevent disease. However, because closely tracking proximity and contagion in an entire community for months was previously impossible, modeling efforts have focused on large populations. As a result, we could say little about how an individual can better welcome good contagion and avoid bad contagion through his immediate social network. This paper describes how a "common" cold spread through a student residence hall community, with information based on daily surveys of symptoms for four months and tracking the locations and proximities of the students every six minutes through their cell phones. This paper also reports how infection occurred – and how infection could have been avoided – based on fitting the susceptible-infectious-susceptible (SIS) epidemic model to symptoms and proximity observations. It combines epidemic models and pervasive sensor data to give individually-tailored suggestions about local contagion, and also demonstrated the necessity of extending the epidemic model to individual-level interactions.

Epidemiologists agree on a framework for describing epidemic dynamics – people in a population can express different epidemic states, and change their states according

S.J. Yang, A.M. Greenberg, and M. Endsley (Eds.): SBP 2012, LNCS 7227, pp. 172–179, 2012.
© Springer-Verlag Berlin Heidelberg 2012

to certain events. Computing event rates requires only knowledge about the overall-population at the present time. The susceptible-infectious-recovered (SIR) model, for example, divides the population into susceptible, infectious, and recovered sub-populations (or "compartments"). A susceptible person will be infected at a rate proportional to how likely the susceptible person is to make contact with an infected disease carrier, and an infected person will recover and gain lifetime immunity at a constant rate. Other compartmental models include the susceptible-infectious-susceptible (SIS) model for the common cold, in which infectious people become susceptible again once recovered, and the susceptible-exposed-infectious-recovered (SEIR) model, in which infected carriers experience an "exposed" period before they become infectious.

However, the availability of new data and computational power has driven model improvements, refining compartmental models that assume homogeneous compartments and temporal dynamics, to develop the Epidemiological Simulation System (EpiSimS) that take land use into account [1], and more recently simulations based on the tracking of face-to-face interactions in different communities [2,3,4,5].

These simulations all show evidence in favor of an epidemic dynamics framework, and against the assumption of homogeneous relationship and temporal dynamics. Using these kinds of algorithms with real-world symptom reports and proximity data could offer a much better understanding of how infection actually transfers from individual to individual, allowing for personalized contagion recommendations.

To understand the infection dynamics in a community at the individual level, we use the data collected in the Social Evolution experiment [6], part of which tracked "common cold" symptoms in a student residence hall from January 2009 to April 2009. The study monitored more than 80% of the residents of the undergraduate residence hall used in the Social Evolution experiment, through their cell phones from October 2008 to May 2009, taking daily surveys and tracking their locations, proximities and phone calls. This residence hall housed approximately 30 freshmen, 20 sophomores, 10 juniors, 10 seniors and 10 graduate student tutors. Researchers conducted monthly surveys on various social relationships, health-related issues, and status and political issues. They captured the locations and proximity of the students by instructing the cell phones to scan nearby Wi-Fi access points and Bluetooth devices every 6 minutes. They then collected the latitudes and longitudes of the Wi-Fi access points and the demographic data of the students to make sense of the data set. The data are protected by MIT COUIS and related laws.

This paper makes the following contributions to the field of human behavior modeling: It is among the first to discuss the spread of flu symptoms, tracked daily with cellphone-conducted surveys over an entire community. It is also among the first to model the spread of flu symptoms by looking at proximity tracked by cell phones, paired with a repository of other cellphone-conducted surveys about activity, status, and demographics. Lastly, this paper introduces a multi-agent model that is compatible with compartmental epidemic models and can infer who infected whom and how to avoid catching the flu. The large quantity of behavioral data generated from pervasive computing technology provides the details necessary to shift social sciences research from the level of large populations to individuals, and to enable social sciences to give more personalized advice.

Table 1. Probability of catching symptom = $1-\alpha \times(-\beta \times$number of contacts with symptom), R^2 and p

Symptom	α	β	R^2	p
Runny nose	1.013	0.024	0.52	0.04
Sadness	0.991	0.016	0.63	0.13
Stress	1.001	0.035	0.85	0.005
Nausea	0.993	0.006	0.94	0.11

The rest of the paper is organized as follows: In section 2 we describe the structure of face-to-face contact in the residence hall community, and the sensor data that captures this structure. In section 3 we introduce a Bayesian, multi-agent model, related to the Markov jump process, that not only simulates contagion but also makes inferences from observations. In section 4 we demonstrate that we can effectively predict new cases of symptoms, identify cases of symptoms even if students do not report them, and determine the students and contacts that are most critical for symptom-spreading. Hence, we show that the multi-agent model captures how symptoms of the common cold and the flu spread in a student dormitory community.

2   Contagion in Social Evolution Experiment

In the Social Evolution experiment, we offered students $1 per day from 01/08/2009 through 04/25/2009 to answer surveys about contracting the flu, regarding the following specific symptoms: (1) runny nose, nasal congestion, and sneezing; (2) nausea, vomiting, and diarrhea; (3) frequent stress; (4) sadness and depression; and (5) fever. Altogether, 65 residents out of 84 answered the flu surveys, each of whom answered for half of the surveyed period. The correlation between stress and sadness is 0.39, while the correlations between other pairs are about 0.10.

The symptom self-reporting in the Social Evolution data seems to be compatible with what the epidemic model would indicate: symptoms other than runny nose are probabilistically dependent on that student's friendship network. The durations of symptoms were about two days, and fit the exponential distribution well ($p \approx 0.6$ in Kolmogorov hypothesis testing). The chance of reporting a symptom is about 0.01, and each individual had a 0.006~0.035 increased chance of reporting a symptom for each additional friend with the same symptom (Table 1). These parameters are useful for epidemic simulation in the residence hall network, and for setting the initial values of fitting an epidemic model to real-world symptom observations and sensor data. The symptom surveys show some repeated infections, several clustered infections, the persistence of infections in larger clusters, and the persistence of infections caused by individuals who took longer to recover.

In this data, a student with a symptom had 3-10 times higher odds of seeing his friends with the same symptom (again, except for runny nose). As such, it makes sense to fit the time-tested infection model with real-world data of symptom reports and proximity observations, and infer how friends infect one another through their contacts. In order to determine whether the higher odds could somehow be due to chance, we conducted the following permutation test to reject the null hypothesis that "the friendship network is unrelated to symptoms," and we can reject that null hypothesis with $p < 0.05$. The permutation test shuffles the mapping between the students and the nodes in the friendship network and estimates the probability distribution of the number of friends with the same symptom among all possible shuffling. If friendship networks were not related to the timing of when a student exhibits a symptom, then all mappings between the students and the nodes would be equally likely, and the number of friends with the same symptom would take the more likely values.

3   Modeling Infection Dynamics

In this section, we propose a discrete-time stochastic multi-agent SIS model, along with a corresponding inference algorithm to fit this multi-agent model to real-world data on proximity and symptom reporting. The inference algorithm does three things. First, it learns the pa-rameters of the multi-agent model, such as rate of infection and rate of recovery. Second, it estimates the likelihood that an individual was infectious from the contact he had with other students, and from whether those others reported symptoms when the individual's symptom report is not available. Finally, it enables us to make useful pre-dictions about contracting infections within the community in general.

Discrete-Time Stochastic Multi-agent SIS Model to Fit Real-World Infection Dynamics

- Input:
  - A dynamic network, $G_t = (N, E_t)$, where nodes representing people, bi-directional edges $E_t = \{(n_1, n_2): n_1 \text{ is near } n_2 \text{ at time } t\}$ representing "nearby" relation, and
  - Hyperparameters which provide prior information about: α — the probability that an infectious person outside of the network makes a susceptible person within the network infectious, β — the probability that infectious person within the network makes a susceptible nearby person infectious, and γ — the probability that an infectious per n becomes susceptible. The above variables are all assumed to be distributed according to beta distributions defined by these given hyperparameters.
  - Hyperparameters which define the prior probability of observing various symptoms depending on whether or not a person currently has a cold.
- Output: a matrix structure indexed by time and node. The state of node at time is either 0 (susceptible) or 1 (infected). The symptoms of node at time is probabilistically dependent on the state of node at time.
- Procedure:
  - Initialize all parameters using their prior distributions, and assume that all people are susceptible at time $t = 1$.

— For each subsequent time $t + 1 = 2, ..., T$ we assume the following generative model:

   o Infectious person becomes susceptible with probability γ, according to a Bernoulli distribution. If the Bernoulli trial is a success (the infectious person is now susceptible), $X_{n,t+1}$ is set, deterministically, accordingly, and the resulting symptoms $Y_{n,t+1}$ are set stochastically, from their probability distribution, conditioned on $X_{n,t+1}$.

   o Infectious persons within and outside of the network contribute to turning a susceptible person infectious, and the contributions happen independently:

     ■ Person n becomes infectious via contact with another infectious person in their network at time t. Each infectious contact, as specified by G_t, infects n with probability β, according to a Bernoulli distribution.

     ■ Person n is infected by someone outside the network, with probability α, according to a Bernoulli distribution.

   Set $X_{n,t+1}$ accordingly if any of the above Bernoulli trials is a success (a susceptible person is now infectious). Also set $Y_{n,t+1}$ stochastically, from its probability distribution, conditioned on $X_{n,t+1}$.

The probability of seeing a state sequence/matrix $\{X_{n,t} : n, t\}$ is therefore

$$
P(\{X_{n,t} : n, t\}, \alpha, \beta, \gamma)
$$
$$
= P(\alpha)P(\beta)P(\gamma) \prod_{n} P(X_{n,1}) \prod_{t,n} P(X_{n,t+1} | (X_{n',t}), \alpha, \beta, \gamma)
$$
$$
= P(\alpha)P(\beta)P(\gamma) \prod_{t,n} \gamma^{1_{X_{n,t}=1} \cdot 1_{X_{n,t+1}=0}} \cdot (1-\gamma)^{1_{X_{n,t}=1} \cdot 1_{X_{n,t+1}=0}}
$$
$$
\cdot \left(\alpha + \beta \cdot \sum_{(n',n,t) \in E} X_{n',t} \right)^{1_{X_{n,t}=0} \cdot 1_{X_{n,t+1}=1}}
$$
$$
\cdot \left(1 - \alpha - \beta \cdot \sum_{(n',n,t) \in E} X_{n',t} \right)^{1_{X_{n,t}=0} \cdot 1_{X_{n,t+1}=0}}
$$

We employ a Gibbs sampler to iteratively sample infectious/susceptible state sequence from Bernoulli distributions, sample events conditioned on state sequence from Bernoulli distributions, and sample parameters from Beta distributions. This provides an algorithm for performing inference in the above generative model. We can infer values of states X, and even missing values in symptoms Y, conditioned on the values of Y which we observe, and the interaction network G. An in depth description of our model and inference algorithm, and further discussion can be found in [7].

The SIS model describes infection dynamics in which the infection doesn't confer long-lasting immunity, and so an individual becomes susceptible again once recovered. The common cold has this infection characteristic.

4   Experimental Result

In this section we model the contagion which existed in the residence hall community. We estimate, at the community level, the parameters of susceptible-infectious-susceptible (SIS) infection dynamics. At the individual level, we describe the results of using the Gibbs sampling algorithm to fit the discrete-time multi-agent SIS infection dynamics to symptom observations.

We took several steps to calibrate the performances of the multi-agent model and support vector classifier on synthetic data. First, we synthesized 50 time series – each 128 days long – from the Bluetooth proximity pattern in the Social Evolution data and different parameterizations. Then, we randomly removed the infectious/susceptible data from 10% of the population, added noise to the remaining data in each time series, and averaged the performances on inferring the held-out data corresponding to each method and parameterization.

We ran Gibbs samplers for 10,000 iterations, got rid of the initial 1000 burn-in iterations, and treated the remaining 9000 iterations as samples from the posterior distribution. We trained the support vector classifier from another 1000-day time series synthesized using the right parameterization, and used the number of infectious contacts yesterday, today, and tomorrow as a feature. We assigned different weights to the "infected" class and the "susceptible" class to balance the true prediction rate and the false prediction rate.

All methods can easily identify 20% of infectious cases in the missing data with little error, but the model-based method using our dynamic multi-agent system consistently performs better than the support vector classifier. Less noise in symptom observation and in the individuals' contact networks significantly improves the performance of inferring missing data, as shown through the ROC (receiver operating characteristic) curves in the left panel of Fig. 1. An ROC curve indicates better performance if it correctly predicts more positive cases and incorrectly predicts fewer negative cases, or equivalently if it is closer to the top-left corner, or it has the larger area below.

The support vector classifier performs worse – especially in identifying the isolated infectious cases in the missing data – because it assumes that its cases are i.i.d (identical and independently distributed) and because including the temporal structure of epidemic dynamics into the features is not an easy task. The support vector classifier also assumes that we either already have enough training data or can synthesize training data. This assumption generally cannot be satisfied for the kinds of problems we are interested in here.

In order to infer latent common cold time series that best fits the multi-agent SIS model from dynamical Bluetooth proximity information and symptom self-report in the Social Evolution data using our Gibbs sampler, we extracted the hour-by-hour proximity snapshot over the 107 days we were monitoring symptoms and interpolated the hourly symptom report as the submitted daily symptom report. We assumed that the symptoms are probabilistically independent given the common cold state. We ran the Gibbs sampler for 10,000 iterations, removed the first 1000 burn-in iterations, and took the rest as samples of the posterior probability distribution of common cold states conditioned on symptom self-reports.

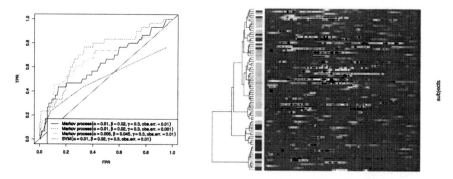

Fig. 1. (Left) Less observation error (obs.err.=0.001) and better knowledge about network ($\beta = 0.045$) lead to better trade-off between true positive rate (TPR) and false positive rate (FPR). The support vector classifier has worse trade-off between TPR and FPR than the multi-agent Markov model. (Right) An agent-based model can infer common cold state, and captures infection from symptom self-report and proximity network. Sizes of black dots represent the number of symptoms reported, ranging from zero symptoms to all symptoms, and no black dot means no self-report.

The right panel of Fig. 1 shows the (marginal) likelihood of the daily common-cold states of individuals. Rows in this heat map are indexed by subjects, arranged so that friends go together, and are placed side by side with a dendrogram that organizes friends hierarchically into groups according to the distance between the individuals and groups. Different colors on the leaves of the dendrogram represent different living sectors in the student dorm. Columns in this heat map are indexed by date in 2009. Brightness of a heat-map entry indicates the likelihood of being infectious. The brighter a cell is, the more likely it is that the corresponding subject is infectious on the corresponding day. Sizes of black dots represent the number of reported symptoms, ranging from zero symptoms to all symptoms. When a black dot doesn't exist on the corresponding table entry, the corresponding person didn't answer the survey on the corresponding day.

This heat map shows clusters of common cold happenings, and in each cluster a few individuals reported symptom. When interpersonal proximities happened in larger social clusters, symptom clusters lasted longer and involved more people. A study of the heat map also tells us what the Gibbs sampler does in fitting the multi-agent SIS model to the symptom report: a subject often submitted flu-symptom surveys daily when he was in a "susceptible" state, but would forget to submit surveys when he was in the "infectious" state. The Gibbs sampler will nonetheless say that he was infectious for these days, because he was in the infectious state before and after, an infectious state normally lasts four days, and many of his contacts were in the infectious state as well. A subject sometimes reported symptoms when none of his friends did in the time frame. The Gibbs sampler will say the he was in the susceptible state, because the duration of the symptom reports didn't agree with the typical duration of a common cold, and because his symptom report was isolated in his contact network.

The inferred infectious state from symptom reports and hourly proximity networks normally lasts four days, but could be as long as two weeks. A student often caught a cold 2 ~ 3 times from the beginning of January to the end of April. The bi-weekly searches of the keyword "flu" from January 2009 to April 2009 in Boston – as reported by Google Trends – explains 30% of variance in the number of (aggregated) bi-weekly common cold cases inferred by the Gibbs sampler, and network size explains another 10%.

The timing of different symptoms with regard to the inferred common cold cases follows interesting patterns. Stress and sadness normally began three days before the onset of a stretch of infectious state, and lasted two weeks. Runny nose and coughing began zero to two days before the onset of a symptom report and ended in about seven days, and they have similar density distributions. Fever normally occurred on the second day after the onset of a stretch of infectious state, and lasted for about two days. Nausea often happened four days before the onset of reaching an infectious state, then disappeared and reappeared again at the onset.

5   Conclusions

The study of infection in a small population has important implications both for refining epidemic models and for advising individuals about their health. The spread of infection in this context is poorly understood because of the difficulty in closely tracking infection in a complete community. This paper showcases the spread of an infection centered on a student dormitory, based on daily symptom surveys over a period of four months and on proximity tracking through resident cellular phones. It also demonstrates that fitting a discrete-time multi-agent model of infection with real-world symptom self-reports and proximity observations give us useful insight in infection paths and infection prevention.

References

1. Eubank, S., Guclu, H., Kumar, V., Marathe, M., Srinivasan, A., Toroczkai, Z., Wang, N.: Modelling disease outbreaks in realistic urban social networks. Nature 429, 180–184 (2004)
2. Isella, L., Stehle, J., Barrat, A., Cattuto, C., Pinton, J., Van den Broeck, W.: What's in a crowd? analysis of face-to-face behavioral networks. J. Theor. Biol. 271, 166–180 (2010)
3. Salathe, M., Kazandjieva, M., Lee, J., Levis, P., Feldman, M., Jones, J.: A high-resolution human contact network for infectious disease transmission. Proc. Natl. Acad. Sci (USA) 107, 22020–22025 (2010)
4. Hufnagel, L., Brockmann, D., Geisel, T.: Forecast and control of epidemics in a globalized world. Proc. Natl. Acad. Sci. USA 101, 15124–15129 (2004)
5. Stehle, J., Voirin, N., et al.: Simulation of an SEIR infectious disease model on the dynamic contact network of conference attendees. BMC Medicine 9(1), 87 (2011)
6. Dong, W., Lepri, B., Pentland, A.: Modeling the coevolution of behaviors and social relationships using mobile phone data. In: Proc. ACM MUM (2011)
7. Dong, W., Heller, K., Pentland, A.: Modeling Infection with Multi-agent Dynamics. arXiv 1201.xxxx [cs.MA, cs.SI]

Dynamic Multi-chain Graphical Model for Psychosocial and Behavioral Profiles in Childhood Obesity

Edward H. Ip[1], Qiang Zhang[1], and Don Williamson[2]

[1] Wake Forest University School of Medicine, Medical Center Blvd., WC23,
NC 27157, USA
{eip,qizhang}@wakehealth.edu
http://www.phs.wfubmc.edu/public/bios/home.cfm
[2] Pennington Biomedical Research Center, 6400 Perkins Road,
Baton Rouge, LA, USA
Donald.Williamson@pbrc.edu

1   Introduction

Childhood Obesity as a System Problem. Childhood obesity is a social epidemic [1],[2] that persists from childhood to adolescence and well into adulthood. Energy-intake and energy-expenditure-related behaviors – diet and physical activities – form the core of the energy-balance equation. These behaviors are heavily influenced by a complex set of interrelated psychosocial, biological, and ecological variables. The recent literature is beginning to recognize the need to study it in terms of multiple chains of causal influences flowing from distal social factors to proximate, individual factors and behaviors, and in terms of drivers of change in trends and patterns over time. This ecological, multilevel perspective calls for the analysis of obesity as a complex system that involves information collected from multiple sources and at multiple time points.

Conceptual System Framework of Glass & McAtee. To achieve this goal, an empirically testable systems approach to childhood obesity must be built upon a meaningful conceptual framework that provides some structure to the complexity and that at the same time is able to delineate the roles of different variables within the system. Glass and McAtee [3] outline such a conceptual framework for an approach that is applicable to the study of childhood obesity. The framework, which we call a multilevel behavioral-social-ecologic model, can be visualized as a system that contains two primary axes: time (as the horizontal axis) and a hierarchy of systems from genes, to organs, to social networks and communities (as the vertical axis). The chain of causal influence flows from distal social factors to proximate individual factors, as the time axis depicts temporal influences, which could be conceptualized as the life course from birth to death as well as societal changes at the population level. On the one hand, social and environmental forces external to individuals are treated along the vertical axis, as rising above the "waterline," where individual behavior occurs. On the other

S.J. Yang, A.M. Greenberg, and M. Endsley (Eds.): SBP 2012, LNCS 7227, pp. 180–187, 2012.

hand, individual genes and biological systems lie beneath the waterline. In systems language, the social conditions, cultural norms, poverty, etc. are treated as control parameters that affect the probability of behaviors that are "causes" of obesity. To distinguish them from direct causal risk factors, each set of variables representing a specific aspect of the relatively stable social and built environments is called a risk regulator. For example, culture is a risk regulator in the sense that it does not directly "cause" any single disease in everyone similarly exposed. Yet, culture operates at a system level to up- or down-regulate the likelihood of key risk factors (behaviors such as overeating) dynamically, and presumably over the entire life course. In summary, the multilevel behavioral-social-ecologic model offers a coherent, multilevel framework for the study of childhood obesity in both the social and biological contexts.

Operationalizing the Glass & McAtee Conceptual Model. The proposed dynamic multichain graphical model (DMGM) is a tool for analytically mapping the multilevel behavioral-social-ecologic modeling framework to an empirically falsifiable dynamic system. Drawing upon prior work on dynamic Bayesian network (DBN) [4], in particular the hidden Markov model – a special case of the DBN, the DMGM employs a three-pronged strategy to operationalize the ecologic conceptual framework. First, it treats relatively stable contextual risk regulators (e.g., environmental factors such as density of fast food restaurants in a neighborhood) and direct (causal) risk factors (e.g., sedentary lifestyle) as two distinct constructs. Operationally, such a distinction allows a potentially daunting number of levels of variables to be broken down into two more manageable pieces, or "spaces" (see paragraph below). The second strategy is to exploit the sparse conditional independence structure and use multichain graphical models to delineate parallel causal mechanisms and their interactions. Finally, the DMGM employs modern statistical techniques, including the random-effects models and robust estimation methods, for handling clustered data within each space.

Acknowledging the risk of oversimplified representation, we present the DMGM as a hybrid graphical model that consists of two spaces: (1) the causal space – a system of interrelated outcome variables, possibly temporal in nature, among which some hypotheses about causal relationships can be properly formed based on substantive knowledge; and (2) the regulatory space – a collection of relatively stable risk regulators that are treated like predictors in regression. They have the capacity to modulate system parameters. The modeling of the causal space is consistent with the graphical model tradition, in which the joint distribution is the goal of inference. The causal space is thus the equivalent of a DBN, whereas the regulatory space variables are treated more or less as fixed and are not fully engaged in inference about the joint distribution within the causal space. In this sense, the current work is an extension of the work on the mixed-effects hidden Markov model in Ip et al. [7] and Zhang et al. [6].

The remainder of this article is organized as follows. First, we describe the architecture of a dynamic system as operationalized by the DMGM. Then we describe the data set. Finally, we report data analysis results to illustrate the DMGM methodology.

2   Multichain Hidden Markov Model

In this specific presentation, the hidden Markov model (HMM) is used as a basic model. For our purpose, one can envision the HMM as an extension of the latent class model for longitudinal data. A hidden (latent) categorical variable represents a specific construct, or a behavioral profile such as dietary behavior, as measured by multiple indicators, such as fat, carbohydrate, and protein contents. The categories of the hidden variables within a specific profile are called hidden states. Besides assuming that indicators are conditionally independent given hidden states, the hidden variables across time are assumed to follow the Markov assumption, namely that the value of the hidden variable at time t depends only on its value at time $t-1$ and not on previous values. Both the latent class structure (e.g, the number of states and the state parameters) and the transition probabilities between states are assumed to remain constant across time. The corresponding Markov chain is thus homogeneous. In DMGM, there exist more than one Markov chain, and each chain represents a trajectory of behavioral changes within a specific domain of behavior. To summarize trajectories of individual changes across multiple domains, a second-level HMM treats categories of the hidden class variable at the domain level as indicators.

The term superstate is used to refer to discrete, latent (hidden) states that represent phenotypes of individuals with characterization in several domains – in our application these are the domains of psychosocial food-intake behavior and physical-activity behavior. Note that if within each domain there are four states, then there are $4^3 = 64$ possible combinations of superstates. However, some of these states would not be empirically supported by the data. The second level of hidden states or superstates is a useful way to capture the heterogeneity across the domains and to summarize the multiple trajectories of change.

We have just outlined the structure of the causal space that represents direct causal factors within the system. Up to this point, the DMGM is still a special case of the dynamic Bayesian network. In Fig. 1, the nodes and arcs (for two chains, not counting the chain of superstates) inside the large box constitute the causal space. Estimation of model parameters can readily be implemented by standard dynamic Bayesian network algorithms. Figure 1 also shows input from the so-called regulatory space. The input variables in the regulatory space are treated like regressors in (conditional) linear models. By design, the separation of the causal and regulatory space limits the number of risk factors (parental nodes) that are present in the causal space. Factors that are less directly linked to childhood obesity outcomes are "moved" into the regulatory space. Indeed, one of the main advantages of segregating the causal and the regulatory spaces is the computational feasibility of handling data with a sizeable number of levels and/or a large number of variables. The technical details of implementation are given in [7] and [6].

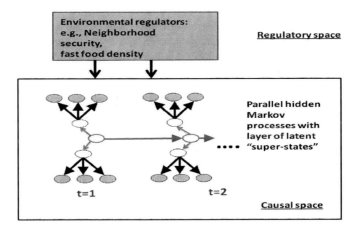

Fig. 1. Architecture of the dynamic multichain Markov graphical model and the causal and regulatory spaces. Two chains together with a chain for superstates are shown within the causal space.

3   LA Health Data and Simulated Data Sets

Using both data from the Louisiana (LA) Health Study [8] and simulated data, we proceed to illustrate the DMGM approach. The LA Health Study was a clustered, randomized, controlled trial that enrolled $N = 2,201$ school students across 17 rural school clusters in the state of Louisiana, which has one of the highest rates of poor health in the nation. Students were randomly assigned to one of three arms – primary prevention (PP), combined primary and secondary prevention (CP), and no-intervention control (C). PP emphasized the modification of environmental cues, enhancement of social support, and promotion of self-efficacy for health behavior change. CP relied on both the PP approach and an Internet-based educational program reinforced with regular classroom instruction and synchronous online counseling and asynchronous (e-mail) communication for children and their parents. In addition to the primary endpoint of body-mass index (BMI), other outcome measures were collected in the LA Health Study: (1) energy and nutrient measurements, (2) physical and sedentary activity measurements, and (3) psychosocial measurements. For the first category, digital photos were used for food selection at school, and dietary measures, including total calories, macro-nutrient content, and total dietary fat, were collected. Physical activities were measured by the Self-Administered Physical Activity Checklist (SAPAC), while sedentary behavior was assessed by questionnaires about the number of hours spent on TV, video, computer, homework, and telephone. The Children's Eating Attitudes Test (ChEAT) and the Child Depression Inventory (CDI) were used to respectively measure eating disorder symptoms and symptoms of childhood depression. Subscales from ChEAT such as social pressure and food preoccupation were used for our analysis. Thus, data from three domains that are most proximate to obesity – diet, physical activities, and psychosocial – were included in the causal space. Measurements were taken at a total of three time-points across a span of 18-24 months.

Currently, almost no data are available at the community/environment level in the LA Health study. In order to illustrate the full DMGM framework, we developed two models for the regulatory space. The first model (Regulatory Model RM1) used intervention status as a summarized measure of the environmental effect, and the second (Regulatory Model RM2) used simulated data on two indexes that indicate community-level factors – a community security index and the density of fast-food restaurants. In the second model RM2, the two indexes were independently sampled from a standard normal distribution with mean values directly related to the outcome variable of health state using a set of pre-specified parameters. For both RM1 and RM2, we used generalized, mixed effects, linear models to describe how students' transitions from one health state to another were regulated by the respective environmental variables.

Table 1. Descriptive Statistics of LA Health Female Participants ($N = 687$)

Characteristics		Mean/Percentage	SD
White		23.7	
Black		74.7	
Age		13.95	1.19
BMI		23.71	6.74
Psycho-social profile (factor loading)			
	Overconcern	11.28	5.34
	Dieting	8.25	3.85
	Social Pressure	6.02	3.86
	Self-control	6.31	3.01
	Depression	46.33	8.33
Food intake profile (percentage)			
	% from Fat	33.05	6.29
	% from Protein	18.34	4.65
	% from Carbohydrate	49.68	8.72
Physical activity profile (min/day)			
	Total TV	174.28	137.71
	Before School PA	7.23	18.18
	During School PA	17.80	23.98
	After School PA	65.26	69.99

4   Results of Data Analysis

Because female and male students exhibit rather different developmental trajectories, we conducted separate analyses for each gender. In the interest of saving space, we only show the results for female students. Table 1 provides a brief description of the (continuous) variables used for creating the three separate profiles - psychosocial (PS), food intake (FI), and physical and sedentary activities (PSA). In this presentation, we used simple models for the first-level hidden states. The number of hidden states were, respectively, 2, 3, and 2 for the three different domains of interest: PS, FI, and PSA. To visualize the profiles across the chains, as generated by the HMM, we depict the results in the form of a dot chart with a bar of length ±2 standard deviations centered at the dot (Fig. 2). For the PS domain, State 1 represents "High-Concerns-About-Weight," and State 2 represents a "Low-Concerns-About-Weight" state (Fig. 2(a)). For the FI domain, the three resulting states were respectively labeled High-Carb, High-Fat, and High-Protein (Fig. 2(b)). For the PA domain, State 1 was labeled Physically-Inactive, and State 2 was labeled Physically-Active (Fig. 2(c)).

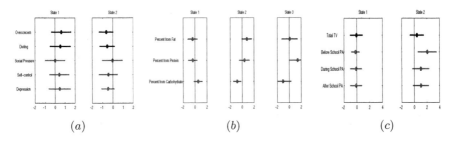

(a) (b) (c)

Fig. 2. Conditional probability of states in profiles for (a) psycho-social (PS), (b) food intake (FI), and (c) physical and sedentary activities (PSA)

The superstates profiles that were derived from domain-level states are shown in Fig. 3. The domains are represented by the panels, and each column within a panel represents the profile of a superstate for the specific domain. For example, the third column in the PS-panel shows that Superstate 3 contains a high proportion of Low-Concerns-About-Weight female students (column 3, panel 1). This superstate also contains a high proportion of High-Carb (column 3, panel 2) and Physically-Inactive (column 3, panel 3) female students. Therefore, this superstate is dominated by female students that are not concerned about their weight, eat a relatively high-carb diet, and are physically not active. Superstate 2 consists of female students who are not highly concerned about weight, consuming a relatively high-fat diet compared with the other two superstates, but staying physically active. Superstate 1 appears to be a state that is characterized by a high level of anxiety about weight and diet, a high-carb dietary habit, and a low-physical-activity lifestyle. The transitions over time across the different superstates are revealing (Table 2). Superstate 2, which is characterized by a high level of physical activity, has a high probability of transitioning into State 1 or State 3, with probabilities 0.40 and 0.39, respectively. This implies that the physically active female students tend to engage in fewer physical activities during the last two time points of the study. The interventions – both primary and combined – do not moderate this trend. From the transition probability

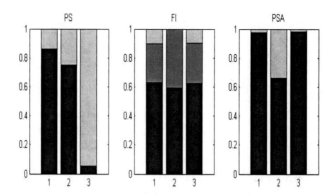

Fig. 3. Conditional probability profile of the super-states 1,2, and 3. The lowest-numbered category of a domain-level state appears at the bottom of the stacked bar chart.

Table 2. Estimated prior state probabilities (α) and transition probabilities (τ) between superstates, in which an entry (i,j) indicates the probability of transition from the i-th superstate to the j-th superstate.

	Superstate 1	Superstate 2	Superstate 3
α:	0.24	0.50	0.25
τ:	0.93	0.00	0.07
	0.40	0.21	0.39
	0.00	0.00	1.00

table, it can be also seen that Superstates 1 and 3 are both relatively stable, implying that regardless of these female students' levels of concern about weight, their high-carb and inactive lifestyle remain unchanged throughout the period of study. For RM1, we examined how transition probabilities were regulated by the environmental variable Intervention Status. We found that CP has a marginally significant effect in reducing the likelihood (OR=0.28, p=0.06) of students moving from Superstate 2 (physically active) to Superstate 1 (high anxiety, not active). CP also has a statistically significant effect of preventing students from moving from Superstate 2 to Superstate 1 (OR=0.13, p=0.002). In this analysis, we included baseline BMI, age, gender, and race, which are individual-level but not environmental variables, as covariates. Only BMI is statistically significant. The intervention PP was not statistically significant in affecting the dynamic of behaviors. For RM2, the model performed quite well in recovering the assigned values of the parameters, as both the parameters for security and fast-food density were located within the 95% confidence limits. We also found that the results were rather sensitive to the variation of the simulated indexes, as measured by their standard deviations. As a next step, we are currently collecting real environmental data for the schools and communities.

5   Discussion and Conclusion

This paper illustrates how the DMGM framework operationalizes a multilevel behavioral-social-ecologic conceptual model. The proposed approach can be expanded to include important system features such as feedback loop and time-varying environmental variables. The limitations of the study include the relatively small number of longitudinal observations, the lack of significant regulatory variables, and the lack of "below-skin" variables such as physiologic and genetic measures. In conclusion, the dynamic approach could be a valuable tool for modeling childhood obesity and other socially significant systems.

Acknowledgment. The study is supported by NHLBI grant 1U01HL101066-01 (PI: Edward Ip).

References

1. World Health Organization. Dietary change and exercise needed to counter chronic diseases worldwide. Rev. Panam Salud Publica 13, 346–348 (2003)
2. Hedley, A.A., Ogden, C.L., Johnson, C.L., et al.: Prevalence of overweight and obesity among US children, adolescents, and adults 1999-2002. Journal of the American Medical Association 291, 2847–2850 (2004)
3. Glass, T.A., McAtee, M.J.: Behavioral science a the crossroads in public healt: Extending horizons, envisioning the future. Social Science and Medicine 62, 1650–1671 (2006)
4. Ghahramani, Z.: Learning Dynamic Bayesian Networks. In: Giles, C.L., Gori, M. (eds.) IIASS-EMFCSC-School 1997. LNCS (LNAI), vol. 1387, pp. 168–197. Springer, Heidelberg (1998)
5. MacDonald, I.L., Zucchini, W.: Hidden Markov and other models for discrete-valued time series. CRC Press (1997)
6. Zhang, Q., Snow Jones, A., Rijmen, F., Ip, E.H.: Multivariate discrete hidden Markov models for domain-based measurements and assessment of risk factors in child development. Journal of Computational and Graphical Statistics 19(3), 746–765 (2010)
7. Ip, E.H., Snow-Jones, A., Heckert, D.A., Zhang, Q., Gondolf, E.: Latent Markov model for analyzing temporal configuration for violence profiles and trajectories in a sample of batterers. Sociological Methods and Research 39, 222–255 (2010)
8. Williamson, D., Champagne, C., Harsha, D., Han, H., Martin, C., Newton Jr., R., Stewart, T., Ryan, D.: Louisiana (LA) Health: Design and methods for a childhood obesity prevention program in rural shcool. Contemporary Clinical Trials 29(5), 783–795 (2008)

The Imperative for Social Competency Prediction

Robert Hubal

RTI International, Research Triangle Park, NC, USA
rhubal@rti.org

Abstract. Some military personnel returning from deployment show social competency deficits: They act impulsively and make risky decisions, misinterpret interaction cues, experience difficulties with personal relationships, and adopt high-risk behaviors. These adverse social skills directly influence, among other important variables, psychological health and quality of life. Social skills deficits are not just a military concern; for instance, at-risk adolescents and reintegrating prisoners must also learn to demonstrate social competence. Meanwhile, today's screening is limited in its ability to assess current—and predict future—social competency; typical neurocognitive assessment is not designed to assess social competence in realistic situations. The author proposes a tool to improve screening by identifying social competency deficits through assessment of behavior in simulated social situations. This is important not only to more accurately assess adverse behaviors, but also to predict future behaviors and their causes, to focus intervention to address social competency deficits before adverse behaviors are ever exhibited.

Keywords: Social competency, interaction skills, situated assessment.

1 Introduction

Social skills are those abilities that underlie individuals' management of their behavior in a variety of challenging or demanding social situations. Social skills *deficits* manifest as poor behaviors in such situations. For example, in a confrontational situation, some individuals may perceive linguistic, gestural, and expressive cues of the dialog partner as indicating hostile intent. Misinterpretation of these cues—the words used during the situation, the gestures used and facial expressions shown by the dialog partner—could lead to an escalation in the potential for violence or an adverse outcome as a result of the confrontation. Other situations where social skills are essential are those involving risky choices, those involving impulsive versus deliberate decisions, those involving familial or personal relationships, and those involving negotiation or the expressing of one's preferences in conflict with peer opinions or influences.

These situations and others are prevalent for U.S. military personnel returning from deployment, and an increasing number of personnel are behaving poorly. Social skills deficits demonstrated by these personnel are manifested in numerous ways, including:

S.J. Yang, A.M. Greenberg, and M. Endsley (Eds.): SBP 2012, LNCS 7227, pp. 188–195, 2012.
© Springer-Verlag Berlin Heidelberg 2012

- *Psychological health.* Some 40% of personnel returning from Iraq and Afghanistan experience high rates of psychological problems such as posttraumatic stress disorder and depression [1]. These mental health conditions are associated with substance abuse disorders and with behavioral problems such as poor anger management and aggression [2]. They may get progressively worse post deployment [3].
- *Substance use.* Anywhere from 12-15% to nearly one-third of returnees screen positive for or experience alcohol misuse [4,5]. It is estimated that, between 2004 and 2006, 25% of young veterans suffered from substance abuse disorders in the preceding year [6]. Certain substance abuse trends for deploying [7] and postdeployed [8] personnel are concerning.
- *Aggression and violence.* Intimate partner violence is a significant concern for spouses of personnel returning from combat [9], as anger problems mediate between stress disorders and intimate partner violence among veteran couples [10]. Other indicators of aggression include risky driving and hypervigilance [11,12]. Additionally, there is a distressing trend of suicide by veterans [13].

The problem is not just one of postdeployed military personnel, however. Social skills deficits are demonstrated as well by formerly incarcerated persons [14], at-risk youth [15], certain individuals along the autism spectrum [16], the mentally ill [17], even pediatric cancer survivors [18]. The intent of this paper is to demonstrate how prediction of social competency is possible through behavior modeling using simulated situations and why such prediction would benefit intervention.

1.1 Improving Screening

In particular the vision is a tool to improve health screening by identifying social skills deficits and helping to focus interventions that address those deficits. It is proposed here to "situate" the assessment using virtual vignettes that the author pioneered with inner-city, at-risk adolescents [15]. The idea is to present social situations adapted for the given population (returning personnel, autistic children) to have individuals *demonstrate* their social skills *rather than describe* their attitudes and beliefs as is typical with conventional assessments. The aim would be for user behaviors exhibited while engaged in the set of virtual vignettes to compare favorably against current best practice conventional clinical measures of social skills. Operationalizing improved screening for this situated assessment would involve, relative to existing screening, identifying more individuals with social skills deficits, identifying more specific skills deficits through detailed behavior models that could be used to focus subsequent clinical intervention, and predicting, with greater accuracy, actual social behaviors during an interval between a baseline and a follow-up assessment.

2   Social Skills Assessment

Typical best practice assessment of social skills involves placing individuals into hypothetical (but considered real life) social situations. Examples include asking individuals what they would do in these situations and measuring responses using observation and validated rubrics, presenting non-interactive videotapes to gauge individuals' reactions, and presenting questionnaires that may involve inaccurate reporting but are still considered best available [19]. Other approaches are usually more focused on specific social constructs such as quality of life and family functioning, but even with these constructs there are not systematic assessment approaches [20].

The author has conducted research that represents a different approach to improving assessment of social skills. These studies evaluate decision making and social skills using situated assessment—implemented using a gaming engine and established artificial intelligence models [17,21]—based on types of decision making and social skills that have been associated with violent behavior and drug use and that are targeted by violence and drug abuse prevention programs. These skills include emotional control, information seeking, expressing preferences, negotiation and willingness to compromise, and using non-provocative language. The skills are assessed by simulating in a virtual environment real social encounters that may have adverse consequences.

In these studies, behavior is assessed using two approaches. First, individuals' engagement is observed while they interact within the virtual environment. Specifically considered are their body language, tone of voice, emotional control, and reaction time, for each vignette, mapping each against a predefined scale. Users have been found to be engaged during the vignettes, quickly suspending any disbelief that they may have in speaking with a virtual character once the character responds appropriately to the first few dialog exchanges [15]. These observations provide one source of data for assessing appropriate social behaviors.

Second, which path to outcome that is taken is evaluated. That is, each vignette may have any of several outcomes—usually one or more good outcomes demonstrating avoidance of risky behavior and one or more poor outcomes demonstrating risky behavior—and any of several paths leading to a given outcome. Different paths (i.e., dialog exchanges, manipulation of objects, and other components of the social interaction) represent more or less appropriate social skills. For example, a vignette may involve a virtual character trying to entice an individual into a confrontation. One good outcome is to defuse the situation using dialog. A poor outcome is to vehemently disagree with the character by raising one's voice toward the character. On the path toward any outcome there are worse and better exchanges, such as those that cause a decrease or an increase in how anxious or angry the character behaves [21] or those that demonstrate desired skills such as seeking additional information or expressing one's preferences. These exchanges are behaviorally based—that is, demonstrated by the individual—hence provide data for assessing appropriate social behaviors.

2.1   The Need for Predictive Ability

There is reason to believe that engagement with virtual vignettes could predict real-world, socially adverse behaviors *before* an individual has had the opportunity to demonstrate any adverse behaviors. To test this conjecture would require monitoring user performance with vignettes at a baseline, then following up after some time (e.g., three months). Individuals' actual behaviors in the interim (captured through a questionnaire) would be mapped against baseline behaviors in an attempt to determine predictive patterns of social skills deficits. Thus, supposing that a given individual was arrested for battery in the three months following a baseline, correlations could be sought in that individual's baseline vignette performance that make theoretical sense, such as demonstration of aggression, hostile intent bias, and/or impulsivity in certain situations. These behaviors would have had to have been shown across the range of vignettes (i.e., consistently and reliably, in appropriately different situations). In contrast, supposing that a given individual attempted suicide in the three months following a baseline (such sensitive information can be captured using computer-assisted self-interviewing), relationships could be sought with baseline behaviors that showed, for instance, lack of engagement or interest in the dialogs, inability to express preferences, or slowed reaction time. Still other mappings would result for additional intervening events, such as arrest for a reason other than battery (perhaps drunk driving), acquisition of a new substance use habit, or separation or divorce. With sufficient breadth of vignettes, predictive models developed initially with a sample of participants could be refined based on each subsequent participant's baseline and interim behaviors. The ultimate intent for predictive behavior models is to benefit clinical intervention and even prevention.

3   Tool Development

3.1   Moving beyond Existing Best Practices for Assessment of Social Skills

All U.S. military personnel returning from deployment are supposed to complete a self-report postdeployment health assessment and a reassessment after three and six months, to determine if they have developed physical or psychological health illnesses. These instruments, though, do not directly assess social competency. Additionally, predeployment neurocognitive assessment [22] is now mandated by the U.S. military, thereby establishing a baseline that could be used for comparison against postdeployment assessment to gauge the extent of any possible injury (but see [23]). This assessment is designed to detect executive functioning, attention, impulse control, working memory, and reasoning and decision-making ability through a battery of cognitive and emotional tasks, but it does not tap into interpersonal relations, negotiation skills, emotional intelligence, empathic ability, or recognition of other's intent.

To augment neurocognitive assessment, virtual vignettes—including all relevant virtual objects, persons, actions, sounds, and settings needed for the designed situations—are proposed to situate individuals and elicit behaviors so that social skills competencies or deficits could be demonstrated. Numerous existing applications involve virtual vignettes for training and assessment of interaction skills. These applications have been geared toward a wide variety of behaviors, including interpersonal skills and emotional control among adolescents, simulated patients for bioterrorism preparedness and for pediatric clinical diagnosis, simulated mentally ill consumers encountered by law enforcement officers, and general research participants or medical patients for training researchers to obtain informed consent (for just one review, see [24]). In these applications, individuals interact with characters rendered on the screen through voice, text, menu selection, and/or cursor activity, and the characters respond appropriately with speech, movement, and expression.

With the proposed tool, not just basic cognitive constructs would be addressed, such as are measured using neurocognitive tests, but also higher-level skills, such as are needed in real-world situations. Behavioral constructs representative of social skills deficits include poor emotion expression recognition, impulsivity, insensitivity to penalties, hostility bias, gender stereotyping, acceptance of dating or partner violence, and risky decision making. Hence, virtual vignettes would need to require complex responses from the individuals. Multiple versions of multiple vignettes would be developed, addressing the range of social skills competencies. Vignettes would have to be tailorable to cover the range of individuals' experiences, backgrounds, working and living environments, and social skills deficits.

3.2   Designing and Developing Virtual Vignettes

Designing and developing virtual vignettes is effortful but not complicated. To inform the design, a developer must consider several factors. First, the developer must ensure that the range of vignettes covers the range of social competency skills that need to be addressed. This effort involves subject-matter expertise and creativity in situations presented, as well as variation in possible responses that are available to the user at any decision point. Second, the developer must determine how to follow and assess the user's behavior within a vignette. This task is accommodated by mapping variable values to the social skills competencies and updating them as the interaction progresses, as the user employs or fails to employ appropriate skills [15]. Third, the developer must allow any dialog flow to branch at decision points based on the individual's input, so that the dialog remains unscripted; this may be accomplished by the use of an augmented transition network [20] that manages the dialog state. Eliciting dialog structures for virtual vignettes requires several steps, including interviews with experts to understand key drivers of the dialog flow that are found to lead to adverse consequences, translation of these data into subtopics and conditional statements that define how and when key drivers cause subtopics to be discussed, and definition of grammars that cover the range of possible utterances by either dialog partner.

A developer should expect there to be many subtopics to discuss in any situation, given initial conditions and a narrative for the situation. Extensive branching between subtopics, an influence of having discussed one subtopic on discussions of another subtopic, considerable emotional content and cognitive dissonance exhibited by the individual, and complicated conditional statements all affect the flow of the dialog. For such complex applications one approach is to use a series of linked tables that developers populate with experts' help that systematically organize models of how each virtual character should behave (i.e., its language, gestures, and emotional responses) in response to all of the predicted inputs from the user and within all of the different simulated conditions. That is, behavior models would specify how the emotional, physiological, cultural, and cognitive states of the virtual characters change based on user input and time course, and how they would be influenced by contextual factors such as the social role played by the virtual character and the setting of the dialog [21,25]. Rendering of the characters and setting could be done via nearly any commercial game engine, and non-dialog interaction (e.g., manipulating virtual objects in the virtual environment) realized via object-based selection maps. The many existing applications employ varying techniques to implement their virtual vignettes but all involve some type of underlying behavior modeling.

3.3   Validating through Baseline Assessment and Follow-Up

No existing application, however, is now used to directly predict future behaviors. To validate a situated assessment tool and gauge its predictive capacity, individuals would engage in the virtual vignettes at baseline, and their interim behaviors between baseline and some time post-baseline would be captured, as described above. Relevant events such as demonstration of psychological illness or risky behavior, as well as important background characteristics, would influence how baseline performance feeds into predictive models. Validity checks against standard practices would use methods that have previously been employed [26], including descriptive statistics for individuals' level of engagement and principal components analysis to identify latent constructs of virtual vignette performance measures. The data from each successive individual would refine predictive models.

Situated assessment using virtual vignettes is expected to have predictive power for at least two reasons. First, it is known how to develop virtual vignettes and assess current skill. It is reasonable to believe current skill would influence future actions. Second, because of the greater realism than neurocognitive assessment and the greater dynamism than static vignette assessment, virtual vignettes may enable the determination of how—or within what bounds—an individual would behave in the future under different conditions, given how that individual is behaving now. Furthermore, based on previous findings [26] that showed differential performance on the virtual vignettes prior to any intervention between adolescents who had or had not previously been diagnosed with conduct disorder, it is not unlikely that different groups of individuals (returning personnel, autistic children) would also perform differentially.

3.4   Developing Clinical Guidance for Use by Clinical Staff

A guide for the effective use of this tool—to tailor intervention—would be useful for clinical staff. The clinical use guide would describe how the outcomes from the set of vignettes presented to an individual indicate specific clinical needs. For instance, if the individual were to demonstrate a hostile intent bias through his/her actions, by misinterpreting gestures and facial expressions or by guiding the dialog toward confrontational branches, but not demonstrate any impulsivity in his/her decisions, then a clinician could use this information to focus subsequent intervention onto that biased behavior. It would be important to ensure that the guide also considered what derives from other assessments, so as to check consistency (e.g., between an assessment of impulsivity by one of the neurocognitive tests and that done through the virtual vignettes). These use factors would be developed in consultation with clinical experts.

4   Summary

A tool as is proposed here is needed during clinical interventions for individuals demonstrating or at risk of social skills deficits caused by psychosocial problems— catching potentially socially adverse behaviors *before* they actually occurred. There is such a need for both military and civilian populations. The main challenge is the development of powerful enough predictive behavior models. Assessment, then, would take place within a safe, replicable, adaptable virtual environment simulating real-world social interactions. The tool would be applicable to clinical personnel in the field or at primary and behavioral health care settings. As a screener, the tool would be meant to provide information that could improve the quality of life, psychological health, and fitness for duty or readiness for reintegration.

References

1. An achievable vision: Report, DoD Task Force on Mental Health. Defense Health Board, Falls Church, VA (2007)
2. Tanielian, T., Jaycox, L.H.: Invisible Wounds of War: Psychological and Cognitive Injuries, Their Consequences, and Services to Assist Recovery. Report MG-720-CCF, RAND Corporation, Santa Monica, CA (2008)
3. Milliken, C.S., Auchterlonie, J.L., Hoge, C.W.: Longitudinal Assessment of Mental Health Problems among Active and Reserve Component Soldiers Returning from the Iraq War. JAMA 298, 2141–2148 (2007)
4. Allison-Aipa, T.S., Ritter, C., Sikes, P., Ball, S.: The Impact of Deployment on the Psychological Health Status, Level of Alcohol Consumption, and Use of Psychological Health Resources of Postdeployed U.S. Army Reserve Soldiers. Mil. Med. 175, 630–637 (2010)
5. Erbes, C., Westermeyer, J., Engdahl, B., Johnsen, E.: Post-Traumatic Stress Disorder and Service Utilization in a Sample of Service Members from Iraq and Afghanistan. Mil. Med. 172, 359–363 (2007)
6. Results from the National Survey on Drug Use and Health: National Findings. DHHS, SAMHSA, Office of Applied Studies, Washington, DC (2007)
7. Report of the U.S. Army Pain Management Task Force. Office of the Army Surgeon General (2010)
8. Bray, R.M., Pemberton, M.R., Lane, M.E., Hourani, L.L., Mattiko, M.J., Babeu, L.A.: Substance Use and Mental Health Trends among U.S. Military Active Duty Personnel: Key Findings from the 2008 DoD Health Behavior Survey. Mil. Med. 175, 390–399 (2010)
9. Karney, B.R., Ramchand, R., Osilla, K.C., Caldarone, L.B., Burns, R.M.: Invisible Wounds: Predicting the Immediate and Long-Term Consequences of Mental Health Problems in Veterans of Operation Enduring Freedom and Operation Iraqi Freedom. Working Paper WR-546-CCF, Calif. Community Foundation (2008)

10. Taft, C.T., Street, A.E., Marshall, A.D., Dowdall, D.J., Riggs, D.S.: Posttraumatic Stress Disorder, Anger, and Partner Abuse among Vietnam Combat Veterans. J. Fam. Psychol. 21, 270–277 (2007)
11. Fear, N.T., Iversen, A.C., Chatterjee, A., Jones, M., Greenberg, N., Hull, L., Rona, R.J., Hotopf, M., Wessely, S.: Risky Driving among Regular Armed Forces Personnel from the United Kingdom. Am. J. Prev. Med. 35, 230–236 (2008)
12. Hoge, C.W., Castro, C.A., Messer, S.C., McGurk, D., Cotting, D.I., Koffman, R.L.: Combat Duty in Iraq and Afghanistan, Mental Health Problems, and Barriers to Care. N. Eng. J. Med. 351, 13–22 (2004)
13. U.S. Army Health Promotion Risk Reduction & Suicide Prevention Report. Army HP/RR/SP Report (2010)
14. Roman, J.K., Visher, C.: Prisoner Reentry Programming. In: Weimer, D.L., Vining, A.R. (eds.) Investing in the Disadvantaged: Assessing the Benefits and Costs of Social Policies, pp. 127–150. Georgetown U. Press, Washington, DC (2009)
15. Hubal, R.C., Fishbein, D.H., Sheppard, M.S., Paschall, M.J., Eldreth, D.L., Hyde, C.T.: How Do Varied Populations Interact with Embodied Conversational Agents? Findings from Inner-City Adolescents and Prisoners. Comput. Human Behav. 24, 1104–1138 (2008)
16. Flynn, L., Healy, O.: A Review of Treatments for Deficits in Social Skills and Self-Help Skills in Autism Spectrum Disorder. Res. Autism Spectr. Disord. 6(1), 431–441 (2012)
17. Hubal, R.C., Frank, G.A., Guinn, C.I.: Lessons Learned in Modeling Schizophrenic and Depressed Responsive Virtual Humans for Training. In: Proc. Intelligent User Interface Conference, pp. 85–92. ACM Press, New York (2003)
18. Bonner, M.J., Hardy, K.K., Willard, V.W., Anthony, K.K., Hood, M., Gururangan, S.: Social Functioning and Facial Expression Recognition in Survivors of Pediatric Brain Tumors. J. Pediatr. Psychol. 33, 1142–1152 (2008)
19. Merrell, K.W.: Assessment of Children's Social Skills: Recent Developments, Best Practices and New Directions. Exceptionality 9, 3–18 (2000)
20. Dijkers, M.P.: Quality of Life After Traumatic Brain Injury: A Review of Research Approaches and Findings. Arch. Phys. Med. Rehabil. 35, S21–S35 (2004)
21. Guinn, C., Hubal, R.: Augmented Transition Networks (ATNs) for Dialog Control: A Longitudinal Study. In: Proc. Int. Conference on Computational Intelligence, pp. 395–400. Acta Press, Calgary (2006)
22. Reeves, D.L., Winter, K.P., Bleiberg, J., Kane, R.L.: ANAM® Genogram: Historical Perspectives, Description, and Current Endeavors. Arch. Clin. Neuropsychol. 22, 15–37 (2007)
23. Sapien, J., Miller, T.C., Zwerdling, D.: Testing Program Fails Soldiers, Leaving Brain Injuries Undetected (2011), http://www.ProPublica.org & http://www.NPR. org
24. Hubal, R., Kizakevich, P., Furberg, R.: Synthetic Characters in Health-Related Applications. Stud. Comput. Intell. 65, 5–26 (2007)
25. Mascarenhas, S., Dias, J., Prada, R., Paiva, A.: A Dimensional Model for Cultural Behavior in Virtual Agents. Appl Artif. Intell. 24, 552–574 (2010)
26. Paschall, M.J., Fishbein, D.H., Hubal, R.C., Eldreth, D.: Psychometric Properties of Virtual Reality Vignette Performance Measures: A Novel Approach for Assessing Adolescents' Social Competency Skills. Health Educ. Res. 20, 61–70 (2005)

Partitioning Signed Bipartite Graphs for Classification of Individuals and Organizations

Sujogya Banerjee, Kaushik Sarkar, Sedat Gokalp,
Arunabha Sen, and Hasan Davulcu

Arizona State University
P.O. Box 87-8809, Tempe, AZ, 85281 USA
{sujogya,kaushik.sarkar,sedat.gokalp,asen,hdavulcu}@asu.edu

Abstract. In this paper, we use signed bipartite graphs to model opinions expressed by one type of entities (e.g., individuals, organizations) about another (e.g., political issues, religious beliefs), and based on the strength of that opinion, partition both types of entities into two clusters. The clustering is done in such a way that support for the second type of entity by the first within a cluster is *high* and across the cluster is *low*. We develop an automated partitioning tool that can be used to classify individuals and/or organizations into two disjoint groups based on their beliefs, practices and expressed opinions.

1 Introduction

The goal of the Minerva[1] project, currently underway at Arizona State University is to increase understanding of movements within Muslim communities actively working to counter violent extremism. As a part of this study, we have collected over 800,000 documents from web sites various organizations in Indonesia. Based on the *support* and *opposition* of certain *beliefs* and *practices*, we can partition the set of organizations $\mathcal{O}$ into two groups $\mathcal{O}_1$ and $\mathcal{O}_2$ and the set of beliefs and practices $\mathcal{B}$ into two groups, $\mathcal{B}_1$ and $\mathcal{B}_2$, such that organizations in $\mathcal{O}_1$ support $\mathcal{B}_1$ and oppose $\mathcal{B}_2$, while the organizations $\mathcal{O}_2$ support $\mathcal{B}_2$ and oppose $\mathcal{B}_1$. With the domain knowledge of the social scientists in our team regarding the beliefs and practices of Indonesian community, we can then *label* one group as being *radical* and other as *counter-radical*.

Although the motivation for our work was driven by Minerva, the the problem that is being addressed in this paper is much broader in nature. In the mathematical sociology community, the problem is known as the *Signed two-mode network partitioning problem* [1]. In its mathematical abstraction, the problem is specified by a *bipartite graph* $G = (U \cup V, E)$ and label function $\sigma : E \rightarrow \{P, N\}$. The node sets U and V may be representing the set of organizations $\mathcal{O}$ and the set of beliefs $\mathcal{B}$ respectively. If the label of an edge from $o_i \in \mathcal{O}$ to $b_j \in \mathcal{B}$ is

---

[1] A project sponsored by the U.S. Department of Defense.

S.J. Yang, A.M. Greenberg, and M. Endsley (Eds.): SBP 2012, LNCS 7227, pp. 196–204, 2012.

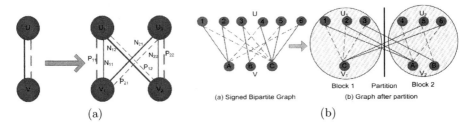

Fig. 1. Partitioning of the node set U and V with the desired goal

P, it implies o_i supports (or has positive opinion) about b_j. If the label of an edge is N, it implies o_i opposes (or has negative opinion) about b_j. The goal of the partitioning problem is to divide the node sets U and V into two subsets (U_1, U_2) and (V_1, V_2) respectively, such that

1. number of P edges (positive opinion or support) between nodes within block 1 (P_{11} between U_1 and V_1) and block 2 (P_{22} between U_2 and V_2) is high,
2. number of P edges between nodes across block 1 and block 2 (edges P_{12} between U_1 and V_2 and P_{21} between U_2 and V_1) is low,
3. number of N edges (negative opinion or opposition) between nodes within block 1 (N_{11} between U_1 and V_1) and block 2 (N_{22} between U_2 and V_2) is low and
4. number N edges between nodes across block 1 and block 2 (edges N_{12} between U_1 and V_2 and N_{21} between U_2 and V_1) is high.

The goal of partitioning is depicted in Fig. 1, where the *green* edges indicate support (i.e, P edges) and the *red* edges indicate opposition (i.e, N edges). We can realize these goals by *maximizing* $[(P11 + P22 + N12 + N21) - (P12 + P21 + N11 + N22)]$.

Signed two-mode network partitioning problem can be applied in a multitude of domains, where the node sets U and V can represent different *entities*. For example, (i) U and V may represent the members of the U.S. Senate/House of Representatives and the bills before the senate/house of representatives where they cast their votes, either supporting or opposing the bill; (ii) U and V may represent the political blogs/bloggers and various issues confronting the nation, where they express their opinions either supporting or opposing issues. Clearly, availability of an automated tool that will co-cluster the entities represented by U and V, will be valuable to individuals and organizations that need a coarse grain (two-modal) partitioning of the data set represented by the node set U and V. This tool can help classify individuals or organizations as *radicals vs. counter-radicals*, or *liberals vs. conservatives* or *violent vs. non-violent*, etc.

The main contribution of this effort is the development of a fast automated tool (and associated algorithms) for co-clustering the entities represented by the node sets U and V. We first compute an optimal solution of the partitioning problem using an integer linear program to be used as a benchmark for our heuristic solution. We then develop a heuristic solution and compare its performance using three real data sets. The real data sets include voting records of the Republican and Democratic members of the 111th US Congress and the opinions expressed in top twenty two liberal and conservative blogs. In all these data sets our partitioning tool produces high quality solution (i.e., with low misclassification) at a low cost (in terms of computation time). To the best of our knowledge, our Minerva research group is the first to present an efficient computational technique for partitioning of signed bipartite graph and apply it to some real data sets.

2   Related Works

As the literature on clustering, classification and partitioning is really vast, due to page limitations, we only refer to the ones that are most relevant to this paper [1,2,3,4,8,7]. The two key features of the partitioning problem addressed in this paper are (i) the graph is *bipartite* and (ii) the weights on the edges are *signed* (i.e., the weights are both positive and negative). Simultaneous clustering of two sets of entities (represented by two sets of nodes in the bipartite graph) was considered in the context of document clustering in [4,8]. In these studies one set of entities are the documents and the other set is terms or words. Although these efforts study the bipartite graph partition problem, they are distinctly different from our study in one respect. In our study, the edge weights are *signed*, whereas the edges weights considered in [4,8] are unsigned. Graph partitioning problem with signed edge weights was studied in [2,3]. However, these studies are also distinctly different from our study in that, while they focus on partitioning general (i.e., arbitrary) graphs, we focus our attention to partitioning bipartite graphs. The study that comes closest to our study is [1,7], where attention is focused on partitioning of a *signed bipartite* graphs. However, neither [1] nor [7] present any efficient algorithm to solve the partitioning problem in signed bipartite graph.

3   Problem Formulation

In this section we formally define the partitioning problem.

Signed Bipartite Graph Partition Problem (SBGPP): An edge labeled weighted bipartite graph $G = (U \cup V, E)$ where $U = \{u_1, u_2, \ldots, u_n\}$ represents entities of type I and $V = \{v_1, v_2, \ldots, v_m\}$ represents entities of type II. Each edge $(u, v) \in E$ has two functions associated with it: (i) label function $\sigma : E \to \{P, N\}$, which indicates the type of opinion (positive or negative), and (ii) weight function $w : E \to \mathbb{Z}^+$, which indicates the strength of that opinion. $A_N = [w_n(u, v)]$ and $A_P = [w_p(u, v)]$ are the weighted adjacency matrix for edges with label N and P respectively. If the node set U is partitioned into U_1 and U_2 and V is partitioned into V_1 and V_2, the strength of the positive and negative opinions of the entities of type I regarding the entities of type II are defined as follows:

For all edges $(u, v) \in E$,

$$P_{11} = \sum_{u \in U_1} \sum_{v \in V_1} w_p(u, v), \ P_{12} = \sum_{u \in U_1} \sum_{v \in V_2} w_p(u, v), \ P_{22} = \sum_{u \in U_2} \sum_{v \in V_2} w_p(u, v)$$

$$P_{21} = \sum_{u \in U_2} \sum_{v \in V_1} w_p(u, v), \ N_{11} = \sum_{u \in U_1} \sum_{v \in V_1} w_n(u, v), \ N_{12} = \sum_{u \in U_1} \sum_{v \in V_2} w_n(u, v)$$

$$N_{22} = \sum_{u \in U_2} \sum_{v \in V_2} w_n(u, v), \ N_{21} = \sum_{u \in U_2} \sum_{v \in V_1} w_n(u, v)$$

Problem: Find a partition of the node set U into U_1 and U_2 and V into V_1 and V_2 such that $[(P11 + P22 + N12 + N21) - (P12 + P21 + N11 + N22)]$ is *maximized.*

4   Computational Techniques

In this section we give a mathematical programming technique to find the optimal solution for the SBGPP. Since computational time for finding optimal solution for large graphs is unacceptably high, we present a heuristic in subsequent section to solve the SBGPP.

4.1   Optimal Solution for SBGPP

The goal of the SBGPP is to partition U into two disjoint sets U_1 and U_2 (similarly V into V_1 and V_2). For each node in $u \in U$ and each partition U_i, $i = 1, 2$, we use a variable b_{ui}. b_{ui} is 1 iff in u is in U_i. Similarly we define variable p_{vi} for all $v \in V$. We will refer $B_1 = U_1 \cup V_1$ and $B_2 = U_2 \cup V_2$ as *blocks* 1 and 2 respectively.

Variables: For each node $u \in U$, $v \in V$ and each partition U_i, V_i, $i = 1, 2$

$$b_{ui} = \begin{cases} 1, & \text{if node } u \text{ is in partition } U_i \\ 0, & \text{otherwise.} \end{cases} \qquad p_{vi} = \begin{cases} 1, & \text{if node } v \text{ is in partition } V_i \\ 0, & \text{otherwise.} \end{cases}$$

The mathematical programming formulation is given as follows:

$$\max \quad L = \sum_{i=1}^{2} \sum_{u \in U_i} \sum_{v \in V_i} (w_p(u, v) - w_n(u, v)) b_{ui} p_{vi}$$

$$+ \sum_{\substack{i,j=1 \\ i \neq j}}^{2} \sum_{u \in U_i} \sum_{v \in V_j} (w_n(u, v) - w_p(u, v)) b_{ui} p_{vj}$$

$$s.t \quad b_{u1} + b_{u2} = 1, \quad \forall u \in U \tag{1}$$

$$p_{u1} + p_{u2} = 1, \quad \forall p \in V \tag{2}$$

The objective function computes the objective value given by the expression L. We want to maximize L. It may be noted that the above quadratic objective function can easily be changed into a linear function by simple variable transformation [6]. Constraint 1 and 2 ensures that each node in U and V belongs to one particular block.

4.2   Move-Based Heuristics

We present a move-based heuristic to find an approximate solution of SBGPP. The move-based heuristic is a variant of well known FM algorithm [5] for partitioning graphs. The algorithm starts with a random initial partition and iteratively moves nodes from one block to another such that the value of the objective function is improved. The *"gain"* of a node is defined as the value by which the objective function increases if the node is moved from one block to the other. In each iteration the node with the highest gain is moved from one block to the other. In case of a tie a node is chosen arbitrarily. After a node is moved, it is locked and is not moved until the next pass. The heuristic is presented in Algorithm 1. It should be noted that original FM algorithm will not work for our problem as SBGPP relates to signed bipartite graphs with a completely different objective function and doesn't have any size constraints. As a result the node gain computation routine Algorithm 2 is considerably different from the original

---

Algorithm 1: Move-based Heuristic (MBH)

> **Input**   : A weighted signed bipartite graph $H = (U \cup V, E)$
> **Output**: A partition of the nodes U_1, U_2 and V_1, V_2 such that objective value L
>   is maximum

1 $L \longleftarrow 0$;
2 **for** $i \longleftarrow 1$ **to** r **do**
3   Generate a random partitioning of the nodes in U into U_1 and U_2 and nodes in V into V_1 and V_2;
4   **repeat**
5   Compute gains of all nodes using Algorithm 2 ;
6   **repeat**
7   Among all the unlocked nodes select the node of highest gain. Move the node to the other block and call it base node. Lock the base node. If the objective function value is best of the all the values seen so far in this iteration then save this partition;
8   Update the node gains of all the free neighbors of the base node;
9   **until** *Until all the nodes are locked*;
10   Change the current partition into a new partition that has the largest value of the objective function in this pass ;
11   Unlock all the nodes;
12   **until** *If the objective value L' improves during the last pass*;
13   **if** $L' \geq L$ **then** $L' \leftarrow L$ and save the current partition
14 **return** L and the final partition of nodes

Algorithm 2: Node Gain Computation

Input : A weighted signed bipartite graph $G = (U \cup V, E)$
Output: Gains of all nodes
1 **foreach** *node* $u \in U \cup V$ **do**
2 $\quad gain(u) \longleftarrow 0$;
   $\quad$ // FBlock = "from block" of node u, ToBlock = "to block" of node
   $\quad$ u, w(e) = weight of edge e and # = number
3 $\quad$ **foreach** *edge* $e \in E$ *with* $l(e) = N$ *of node* u **do**
4 $\quad\quad$ **if** *# nodes of e in ToBlock is* 0 **then** $gain(u) \leftarrow gain(u) + 2 * w(e)$;
5 $\quad\quad$ **if** *# nodes of e in FBlock is* 1 **then** $gain(u) \leftarrow gain(u) - 2 * w(e)$;
6 $\quad$ **foreach** *edge* $e \in E$ *with* $l(e) = P$ *of node* u **do**
7 $\quad\quad$ **if** *# nodes of e in ToBlock is* 0 **then** $gain(u) \leftarrow gain(u) - 2 * w(e)$;
8 $\quad\quad$ **if** *# nodes of e in FBlock is* 1 **then** $gain(u) \leftarrow gain(u) + 2 * w(e)$;

FM algorithm. Algorithm 1 runs for r different initial random partition of the nodes to avoid the possibility of being stuck at a local maxima. In practice the heuristic converges very fast, mostly in 2 to 3 passes.

5   Experimental Results and Discussions

To validate the effectiveness of our heuristic and benchmark its performance we tested the heuristic on real world data. The real world data consists of *US Congress* (SENATE, REP) and *political blogosphere* (BLOG) data sets.

5.1   US Congress Data [SENATE, REP]

The US Congress has been collecting data since the very first congress of the US history. This data has been encoded as XML files and publicly shared through the govtrack.us project[2]. From various types of data available at the project site, we collected the *roll call votes* for the 111th US Congress which includes The Senate and The House of Representatives and covers the years 2009-2010. The 111th Senate data contains information about 108 senators and their votes on 696 bills[3]. The 111th Congress has 451 representatives and the data contains their vote on 1655 bills.

We extracted the SENATE and REP data in adjacency matrices $A_{|U| \times |V|}$, with U vertices representing the congressmen, and the V vertices representing the bills. The edge (u_i, v_j), $u_i \in U, v_j \in V$ has weight 1 if the congressman u_i votes 'Yea' for the bill v_j , -1 if the congressman votes 'Nay', and 0 if he did not attend the session. We have the original classification vector for both the congressmen and the bills in terms of which party they represent (or which

---

[2] http://www.govtrack.us/data
[3] Normally, each congress has 100 senators (2 from each state), however in many of the congresses, there are unexpected changes on the seats caused by displacements or deaths.

party sponsored the bill). The first two columns of Table 1 provide information about this data as well as the partitioning accuracies of the algorithms. Figure 2 depicts the partitioned vote matrices of the 111th US Congress data, where rows representing the congressmen and the columns representing the bills. Also, the light green color represents 'Yea' votes, and dark red represents 'Nay' votes.

5.2 Blog Data [BLOG]

As Web 2.0 platforms gained popularity, it became easy for web users to be a part of the web and express their opinions, mostly through blogs. Most blogs are maintained by individuals, whereas there are also professional blogs with a group of authors. In this study, we focus on a set of popular political liberal or conservative blogs that have a clearly declared positions. These blogs contain discussions about social, political, economic issues and related key individuals. They express positive sentiment towards individuals whom they share ideologies with, and negative sentiment towards the others. In these blogs, it is also common to see criticism of people within the same camp, and also support for people from the other camp.

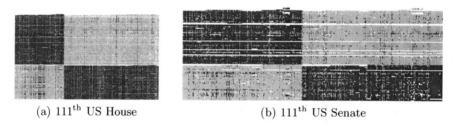

(a) 111th US House   (b) 111th US Senate

Fig. 2. Vote matrix of US Congress after partitioning

Table 1. Descriptive summaries of the graphs for each dataset with the Heuristic accuracy

	111th US Senate	111th US House	Political Blogosphere
Vertices in U	64 Democrat 42 Republican Senator	268 Democrat 183 Republican Representatives	13 Liberal 9 Conservative Blogs
Vertices in V	696 Bills	1655 Bills	20 Liberal 14 Conservative People
Graph Density	88.36 %	91.23 %	39.04 %
Heuristic accuracy	100.00%	99.56%	98.21%

In this experiment, we collected a list of 22 most popular liberal and conservative blogs from the Technorati[4] rankings. For each blog, we fetched the posts for the period of 6 months before the 2008 US presidential elections (May - October, 2008). We expected to have high intensity of the debates and discussions

---

[4] http://technorati.com

and resulting in a bipolar clustering in the data. Table 2 shows the partial list of blogs with their URLs, political camps and the number of posts for the given period.

We use AlchemyAPI[5] to run a named entity tagger to extract the people names mentioned in the posts, and an entity-level sentiment analysis which provided us with weighted and signed sentiment (positive values indicating support, and negative indicating opposition) for each person. This information was used to synthesize a signed bipartite graph (the BLOG data), where the blogs and people correspond to the two sets of vertices U and V. The a_{ij} values of the adjacency matrix A are the cumulative sum of sentiment values for each mention of the person v_j by the blog u_i.

To get a gold standard list of the most influential liberal and conservative people, we used The Telegraph List[6] for 2007. The third column of Table 1 provides information about this data as well as the partitioning accuracies of the algorithm.

Table 2. Political Blogs

Blog name	URL	Political view	Size
Huffington Post	http://www.huffingtonpost.com/	Liberal	3959
Daily Kos	http://www.dailykos.com/	Liberal	1957
Boing Boing	http://www.boingboing.net/	Liberal	1576
Crooks and Liars	http://www.crooksandliars.com/	Liberal	1497
Firedoglake	http://www.firedoglake.com/	Liberal	1354
Hot Air	http://hotair.com/	Conservative	1579
Reason - Hit and Run	http://reason.com/blog	Conservative	1563
Little green footballs	http://littlegreenfootballs.com/	Conservative	787
Atlas shrugs	http://atlasshrugs2000.typepad.com/	Conservative	773
Stop the ACLU	http://www.stoptheaclu.com/	Conservative	741
Wizbangblog	http://wizbangblog.com/	Conservative	621

6 Conclusion

In this paper we study the problem of partitioning *signed bipartite graph* with relevant application in political, religious and social domains. We provided a fast heuristic to find the solution for this problem. We tested the high accuracy of our heuristic on three sets of real data collected from political domain.

References

1. Andrej, M., Doreian, P.: Partitioning signed two-mode networks. Journal of Mathematical Sociology 33, 196–221 (2009)
2. Bansal, N., Blum, A., Chawla, S.: Correlation clustering. Machine Learning, 238–247 (2002)

---

[5] http://www.alchemyapi.com
[6] The-top-US-conservatives-and-liberals.html

3. Charikar, M., Guruswami, V., Wirth, A.: Clustering with qualitative information. In: Proceedings of the 44th Annual IEEE FOCS (2003)
4. Dhillon, I.S.: Co-clustering documents and word using bipartite spectral graph partitioning. In: Proceedings of the KDD. IEEE (2001)
5. Fiduccia, C., Mattheyses, R.: A linear-time heuristic for improving network partitions. Papers on Twenty-Five Years of Electronic Design Automation, pp. 241–247. ACM (1988)
6. Sen, A., Deng, H., Guha, S.: On a graph partition problem with application to vlsi layout. Inf. Process. Lett. 43(2), 87–94 (1992)
7. Zaslavsky, T.: Frustration vs. clusterability in two-mode signed networks (signed bipartite graphs) (2010)
8. Zha, H., He, X., Ding, C., Simon, H., Gu, M.: Bipartite graph partitioning and data clustering. In: Proceedings of the 10th International Conference on Information and Knowledge Management, pp. 25–32. ACM (2001)

Information and Attitude Diffusion in Networks

Wai-Tat Fu and Q. Vera Liao

University of Illinois at Urbana-Champaign, Urbana IL 61822, USA
wfu@illinois.edu

Abstract. The availability of diverse information does not guarantee that
a person's views will be equally diverse. Research has consistently shown
that attitudinal positions on an issue will lead to selective reception and
dissemination of information, which in turn have reciprocal effects on those
attitudes. The current paper aims at understanding how the diffusion of information and attitudes in networks are dynamically related to each other
from a computational perspective. Simulations from an agent-based model
show that selective exposure of information and social reinforcement in active dissemination of information can lead to polarization of attitudes in
the network. Network structures are shown to have significant effects on
information and attitude diffusion. While simple contagion models of information diffusion predicts that hub nodes in a small-world network can
facilitate propagation of information, our model shows that hub nodes can
induce stronger polarization of attitudes when information and attitude
diffusion can mutually influence each other. Results highlight the importance of incorporating social science research in network models to better
establish the micro-to-macro links.

Keywords: Attitude diffusion, selective exposure of information, active
dissemination of information, information cascades.

1 Introduction

Research on information diffusion shows that a small number of bridge links in a
network can lead to wide spread propagation of information [4, 12]. However, to
predict the spread of more complex social behavior, such as political, religious, or
cultural movements, it is important to understand how information may impact
attitudes of individuals, such that one can predict how their behavior will be
changed by the information. There is, however, still a lack of research on how
information diffusion is related to attitude diffusion, and how their relations are
impacted by different network structures. Indeed, observations of most groups
seem to suggest that there is almost always some diversity of opinion even after
extensive exchanges of information, and our world remains incredibly divided
even on basic factual issues[9].

2 Background

Research on information diffusion show that propagation of information can
be surprisingly efficient in networks that have few bridging or weak links [4]).

S.J. Yang, A.M. Greenberg, and M. Endsley (Eds.): SBP 2012, LNCS 7227, pp. 205–213, 2012.
© Springer-Verlag Berlin Heidelberg 2012

Assuming that information from one node can trigger activation (propagation of information) of its neighboring nodes, these weak links act as shortcuts in the network such that information can quickly spread through the network in ways that are similar to the spreading of diseases. However, while a single exposure is enough to propagate information (e.g., a piece of news) from one node to another as assumed in these epidemiological models of information diffusion, the spread of the impact of the information, such as changes in attitudes or beliefs, goes beyond a single exposure to the information and does not occur *tabula rasa*. Decades of research in social psychology have revealed a large number of factors (e.g., informative vs normative influence, compliance, power, involvement, ego and identity expression [10], etc) that impact attitude formation and change. As a first step, the current goal is to abstract away these important details to investigate the effects of structural relations between information and attitude diffusion. Specifically, we will focus on the interactions among the processes of the: (1) impact of information on individual attitude change [8], (2) selective exposure to information [5], and (3) active dissemination of information [1]. The current goal is to quantify these relations and computationally examine how they impact information and attitude diffusion in different networks.

2.1 Individual Attitude Change

Attitude can be defined generally as an evaluation of an object that ranges from people, groups, and ideas. Changes in attitude towards an object may spread through a network. While propagations of information between pairs of individuals are assumed to be independent in simple contagion models, attitude diffusion has local dependencies - that is, the state of a single node depends critically on the states of the node's neighbors. This is consistent with the large number of studies in social psychology that show that evaluation of information is subject to social influence (e.g., [1]) such that the more people who oppose one's position (relative to those who share it), the more likely one will change their attitudes. The impact of consensus information on attitudes implies that the level of heterogeneity of the network in terms of different number of neighbors and diversity of attitudes – is important for attitude change. In the current model, we assume that as an agent receives a piece of information that argues for or against an idea i, the attitude towards the idea A_i will be changed according to a threshold model [4]:

$$A_i = \frac{A_m}{1 + exp[-c(S_i - \tau_A)]} \tag{1}$$

A_m is the maximum attitude value, S_i represents the strength of social influence, c is a slope parameter, and τ_A is a threshold parameter. The sigmoid function is traditionally used in threshold model. We set $S_i = P_i/(P_i + N_i)$, in which P_i and N_i are the frequencies of information supporting and opposing idea i respectively. S_i represents the amount of information the agent receives that may change her attitude, and it is expressed as a fraction of relative information that supports the attitude. The frequencies are calculated by summing up the information sent by its neighbors (local persuasion) as well as information broadcasted throughout the network (global information, e.g., from a mass media such as TV). In the current model, local persuasion will always be received (e.g., conversation with a friend), but global information may or may not be received (see the next subsection on selective exposure of information).

It is important to note that, unlike simple contagion models (e.g., [8]), we assume that attitude change is mediated by information exchanges. This is important because we will model how likely people with different attitudes may decide to spread information in the network, and we assume that even when two neighbors are connected, their attitude changes depend on how frequently they exchange information. We justify this assumption by considering the fact that people connect for many reasons (sharing an office, cooperating on a task, etc) but unless they start to talk and exchange information about an issue, their attitudes may not influence each other.

Similar to previous threshold models, τ can be interpreted as the resistance to persuasion, and c can be interpreted as the sensitivity to impact of new information on attitudinal positions. Due to space limitation, we will not show their effects in this paper and will fix the values at c=1 and τ=0.1 throughout the simulations.

2.2 Selective Exposure of Information

The availability of diverse information does not guarantee that opinions will be equally diverse. For example, individuals strongly committed to certain religions often avoid contact with information or people that can tempt them away from their doctrine. Research shows that, unless people are guided by rather rare accuracy motives, they actively choose information that confirms their prior attitudes. This congeniality bias is found to be robust across a diverse range of situations [5]. One general explanation for this bias is based on the need to defend ego-relevant beliefs and reduce the discomfort of disconfirmation. In fact, research has consistently shown that the greater the ego threat (e.g., the stronger the attitude one has), the stronger the motivation to defend prior attitudes by shunning inconsistent information [5]. In the model, we assume that attitude and the propensity to select information has a reciprocal relationship. Specifically, the log odds O_i of selecting a supporting information to idea i is represented as

$$O_i = k * log(\frac{A_i}{A_m - A_i}) \tag{2}$$

In eqn 2, k stands for the strength of the congeniality bias - the larger the value of k, the more likely the agent will select information that is consistent with their attitude. For example, people who have high involvement in a particular idea (e.g., religion) tend to have stronger ego threat when exposed to opposing views, and thus have a stronger tendency to avoid (or not pay attention to) inconsistent information.

2.3   Active Dissemination of Information

Forming an attitude also alters the course of information dissemination. In the beginning, people who strengthen their attitudes may be mere receptors and seekers of information, but as their attitudes are integrated with their identity, many adopt an activist role. Strengthening confidence in an initial attitude can over time trigger feelings of invincibility, which result in communicating one's attitudes to others and actively debating individuals with opposite points of view. For example, many argue that people who seek to express their views on social media are often those who want to argue with others and to influence people to agree with their views[9].

We assume that the change from a passive to active role depends on the strength of the attitude. However, maintaining an activist's role in disseminating information is subject to social reinforcement i.e., when an agent perceives that the information is received by others, the act of dissemination is reinforced. This reinforcement accumulates as more people show positive reception to the information [6]. Specifically, we assume that there is an attitude threshold λ for information dissemination, and unless the magnitude of the agents attitude is above λ, the agent will not disseminate information. Once the threshold is reached, the probability P_D that an agent will disseminate information again after n attempts is determined by its utility $U_D(n+1)$, which are calculated by

$$U_D(n+1) = U_D(n) + \alpha * [R(n) - U_D(n)] \tag{3}$$

$$P_D = \frac{1}{1 + exp(\beta - U_D)} \tag{4}$$

In the above equations, $R(n)$ is the social reinforcement received, which is operationalized here as the number of times the information is received [6], and β represents the utility of not disseminating information, and is set to 1. For the current purpose, it is assumed that the agent knows how many of the agents received the information (e.g., the number of times a video is watched on YouTube). We set $\lambda=0.1$ and $\alpha=0.5$. Due to space limitation we cannot demonstrate the effects of these parameters, but these values are chosen to demonstrate the typical effect of active dissemination of information on information and attitude diffusion (i.e., they are in a range that does not radically change the behavior of the model).

3   The Model Simulation

We aim at simulating how information and attitude diffuse in networks to explore how network structures moderate the impact of information that supports or opposes a particular idea on the distribution of attitudinal positions among the agents. Agents are characterized by a continuous function of attitudinal positions with mechanisms of attitude change, selective information exposure, and information dissemination as discussed above. We assume that the processes of selective information exposure and dissemination are functions of the strength of agents attitudinal positions, and thus can explain how information and attitude diffusion interacts in different networks.

3.1   The Network

We created networks with 1000 agents (N), with each agent initially having n neighbors (n=5, 10, and 20). Following Watts and Strogatz [11], we use a perturbation algorithm to study the effects of changing connections on information and attitude diffusion. For each agent, with probability p, one of its neighbors is removed and a randomly selected agent is added as its neighbor. As p increases from 0 to 1, the network transition from a regular to a small-world to a random network. We then use a uniform random distribution [-1, 1] to assign prior attitude strength to each agent ($A_m = 1$).

3.2   Spread of Information and Attitude

Global information is broadcasted to all agents in each simulation cycle, and the information is randomly chosen to be supporting or opposing the idea. However, the probability that each agent will pay attention to (i.e., receive) the global information is determined by eqn 2. At the end of each cycle, attitude strength is updated according to eqn 1. When the magnitude of the attitude strength exceeds the dissemination threshold λ, the agent will attempt to disseminate the last received information to its neighbors in the next cycle. Spread of information will be measured by the percentage of agents who received the information over 10,000 simulation cycles (all networks stabilized at that point).

We measure attitude distribution by calculating the relative mean difference in the attitude strengths $|A_i - A_j|$ between any two agents i and j at the end of the simulation cycles. This is mathematically equivalent to the *Gini coefficient* G. A higher G value indicates that the network has more polarized attitudes. For N agents, G can be calculated as:

$$G = \frac{\sum_i^N \sum_j^N |A_i - A_j|}{2N \sum_i^N A_i} \tag{5}$$

4   Results

The top panel of figure 1 shows the Gini coefficients (G) in networks with different values of k (which represents the strength of congeniality bias in selecting information) as a function of p. A higher value of G indicates stronger polarization of attitudes. One can clearly see that network structures have significant effects on attitude diffusion. In general, polarization increases as p increases from 0.001 to 0.1, which reflects a transition from a regular to a small-world network; but as p is larger than 0.1, polarization quickly drops. Polarization also becomes stronger as k increases. In addition, as the initial number of connections (n) increases, the effects of network structures on attitude diffusion are strengthened.

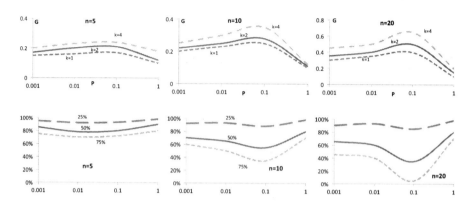

Fig. 1. [Top]:Gini coefficient calculated based on distributions of attitude strengths in the network (note that scales of y-axis have changed).[Bottom]:Percentage of agents who received 25%, 50%, and 75% of the global information.

To understand these counterintuitive results, one needs to note that as p increases, there are a number of changes in the network that impact information and attitude diffusion. When p increases from 0 to 0.1, the average path length dramatically decreases as links are re-created, which changes the network from a regular to a small-world network [11]. In other words, as p increases there are more hub nodes that reduce the average path length in the network.

While hub nodes are typically associated with more efficient information diffusion, they are more likely exposed to heterogeneous information in our simulation. In contrast, nodes that are connected locally tend to be less connected to the rest of the network (except through the hub nodes). Initially, these local neighboring nodes are exposed to only global information, which drives their attitudes randomly to either position. However, when any of these neighboring nodes reach the attitude threshold to disseminate information, they will have a large influence on its neighbors. As neighboring nodes are persuaded and shifted their attitude to one side, they become more selective in receiving global information, triggering a form of local information cascades of attitude change [3]. Given that active dissemination of information is subject to social reinforcement, denser local connections tend to lead to more local coherence (i.e., similar attitudes) among its nodes. This kind of local subnetwork is commonly observed even in large social networks [7].

So why do small-world networks lead to more polarized attitudes than regular networks? The main reason is that, unlike epidemiological models of information diffusion, hub nodes are actually more difficult to be "activated" in our model (see also [2]). Because hub nodes are exposed to heterogeneous information, they will less likely develop extreme attitudes. Thus, they will less likely become activists and disseminate information to their neighboring nodes. Rather than providing bridging links to its neighbors, a hub node in our model is actually shielding the local nodes from receiving opposing information from other local

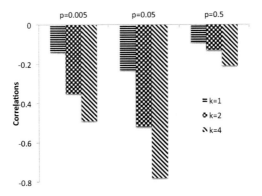

Fig. 2. The correlation between node degrees and magnitudes of attitude strength for representative values of p in the network with n=5

networks. To test this idea, we calculated the correlation between node degrees and their absolute attitude strengths in the networks. If hub nodes are inhibiting attitude diffusion, one will expect a negative correlation between node degree and absolute attitude strength. Figure 2 shows that the negative correlations are strongest when p=0.05 (in the small-world network range) than when p=0.005; and it is weakest when p=0.5. When p is larger than 0.1, the network loses local coherence and thus there is a much smaller chance for information cascades to occur among its neighbors. However, when propensity to selectively receive information is strong (k is high), global influence may still induce smaller scale polarization in the random network. However, in a small-world network, both local and global information cascades may contribute to local subgroups that support each other's attitudinal position while shunning inconsistent information, leading to stronger polarization. Figure 2 also shows that when the number of initial connections increases, the hub becomes larger. However, hub nodes do not promote attitude diffusion. They actually lead to fewer local polarized subgroups.

Finally, we will analyze how attitude diffusion is related to information diffusion. The bottom panel of figure 1 shows the percentages of agents who receive 25%, 50%, and 75% of the information broadcasted to the network. Consistent with the patterns of attitude diffusion, when p increases from 0.001 to 0.1, information diffusion decreases rather than increases. When p further increases from 0.1 to 1, however, there is less polarization of attitudes in the network as local coherence is lost, and increases information diffusion in the network. Similarly, as the number of initial connections increases, the effect of network structures is strengthened because the networks become less polarized. In other words, polarization is limiting the general exposures to information in the network.

5 Conclusions

Using an agent-based model, we have investigated the dynamic relations between information and attitude diffusion, and how the relations are influenced by network structures. We show that as connections are rewired to change from a regular to a small-world network (while keeping density constant), polarization of attitudes tends to intensify. Strong local connections and coherence are critical for the development and maintenance of extreme attitudes among members of a subnetwork, as members receive social reinforcement as they exchange information and begin to shun themselves from information inconsistent with their attitudes. This is similar to recent findings on the effects of hub nodes on complex contagion [2]. Counterintuitively, hub nodes may inhibit information diffusion, as they are difficult to be "activated" by its neighbors because of their exposure to heterogeneous information from different subnetworks. While the model is a general attempt to demonstrate the relations between information and attitude diffusion, it does show clearly that epidemiological models of information diffusion is too simple to capture the spread of social behavior that involves complex interactions among external information and prior beliefs, attitudes, or ideas of individuals. More generally, our results demonstrate the importance to take into account results and theories from cognitive and social behavioral studies to enrich the complex representations and processes of a node in a network, so as to establish a stronger micro-to-macro link between individual and network behavior.

References

[1] Albarracín, D., Johnson, B.T., Zanna, M.P.: The handbook of attitudes. Erlbaum, Mahwah (2005)

[2] Centola, D., Macy, M.: Complex contagions and the weakness of long ties. American Journal of Sociology 113(3), 702–734 (2007)

[3] Fu, W.T., Liao, Q.V.: Quality control of online information: A quality-based cascade model. In: International Conference on Social Computing, Behavioral-Cultural Modeling, and Prediction (2011)

[4] Granovetter, M.: Threshold models of collective behavior. The American Journal of Sociology 83(6), 1420–1443 (1978)

[5] Hart, W., Albarracín, D., Eagly, A., Brechan, I., Lindberg, M., Merrill, L.: Feeling validated versus being correct: a meta-analysis of selective exposure to information. Psychological Bulletin 135(4), 555–588 (2009)

[6] Huberman, B.A., Romero, D.M., Wu, F.: Crowdsourcing, attention and productivity. Journal of Information Science 35(6), 758–765 (2009)

[7] Huberman, B.A., Romero, D.M., Wu, F.: Social networks that matter: Twitter under the microscope. First Monday 14(1) (2009)

[8] Nowak, A., Szamrej, J., Latan, B.: From private attitude to public opinion: A dynamic theory of social impact. Psychological Review 97, 362–376 (1990)

[9] Sunstein, C.R.: The law of group polarization. Journal of Political Philosophy 10(2), 175–195 (2002)

[10] Visser, P.S., Mirabile, R.R.: Attitudes in the social context: The impact of social network composition on individual-level attitude strength. Journal of Personality and Social Psychology 87(6), 779–795 (2004)

[11] Watts, D., Strogatz, S.: Collective dynamics of 'small-world' networks. Nature 393(6684), 440–442 (1998)

[12] Watts, D.J.: A simple model of global cascades on random networks. Proceedings of the National Academy of Sciences of the United States of America 99(9), 5766–5771 (2002)

Using Mixed-Mode Networks to Disentangle Multiple Sources of Social Influence

Kayo Fujimoto

Division of Health Promotion & Behavioral Sciences, School of Public Health,
University of Texas, Health Science Center at Houston
Kayo.Fujimoto@uth.tmc.edu

Abstract. Social network analysis has been widely used to model various forms of network influence on individual behavior. In network studies into peer influence on adolescent risk-taking health behavior, either friendship influence (using one-mode networks) or affiliation-based peer influence (using two-mode networks) have been employed to examine potential sources of network influence on adolescent substance use behavior. However, these sources of peer influence may be potentially confounded with each other. This paper introduces a new network method of incorporating one network's influence into another network's influence patterns, enabling us to decompose the different sources of network influences though the use of a combination of the networks. Specifically, the method incorporates affiliation-based peer influence (operationalized by a two-mode network) with either a second two-mode network, or the peer influence (operationalized by a one-mode network). Empirical examples are included to demonstrate the utility of this new approach.

Keywords: Social network analysis, social influence, one-mode network, two-mode affiliation network, adolescent substance use, health behavior.

1 Introduction

Social Network Analysis (SNA) has been widely employed to understand various kinds of health behavior. In SNA, structural variables are measured based on a distinct set of entities, which are referred to as "mode" [1]. In the one-mode network data, structural variables are measured on a single set of nodes. For instance, such nodes would represent students (N) in the network boundary of school, and the one-mode network data are represented as a friendship matrix X, composed of N rows (nominating students) and N columns (receiving students), with elements of $X=x_{ij}$ equal to 1 if a student n_i (row node) nominated a student n_j (column node) as a friend, and 0 otherwise. In the two-mode network data, structural variables are measured on two distinct sets of nodes. For instance, the first mode is a set of actors, and the second mode is a set of events with which the actors in the first set are affiliated.

S.J. Yang, A.M. Greenberg, and M. Endsley (Eds.): SBP 2012, LNCS 7227, pp. 214–221, 2012.

Such two-mode networks are referred to as an "affiliation network" that records membership for each actor (indexed in row) with each event (indexed in column) in the $N \times K$ matrix form A_{ij}, with the entry of $A = a_{ij}$ equal to 1 if row actor n_i ($i=1, ..., N$) is affiliated with column event v_j ($j=1, ..., K$), and 0 otherwise.

In the studies on the diffusion of adolescent's risk-taking behavior (such as alcohol use, smoking, use of illegal drugs, and risk-taking sexual behavior), SNA has been used to model peer influences by assessing the extent to which adolescents are exposed to peers who engage in risk-taking health behavior. A majority of existing network studies of adolescent health behavioral studies has demonstrated a significant positive correlation between exposure to friends' risk-taking behavior and the likelihood of self risk-taking behavior [2-6].

Recently, a new network method of measuring the level of social influence using two-mode networks has been introduced. In this method, social influence is conceptualized as influence from co-membership with activities/events (affiliation-based social influence), and network model that measures it is termed as "affiliation exposure" model [7, 8]. For instance, it has been reported that affiliation-based peer influence through co-participating with school-sponsored organized sport activities with smokers increased the risk of smoking and drinking among adolescents [8, 9], and such influence was especially stronger when activity members are also close friends [9].

Methodologically, the network exposure model has been widely used to model and estimate network effects [10-12] in diffusion of innovation studies. Network exposure model is designed to measure a given form of network influence (either one-mode or two-mode network) and test its effect on individual behavior to explain how new ideas and practices spread through social networks. Within the framework of network exposure model, various specifications of influence matrix are possible.

This paper introduces a new network method of decomposing the various kinds of network influences though the use of a combination of one-mode and two-mode networks. The method incorporates affiliation-based peer influence operationalized by a two-mode network with either a second two-mode network or the peer influence operationalized by a one-mode network. In the context of peer influence on adolescent risk-taking health behavior, this allows the parsing of affiliation-based peer influence from the co-participation of organized activities into affiliation influence from members who are also friends and from members who are not friends in an attempt to address the potential confounding of these sources of peer influence. Likewise, in the case of a second two-mode network, the new method similarly incorporates two different kinds of affiliation-based peer influence: one affiliation-based peer influence (operationalized by one two-mode network) into another peer influence (operationalized by a second two-mode network). For example, it parses affiliation-based peer influence through organized activity participation into the affiliation influence from those members who share the same crowd identity (such as jock, brains, etc.), and from members who do not share the same crowd identity.

2   Network Method of Modeling Social Influence

2.1   Network Exposure Model

Network exposure model is designed to measure the extent to which an ego is exposed to alters with specific behavioral attributes. Exposure is calculated by multiplying the social network influence weight matrix by a vector indicating whether each actor engages in that behavior. The general formula of the network exposure $\underline{E}$ [12] is defined as:

$$\underline{E} = \frac{\sum_{j=1} W_{ij} y_j}{\sum_{j=1} W_{ij}} \qquad \text{for } i = 1, \cdots, N, \; i \neq j \qquad (1)$$

where E is the exposure vector, W_{ij} is a social network weight matrix, and y_j is a vector of j's behavioral attribute (j=1, $\cdots$, N). By specifying different weight matrices W, this model measures the levels of the given forms of peer influence. The current paper focuses on measuring interpersonal proximity, "cohesion" which measure connectivity among actors as a case of using one-mode network. Cohesion exposure is computed by using a one-mode adjacency matrix, X_{ij}, as a specification of weight matrix W_{ij} By multiplying X_{ij} by nominated actors' behavior y_j and normalizing by the row-sum of X_{ij} (i.e., actor i's outdegree, X_{i+}), the resulting normalized relational exposure of $\underline{E}$ measures the level of nominated actors js who have behavioral attribute of y_j in an ego network.

2.2   Affiliation Exposure Model

Affiliation exposure model [7, 8] uses a two-mode affiliation matrix of A_{ij}. By multiplying an affiliation matrix A_{ij} with its transpose ($A_{ij}\acute{}$), the resulting matrix C_{ij} (= $A_{ij}A_{ij}\acute{}$) is a symmetric matrix with integer values ranging from 0 to K that records the co-membership relation in a network [1]. There are two components in C_{ij}, which is the off-diagonal values that represent the number of activities each pair of actors attended in common, and the diagonal values that represent the number of activities each actor participated. Diagonal values are ignored in calculation. By multiplying C_{ij} by each co-participant's behavior y_j and standardize by it row-sum C_{i+}, the resulting affiliation exposure vector of $\underline{F}$ measures the extent to which an individual co-participated activities with alters who have behavioral attribute of y_j in an affiliation network, which is defined in the following:

$$\underline{F} = \frac{\sum_{j=1} C_{ij} y_j}{\sum_{j=1} C_{ij}} \qquad \text{for } i, j = 1, \cdots, N, \; i \neq j \qquad (2)$$

Affiliation-based social influence is conceptually different from one-mode network influence (such as friendship) since it assumes that co-membership is a structural feature that defines individual social identities and increases the probability of forming acquaintances [13]. However, it may be possible that affiliation exposure is confounded with cohesion exposures, or vice versa. Additionally, one type of affiliation-exposure may be confounded with another type of affiliation exposure. The following method introduces a way to incorporate one network influence into another network's influence patterns, allowing us to decompose the sources of social influences.

2.3   Model Decomposition

To deal with potential confounding of the affiliation exposure with another network exposure, the model was partitioned into two separate models. One example of this would be to measure affiliation exposure based on (a) activity members who are also friends and (b) activity members who are not friends. Another example would be to measure affiliation exposure based on (c) activity members who share the same crowd identity, and (d) activity members who do not share the same crowd identity. Affiliation exposure model that is designed to decompose different components of network influence is defined as follows:

$$\underline{D}_{(a)} = \frac{\sum_{j=1} X_{ij} C_{ij} y_j}{\sum_{j=1} X_{ij} C_{ij}} \qquad \text{for } i, j=1, \cdots, N, \ i\neq \qquad (3)$$

Mathematically, (a) was computed by element-wise product of the co-participation matrix C_{ij} by an adjacency matrix X_{ij} (representing friendship network), and then row-normalized it and matrix-multiplied it by the behavioral vector of alters y_j. To compute (b), the adjacency matrix X_{ij} was subtracted from the unity matrix, and everything else being identical to the computation of $\underline{D}_{(a)}$, which is defined as:

$$\underline{D}_{(b)} = \frac{\sum_{j=1}(1-X_{ij})C_{ij} y_j}{\sum_{j=1}(1-X_{ij})C_{ij}} \qquad \text{for } i, j=1, \cdots, N, \ i\neq j \qquad (4)$$

Essentially, the original C_{ij} matrix has been partitioned based on the (binary) friend network X, such that $C_{ij}=C_{(X)} + C_{(1-X)}$.

Similarly, (c) is computed by element-wise product of the co-participation matrix C_1 with a second two-mode matrix of crowd identity affiliation matrix C_2 that is binarized (representing at least one crowd identities shared with alters, and denoted by the signum in the equations which follow), and then row-normalized it and matrix-multiplied it by the behavioral vector of alters y_j, which is defined by the following formula:

$$\underline{D}_{(c)} = \frac{\sum_{j=1} \text{sgn}(C_{2ij})C_{1ij} y_j}{\sum_{j=1} \text{sgn}(C_{2ij})C_{1ij}} \qquad \text{for } i, j=1, \cdots, N, \ i\neq j \qquad (5)$$

To compute (d), the binarized crowd identity affiliation matrix C_2 was subtracted from the unit matrix, and everything else being identical to the computation of $\underline{D}_{(c)}$, which is defined in the following formula:

$$D_{(d)} = \frac{\sum_{j=1}(1-\mathrm{sgn}(C_{2ij}))C_{1ij}y_j}{\sum_{j=1}(1-\mathrm{sgn}(C_{2ij}))C_{1ij}} \qquad \text{for } i,j=1,\cdots,N,\ i\neq j \qquad (6)$$

2.4   Modeling Network Influence on Behavior

To statistically modeling network influence on individual behavior, one or more network exposure terms defined above can be included as covariates in standard regression analysis to estimate and test their effects on individual behavior. In this model, network exposure variable(s) and the error term are assumed to be independent. Network effects model [14], one variant of network autocorrelation model [15-17], differently estimate network influence under the assumption that endogenous network effect variable is correlated with the error term. Network effects model is based on the assumption that social influence occurs through the dependency among responses to the dependent variable [18]. It treats the network autocorrelation as a parameter and models the autoregressive effects on outcome variable, which is defined in the following formula:

$$y = \rho Wy + X\beta + \varepsilon, \qquad \varepsilon \sim N(0, \sigma^2 I) \qquad (7)$$

where y is a behavioral vector of dependent variable, X is an (N x h) matrix of values for the N actors on h independent variables with unit row vector for the intercept term, β is a (h x 1) vector of regression coefficients, ρ is a scalar estimate of autocorrelation parameter, and W is a (N x N) influence weight matrix. The W matrix can be row or column normalized to unity. In relation to the network exposure model, Wy term with row-normalized W matrix corresponds to the values of network exposure term ($\underline{E}$) defined in the equation (1) or affiliation exposure term ($\underline{F}$) defined in equation (2). When modeling simultaneous decomposed effects of multiple network influences, the decomposed network exposure terms $\underline{D}_{(a)}$ and $\underline{D}_{(b)}$ or $\underline{D}_{(c)}$ and $\underline{D}_{(d)}$ can also be included in the same model.   Furthermore, depending on what the substantive interests are, the main effects of the original exposure terms (i.e., cohesion exposure or affiliation exposure or both) along with the decomposed exposure terms can be included in the model.

3   Disentangling Some Empirical Applications

The following empirical studies illustrate two applications of the new network method to modeling adolescent substance use behavior to combine one network influence patterns with another network's influence patterns, and this will allow us to decompose the sources of social influences from a Social Theory perspective. The first example examines the effect of affiliation exposure based on school sponsored organized activity members who are also friends (versus those who are not friends) on adolescent past-year drinking and current smoking behavior [9]. The second example examines the effect of affiliation exposure based on activity members who share the same crowd identity (versus those who do not share the same crowd identity) on adolescent drinking behavior [19].

The first study [9] consisted of a nationally representative sample of 12,551 adolescents in Grades 7-12 within 106 schools from Add Health data [20]. An ordinal logistic regression analysis was conducted to estimate the effects of decomposition of affiliation influence through organized activities from friends versus non-friends on any drinking or frequent drinking. This study defined friends as reciprocated friendships. Then the study fit cumulative logit models with decomposed affiliation exposure terms (computed separately for sports and clubs activities) through fellow members who are also friends versus who are not friends. The analytic sample was restricted to the 8,433 adolescents (67 % of the sample) who had at least one reciprocated friend. The models included friends' influence, the number of activities participated, and other demographic and socio-economic controlled variables. Results are reported in terms of adjusted odds ratios (AOR).

The effects of decomposed alcohol exposure through fellow sports members who were also friends were significant for both any drinking and frequent drinking (AOR=1.16; p<0.01). Additionally, the magnitude of the effect of decomposed exposure through club members who were also friends was significant (AOR=1.23; p<0.01). Conversely, the decomposed exposure through non-friends sports members was not significant for any drinking and frequent drinking, whereas decomposed exposure through non-friend club members was significant (AOR=1.28; p<0.01). To summarize, the influence from sports member drinkers affects adolescent's drinking behavior only if they are also friends, while the influence from drinking club members affects an adolescent's drinking behavior regardless of friendship.

The second study [19] consisted of five southern California high schools district in Los Angeles County, which is a predominately Hispanic/Latino district. Analytic sample consisted of N=1,716 10th grade adolescents collected in September-November 2010. Ordinal regression was used to model the association between decomposed drinking exposures and various stages of drinking (i.e., susceptible to drinking, lifetime drinking, past-month drinking, and binge drinking). This study modeled cumulative probabilities of respective drinking stages as a function of the decomposed exposure to sports members (1) who shared at least one crowd identification, and (2) who did not share any crowd identification, controlling for drinking exposure to friends and other variables. The results showed that decomposed drinking exposure through sports with shared identification had significant effects on any drinking (i.e., lifetime drinking or more) (OR=1.25; p<0.05), on past-month drinking or more (OR=1.34; p<0.01), and on binge drinking (OR=1.32; p<0.05), but not on being susceptible to drinking. This indicates that as adolescents were exposed more to activity members who shared crowd identity, they were more likely to drink frequently. On the other hand, drinking exposure through sports without shared identification had a significant effect only on the dividing point of being susceptible to drinking, with the odds of intending to drink in the next year, as opposed to not to intending to drink, being 1.28 (p<0.05). These results indicate that the drinking influence from members who do not share crowd identify only matter in an adolescent's cognitive-level determination of an intention to drink, but not the actual drinking.

4   Conclusion

The current paper has introduced a network concept of incorporating one network's influence into another network's influence patterns, and has also introduced a new network method that decomposes affiliation-based influence into two sources of peer influence. This study demonstrates the utility of the method by showing two empirical studies in the context of diffusion of adolescent alcohol use. However, this paper is limited in evaluating the performance, validity or effectiveness of the decomposition process. In social settings the "ground truth" is very elusive (especially with adolescents), and while the theory seems to hold for our examples, future research should explore some statistical properties (such as bias and stability) of the estimation of structural parameter of the decomposed affiliation exposure terms, and continue to evaluate the feasibility of the model in the application of various behavioral studies.

Finally, given the practical difficulties in acquiring the single-mode network data and the ease of acquiring two-mode data, the fact that the proposed method can be applied to both equally well is an another strength of the approach. Although these examples are limited in the context of diffusion of adolescent substance use behavior, the proposed decomposition method can be applied to other fields of public health research.

Acknowledgements. This study was supported by Award Number K99AA019699 (PI: Kayo Fujimoto) from the National Institute On Alcohol Abuse And Alcoholism. The content is solely the responsibility of the author and does not necessarily represent the official views of the National Institute On Alcohol Abuse And Alcoholism or the National Institutes of Health. The first empirical demonstration was based on the analysis of Add Health data, a program project directed by Kathleen Mullan Harris and designed by J. Richard Udry, Peter S. Bearman, and Kathleen Mullan Harris at the University of North Carolina at Chapel Hill, and funded by grant P01-HD31921 from the Eunice Kennedy Shriver National Institute of Child Health and Human Development, with cooperative funding from 23 other federal agencies and foundations. Special acknowledgment is due Ronald R. Rindfuss and Barbara Entwisle for assistance in the original design. Information on how to obtain the Add Health data files is available on the Add Health website (http://www.cpc.unc.edu/addhealth). No direct support was received from grant P01-HD31921 for this analysis. The second empirical demonstration comes from the part of the analysis using the data collected and funded by NIAAA grant 1RC1AA019239-01 (PI: Thomas W. Valente). I also acknowledge Thomas W. Valente and Erik Lindsley for help in completing this paper.

References

1. Wasserman, S., Faust, K.: Social Network Analysis: Methods and Applications. Cambridge University Press, New York (1994)
2. Alexander, C., Piazza, M., Mekos, D., Valente, T.W.: Peers, schools, and adolescent cigarette smoking. Journal of Adolescent Health 29(1), 22–30 (2001)

3. Cleveland, H.H., Wiebe, R.P.: The moderation of adolescent-to-peer similarity in tobacco and alcohol use by school levels of substance use. Child Development 74(1), 279–291 (2003)

4. Crosnoe, R., Muller, C., Frank, K.A.: Peer context and the consequences of adolescent drinking. Social Problems 51(2), 288–304 (2004)

5. Ennett, S.T., Bauman, K.E., Hussong, A., Faris, R., Foshee, V.A., Cai, L., et al.: The peer context of adolescent substance use: Findings from social network analysis. Journal of Research on Adolescence 16(2), 159–186 (2006)

6. Urberg, K.A., Degirmencioglu, S.M., Pilgrim, C.: Close friend and group influence on adolescent cigarette smoking and alcohol use. Developmental Psychology 33(5), 834–844 (1997)

7. Fujimoto, K., Chih-Ping, C., Valente, T.W.: The network autocorrelation model using two-mode network Data: Affiliation exposure and biasness in ρ. Social Networks 33(3), 231–243 (2011)

8. Fujimoto, K., Unger, J., Valente, T.: Network method of measuring affiliation-based peer influence: Assessing the influences on teammates smokers on adolescent smoking. Child Development (in press)

9. Fujimoto, K., Valente, T.W.: Alcohol peer influence from participating in organized school activities among U.S. adolescents: A network approach (manuscript)

10. Marsden, P., Friedkin, N.E.: Network studies of social influence. Sociological Method & Re-search 22(1), 127–151 (1993)

11. Burt, R.S.: Social contagion and innovation: Cohesion versus structural equivalence. American Journal of Sociology 92, 1287–1335 (1987)

12. Valente, T.W.: Network models of the diffusion of innovations. Hampton Press, Cresskill (1995)

13. McPherson, J.M.: Hypernetwork sampling: duality and differentiation among voluntary organizations. Social Networks 3, 225–249 (1982)

14. Doreian, P., Teuter, K., Wang, C.-S.: Network autocorrelation models: Some Monte Carlo re-sults. Sociological Methods and Research 13, 155–200 (1984)

15. Doreian, P.: Linear models with spatially distributed data. Sociological Methods and Research 9, 29–60 (1980)

16. Dow, M.M.: A bi-parametric approach to network autocorrelation: Galton's problem. Sociological Methods and Research 13, 201–217 (1984)

17. Ord, K.: Estimation methods for models of spatial interaction. Journal of the American Statistical Association 70, 120–126 (1975)

18. Leenders, R.A.J.: Modeling social influence through network autocorrelation: Constructing the weight matrix. Social Networks 24, 21–47 (2002)

19. Fujimoto, K., Soto, D., Valente, T.W.: A network approach to decompose peer influences from crowds and organized sports (manuscript)

20. Harris, K.M.: The National Longitudinal Study of Adolescent Health (Add Health), Wave I & II, 1994-1996; Wave III, 2001-2002; Wave IV, 2007-2009 [machine-readable data file and documentation]. Carolina Population Center, University of North Carolina at Chapel Hill, Chapel Hill, NC (2009)

Implicit Group Membership Detection in Online Text: Analysis and Applications

Jeffrey Ellen, Joan Kaina, and Shibin Parameswaran

Space and Naval Warfare Systems Center Pacific
United States Navy, San Diego, CA, USA
{jeffrey.ellen,joan.kaina,shibin.parameswaran}@navy.mil

Abstract. Our thesis is that members of the same group have shared tendencies and nuances in communication style and substance, particularly online. In this paper, we dicuss some potential applications of accuarate authorship affiliation technology. We also discuss related work in similar author identification efforts and the research issues that currently exist when trying to perform automated authorship affiliation. We provide quantitative results from our recent Machine Learning experimenation using Support Vector Machines as some initial validation of our theory. In this paper, we applied our work towards the task of classifying website forum posts by the affiliation of their author. We discuss in detail the stylometric features we used to perform the automated classification and split the original features into individual groups to isolate their respective contributions and/or discriminating capability. Our results show promise towards automating group representation, an important first step in studying group formation.

Keywords: Stylometrics, Text classification, Group detection, Authorship affiliation, Group Membership, Deception detection, Natural Language Processing, Machine Learning.

1 Introduction

As the mechanisms permitting social interactions via computers have become more sophisticated, so has community behavior. And as the interfaces and mechanisms have become more intuitive and familiar, aspects of online social behavior mimic those in offline interactions. According to Communication Accommodation Theory formulated by Giles, et al. in 1973[1], aspects of offline communications and social interactions include mimicry and adoption of language customs and norms within groups, and research shows that this holds for online communications [2]. It should therefore be possible to classify online communications according to the author's affiliation with one or more groups. One social pattern of particular interest to security and financial applications occurs when individuals attempt to infiltrate a group in order to alter the opinion of other group's members, or simply gain credibility from posing as a group member.

S.J. Yang, A.M. Greenberg, and M. Endsley (Eds.): SBP 2012, LNCS 7227, pp. 222–230, 2012.

In this paper, we explore the current state of this research (both from the machine learning as well as the social science perspective) and potential applications of this general technology. For our specific application, we worked towards classifying website forum posts based on their authorship affiliation to a particular group.

1.1   Background and Motivation

There are many domains where knowing the affiliation of an anonymous/pseudonymous author would be of interest.

For instance, the knowledge of an author's affiliations (implicit/explicit group memberships) can help detect bias in financial opinions and reviews thereby ferreting out possible shills. Applying this idea to movie reviews and product reviews could result in more effective advertising and better public relations campaigns. Market researchers, in particular, could use authorship affiliation, along with sentiment analysis, to better ascertain group-wise demographic information of customer reviews. Conversely, consumers could use affiliation to help detect spam, when the communication belies the author's affiliation as a marketer. A more effective and efficient marketplace is of benefit to all consumers.

Another potential application is clustering friends in social networking sites based on their stylometric similarities in statuses, comments, etc. Identifying an individual as a member of a group adds a richer set of features to that individual. Membership in multiple groups can also be used to determine an individual's influence.

In developing technology for the United States Department of Defense, our main goal is assisting intelligence analysis at identifying those who wish to bring harm against others. In our first application of this technology, we are seeking to perform group affiliation for purposes such as determining whether posts of an inflammatory nature within a certain community are being authored by actual members of the community, or by infiltrators and propagandists trying to sway a stable population or political group. It may then be possible to measure the author's influence, the frequency of the topic discussion, the sentiment, etc. to predict a group tipping point that indicates a change in group behavior. Similar to the market research example earlier, in combination with sentiment analysis, improved group-wise demographics might better guide foreign policy decisions.

The work in this paper is an important first step towards achieving some of these goals, more broadly referred to as "Cognitive Information Operations" [3].

1.2   Related Work

Detection of group membership implicit in an anonymous author's authorship style and word usage is a novel application. To the best of our knowledge, there is a shortage of research studies in this particular NLP application. However, there are other NLP applications that employ related methods with different objectives. One such field that is closely related is authorship *attribution* [4], which is an attempt to identify patterns ('signatures' or 'fingerprints') to identify a *particular* author [5][6]. There are numerous studies in this area including one which attempts to identify a particular forum post author from a pre-determined group of 20 authors [7]. Although similar in application area, we feel group affiliation is a more generalized (and hence complex) case of where group level stylometric properties and patterns are identified instead of individual characteristics to be applied to an individual impostor or impersonator.

Our research has not found many other researchers attempting to experimentally validate or apply group stylometrics. A recent survey by Juola suggests that authorship properties can be extended to many groups such as native speakers and gender [8]. However, Juola only specifically references one study identifying author gender [10] and that study specifically targets "formal written documents".

Other techniques researched heavily by the Natural Language Processing community are written text sentiment analysis, topic discovery, and automatic summarization. These topics, although popular with substantial research activity over the past 10 years, are hampered if applied to data written to deceive the audience or otherwise inauthentic. Group affiliation is tangential to these areas and attempts to identify patterns in authorship style that may or may not be a conscious choice of the author, which makes our approach potentially more robust to deception. A recent study [9] used topic analysis to group scientific papers based on the originating conference but in our application area we are specifically trying to differentiate between the affiliations of authors discussing *unconstrained* topics in *informal* blog postings. In addition, the referenced study uses exclusively bag-of-word topics generated by LDA, while we are interested in investigating the applicability of other stylistic features in addition to term selection. We are investigating not just the tangible output text, but the author's decision process in creating that text.

1.3   Communication Modeling Research Issues

Our intuition that linguistic styles and grammatical nuances are shared amongst people with a common background (group) and can be detected via textual media is shared by social scientists and linguists regarding communication [1] and style [19], and algorithms can be developed based on quantifiable aspects of communication model and stylometrics. Other researchers have attempted to classify authors based on gender [10], age [11] and both [9][12]. The right feature can be very powerful: The latter study reported accuracies in the range ~80-90% using only two simple features: sentence length and slang word usage.

It is interesting to note that all previous studies focus on inherent traits that are not a voluntary choice of the author, such as age or native language [13], unlike the group affiliation which we are targeting. One of the most interesting aspects of group affiliation detection is the mixed nature of conscious and unconscious choices in style an author makes as a member of a group. There has been success in individual authorship identification based on a variety of linguistic features [20][21], but regardless of whether or not there is currently a socio-behavioral rationale for why a particular linguistic feature is significant, ultimately performance in automated algorithms is the most significant criteria for our usage. The problem can be attacked from both directions. Some features are trivial to experimentally validate using Machine Learning, others such as respect or eccentricity [3] have solid socio-psychological rationale but are difficult to quantify and utilize algorithmically. Comparison of both theory and results can help determine the most worthwhile features and achieve the best results.

In this investigation we have focused on extracting and testing features that are capable of capturing this mixture of conscious and unconscious communication decisions. We hypothesize that group affiliations of authors can be predicted through their language; their stylistic, structural, and thematic choices, which are manifestations of their cognitive model [3] and will have implicit signatures representative of their natural inclinations and affiliations.

2   Feature Extraction

In our previous publication [22], we started our analysis of group affiliation detection which focused on feasibility and reported performance measures on a composite set of features. This paper continues our research by decomposing and analyzing the different features we used and their contributions as stand-alone representations of the document in question.

- Raw Term Frequency: Much current work on document classification focuses on sorting documents by subject, which naturally has a heavy reliance on term frequency (TF) or a variant thereof. Usually when using a TF feature set, it is conventional to *stem* or *lemmatize* the words. We have chosen to use the raw TF without these modifications to keep the stylistic features intact. Stop words were removed using the list in Natural Language Toolkit (NLTK).
- Exclusive terms: Usage statistic of terms of exclusion. E.g. but, except, without, exclude.
- Negation terms: Usage statistic of terms of negation. E.g. no, never, not, *n't etc.
- Causation terms: Usage statistic of terms showing causation. E.g. due, because, as, since, consequently, hence, so, therefore, accordingly, thus, if, unless, lest etc.
- Function terms: Usage statistic of 303 specific terms that signify function that were reported in [14].
- Noun referencing: Different types of noun referencing employed. E.g. definite/indefinite, article/demonstrative, pronoun immediately preceding noun etc. This is an effort to characterize cultural or group styles e.g. respect to an individual.
- Sentence lengths and closures: Percent of sentences falling in a discrete number of length ranges and types of sentence closures used (normal/questions/exclamations etc.). An attempt to capture and track author-group's structural decisions and syntactic and semantic choices [6].
- Parts Of Speech (POS):   Ratio of terms in each of 50 different POS tags provided in python NLTK package which is a maximum entropy tagger trained against the University of Pennsylvania Treebank data set [15].
- Pronoun usage: Ratio of pronouns in each of 9 different categories: 2nd person, 3rd person, demonstrative, possessive, etc. Along with stylistic choices, this is a simplistic and straightforward effort to capture cultural and other group related word preferences in addressing peers, older, higher stature etc.

The straightforward nature of most of these features (ratios, term-counts etc.) is an attempt to give simple mathematical representations to complex properties of language, and social norms and interactions in a group which will in turn facilitate automation in feature extraction. Features similar to these were also used in a recent work by Khosmood which attempted to disguise authorship through camouflage [16].

3   Experimental Analysis

3.1   Data Sets

We used two different data sets for our experiments, one corpus as our control, and a second corpus representative of our problem domain. The first is the BitterLemons.org corpus. This corpus was released in 2006 [17] and has been used in various linguistic studies since, making it a good baseline. The corpus consists of 594 essays from an Israeli/Palestinian political discussion website, and is balanced with equal number of entries from both sides.

The second corpus we used consisted of two groups of forum posts. The first group consisted of 2636 posts from 9 different websites. These posts were written in Arabic, but we translated the posts to English for our experiments. Most posts in this set would likely be considered extremist from the current United States Government point of view. The other group contained 537 posts from 9 other websites that would most likely be considered 'moderate' from the current United States Government perspective. Most were written in Arabic, with a small number written in Vietnamese, Pashtun, Russian or other languages; we experimented on the English translations. For our experiments we assigned all entries from a particular forum the same label. We selected distinct communities that have some overlap in vocabulary and structure, and trained our machine learning algorithms to discriminate between them.

3.2   Performance Metrics

Commonly used metrics like accuracy, precision, recall and F-measure are biased by the size individual classes and provide reliable evaluations only on balanced datasets. To reduce the effect of an unbalanced dataset on our evaluations and to present a comparison that is independent of specific performance evaluators, we have compared our results under 2 different metrics: Matthew's Correlation Coefficient (MCC) [18], and Balanced Accuracy (BAc). These metrics are frequently used in situations where extremely unbalanced datasets are common, such as medical studies of cancer incidence rates. BAC values are in the range of [0, 1] whereas MCC scores range from -1 to 1.

3.3   Machine Learning Experimentation

We evaluated the group detection and discrimination capability of term frequency (unaltered), and other linguistic features listed in section 2 as individual feature-sets and in combination with each other. For simplicity and clarity, we have chosen linear Support Vector Machines (SVM) for classification. It is trivial and reasonable to substitute this algorithm with kernelized SVM (RBF, polynomial etc.) or any other classification algorithm; we experimented with other algorithms in our previous work [22].

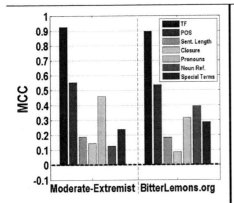

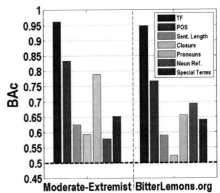

Fig. 1. Classificatio.n performance of different feature-types under Matthew's Correlation Coefficient (MCC). For a two-class problem, random classification corresponds to a MCC value of 0 which is indicated by the dotted line

Fig. 2. Classification performance of different feature-types under Balanced Accuracy (BAc). For a two-class problem, random classification corresponds to a BAc value of 0.5 which is indicated by the dotted line

The performance results obtained by the different feature types on Moderate-Extreme forum data and BitterLemons.org dataset are presented in Figures 1 and 2 respectively. Term frequency, ratio of different POS tags, and ratio of pronoun types provide consistent and considerable discrimination across both data sets demonstrating suitability to identify distinctive styles of groups. The other four categories of features show good performance in our control dataset, while providing only marginal improvement over random classification on the Moderate-Extremist dataset. This can be explained due to the controlled nature of the posts, especially with respect to topic, in BitterLemons.org corpus.

Finally, the results obtained from the combined feature set on both these datasets reveal interesting trends. The fact that the concatenated feature set is able to consistently extract improvements under all performance metric indicates gains to be realized by improving our feature combination method. Also, the high performance of the classification algorithms on both the datasets despite completely different group biases is promising. This indicates that more reliable group detection can be achieved by developing more linguistically oriented features and intelligently combining them with conventional TF-based features.

4   Conclusion

We have introduced the use of stylistic features along with term frequency for implicit author group affiliation detection from written text, specifically informal and semi-formal online text. We were pleased with the high accuracy that we were able to achieve with our straightforward stylistic and TF features on our primary dataset on our target forum data. In some cases this was higher than our benchmark

BitterLemons dataset (Table 1). Our simplified feature combination scheme (concatenation), our results, both TF and augmented TF (TF+LIN) are extremely encouraging. In addition to the classification results obtained from the complete feature set, we have also analyzed the contributions of

Table 1. Group affiliation performance obtained from TF, and combination of TF and our 8 classes of Linguistic Features (TF+LIN)

Metrics →	MCC		BAc	
Features →	TF	TF+LIN	TF	TF+LIN
Moderate-Extreme	0.92	**0.93**	0.96	**0.97**
BitterLemons.org	0.90	**0.91**	0.95	**0.95**

individual feature types and their discriminating capability to gain better understanding of robustness of feature contributions, if any, in different datasets. Based on the small number of datasets analyzed, and the infancy of our linguistic features, it is possible our results are in fact classifying based on a confounding factor such as sentiment analysis or topic modeling instead of the intended stylistic analysis. With respect to our application, final accuracy is the only criteria, and results are extremely promising. With respect to advancing the science and providing understanding and insight, we feel we have started in the right direction.

We feel there is untapped potential for even higher performance. There are other identifying and quantifying properties of group behaviors known to researchers in Social Science which we were not able to evaluate in this round of experimentation. Other features we plan to explore include the measurement of topic emphasis, the use of hedging words, and the use of elusive, noncommittal, and non-specific descriptions. We hope to continue to locate additional Social Science inspired features to try, as well as the theoretical underpinnings of the best performing features. Just as the medical community attempts to categorize groups based on behavioral informatics, we look for predictive features that point to group affiliations and authors' influence on those groups. The ability to quantify or discretize new linguistic properties into features will allow researchers to utilize machine learning algorithms to automatically detect or classify groups. Large scale execution is possible by capitalizing on newly available representations and evolutions of behavior and culture online. We outlined numerous potential applications of authorship affiliation besides our own Department of Defense focus, all of which we feel would benefit society.

Acknowledgements. The authors thank the Office of Naval Research and Martin Krueger for their support of this work. The authors also thank Dr. Marion Ceruti for her tireless contributions. This paper is the work of U.S. Government employees performed in the course of employment and no copyright subsists therein. This paper is approved for public release with an unlimited distribution.

References

1. Giles, H., Taylor, D., Bourhis, R.: Towards a theory of interpersonal accomodation through language. In: Language in Society, vol. 2, pp. 177–192. Cambridge University Press (1973)
2. Postmes, T., Spears, R., Lea, M.: The Formation of Group Norms in Computer-Mediated Communication. In: Human Communication Research, vol. 26, pp. 341–371. Sage Publications (2000)
3. Ceruti, M.G., McGirr, S.C., Kaina, J.L.: Interaction of Language, Culture and Cognition in Group Dynamics for Understanding the Adversary. In: Proceedings of the National Symposium on Sensor and Data Fusion (NSSDF, Nellis AFB, Las Vegas, NV (2010)
4. Holmes, D.I.: Authorship Attribution. Computers and the Humanities 28(2), 87–106 (1994)
5. Zheng, R., Li, J., Chen, H., Huang, Z.: A framework for authorship identification of online messages: Writing-style features and classification techniques. Journal of the American Society for Information Science and Technology 57(3), 378–393 (2006)
6. Stamatatos, E.: A survey of modern authorship attribution methods. Journal of the American Society for Information Science and Technology, 538–556 (2009)
7. Abbasi, A., Chen, H.: Applying Authorship Analysis to Extremist-Group Web Forum Messages. IEEE Intelligent Systems, 67–75 (2005)
8. Juola, P.: Authorship attribution. Foundations and Trends in Information Retrieval 1(3), 233–334 (2006)
9. Booker, L., Strong, G.: Using Topic Analysis to Compute Identity Group Attributes. In: Social Computing, Behavioral Modeling, and Prediction, pp. 249–258 (2008)
10. Koppel, M., Argamon, S., Shimoni, A.: Automatically Categorizing Written Texts by Author Gender. Literary and Linguistic Computing 17(3) (2002)
11. Izumi, M., Miura, T., Shioya, I.: Estimating the date of blog authors by CRF. In: IEEE Pacific Rim Conference on Communications, Computers and Signal Processing, pp. 249–252 (2007)
12. Goswami, S., Sarkar, S., Rustagi, M.: Stylometric Analysis of Bloggers' Age and Gender. In: Proceedings of the AAAI International Conference on Weblogs and Social Media (2009)
13. Koppel, M., Schler, J., Zigdon, K.: Determining an author's native language by mining a text for errors. In: Proceedings of Knowledge Discovery in Data Mining, pp. 624–628 (2005)
14. Argamon, S., Saric, M., Stein, S.S.: Style mining of electronic messages for multiple authorship discrimination: first results. In: Proceedings of Knowledge Discovery in Data Mining, pp. 475–480 (2003)
15. Ratnaparkhi, A.: A Maximum Entropy Model for Part-Of-Speech Tagging. In: Proceedings of the Emperical Methods in Natural Language Processing, pp. 133–142 (1996)
16. Khosmood, F., Levinson, R.: Automatic Synonym and Phrase Replacement Show Promise for Style Transformation. In: Proceedings of the IEEE Ninth International Conference on Machine Learning and Applications, pp. 958–961 (2010)
17. Lin, W.H., Wilson, T., Wiebe, J., Hauptmann, A.: Which side are you on? Identifying perspectives at the document and sentence levels. In: Proceedings of the Tenth Conference on Natural Language Learning, pp. 109–116 (2006)
18. Matthews, B.W.: Comparison of the predicted and observed secondary structure of T4 phage lysozyme. Biochim. Biophys. Acta 405, 442–451 (1975)

19. Burrows, J.F.: Word patterns and story shapes: The statistical analysis of narrative style. Literary and Linguistic Computing 2, 61–70 (1987)
20. Stamatatos, E., Fakotakis, N., Kokkinakis, G.K.: Automatic Text Categorization in Terms of Genre, Author. Computational Linguist 26(4), 471–495 (2000)
21. Argamon-Engelson, S., Koppel, M., Avneri, G.: Style-based text categorization: What newspaper am I reading? In: Proceedings of AAAI Workshop on Learning for Text Categorization, pp. 1–4 (1998)
22. Ellen, J., Parameswaran, S.: Machine Learning for Author Affiliation within Web Forums. In: Proceedings of the IEEE Tenth International Conference on Machine Learning, pp. 100–106 (2011)

Automatic Crime Prediction Using Events Extracted from Twitter Posts

Xiaofeng Wang, Matthew S. Gerber, and Donald E. Brown

Department of Systems and Information Engineering, University of Virginia
{xw4u,msg8u,brown}@virginia.edu

Abstract. Prior work on criminal incident prediction has relied primarily on the historical crime record and various geospatial and demographic information sources. Although promising, these models do not take into account the rich and rapidly expanding social media context that surrounds incidents of interest. This paper presents a preliminary investigation of Twitter-based criminal incident prediction. Our approach is based on the automatic semantic analysis and understanding of natural language Twitter posts, combined with dimensionality reduction via latent Dirichlet allocation and prediction via linear modeling. We tested our model on the task of predicting future hit-and-run crimes. Evaluation results indicate that the model comfortably outperforms a baseline model that predicts hit-and-run incidents uniformly across all days.

1 Introduction

Traditional crime prediction systems (e.g., the one described by Wang and Brown [14]) make extensive use of historical incident patterns as well as layers of information provided by geographic information systems (GISs) and demographic information repositories. Although crucial, these information sources do not account for the rich and rapidly expanding social media context that surrounds incidents of interest. An essential part of this context is the stream of information created by users of services such as Facebook[1] and Twitter[2]. These services allow users to instantly create, disseminate, and consume information from any location with access to the Internet. Recently, Howard et al. argued that such services played a key role in the development and perpetuation of the "Arab Spring" uprisings that took place across North Africa and the Middle East beginning in December of 2010 [10]. The authors found, among other things, evidence that social media activity of particular types preceded mass protests and other incidents.

Whereas the study conducted by Howard et al. was retrospective, this paper presents a preliminary investigation of the *predictive* power of social media information, in particular information produced by the Twitter service. We hypothesized that information extracted from the Twitter service would - if properly structured and modeled - provide indicators about the likelihood of future

---

[1] http://www.facebook.com
[2] http://www.twitter.com

S.J. Yang, A.M. Greenberg, and M. Endsley (Eds.): SBP 2012, LNCS 7227, pp. 231–238, 2012.
© Springer-Verlag Berlin Heidelberg 2012

incidents. Our investigation did not attempt to acquire and digest all "tweets" (short messages created by Twitter users); rather, we pulled tweets from the Twitter feed of a news agency covering the area of Charlottesville, Virginia. Consider the following tweets:

(1) TRAFFIC ALERT: Rt. 20 closed due to a wreck.
(2) Road closed at JPA and Shamrock due to tree falling over road.

Intuitively, these tweets provide evidence of an increased hazard level along roadways, which, in turn, might lead to an increased number of accidents or hit-and-run crimes. The goal of our investigation was to build a predictive model of criminal incidents that leverages this type of evidence. We used state-of-the-art natural language processing (NLP) techniques to extract the semantic event content of the tweets. We then identified event-based topics using unsupervised topic modeling, and used these topics to predict future occurrences of criminal incidents. With the tweet information alone, our predictive model comfortably outperformed a uniform prediction baseline on held-out data.

2   Related Work

2.1   Crime Mapping and Prediction

The task of crime prevention is constrained by scarce resources (e.g., time, patrol units, and finances). Analysts, therefore, often employ a variety of computer systems to identify and visualize areas of high crime, otherwise known as "hot-spots" [7]. Crime hot-spots indicate spatial areas of relatively high threat according to some underlying model. A common model - one promoted by Eck et al. - relies on kernel density estimation (KDE) from the criminal history record of an area [7]. KDE is an efficient method of computing a continuous surface, where the relative threat (i.e., "hotness") of an area is indicated by its color and/or vertical height. Chainey et al. investigated the use of KDE for crime prediction [6]; however, KDE as a predictive model suffers from (1) a lack of portability (crimes cannot be predicted for previously unseen regions), and (2) a lack of contextual information such as that coming from social media services. In a similar vein, Mohler et al. applied the self-exciting point process model (previously developed for earthquake prediction) as a model of crime [12]. This model, like ones based on KDE, relies on the prior occurrence of crimes in a particular area and thus cannot generalize to previously unobserved areas.

Wang and Brown proposed a different approach to crime modeling [14]. In their approach, prior criminal incidents are used as supervised training data within the predictive model. Geographic locations are characterized by a rich set of spatial and demographic features instead of the simple geographic coordinates used by KDE-based approaches. Example features include the distance to the nearest business and the number of divorced individuals in the region. This representation permits crime prediction in previously unseen places, as the correlation between, for example, burglary and business proximity can often be generalized from one area of a city or country to another area. Furthermore, the

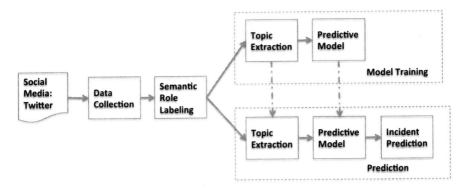

Fig. 1. Overall process of criminal incident prediction using tweets

feature-based representation permits the addition of information such as that extracted from social media services. Thus, the work presented in this paper should be viewed as complementary to the work reported by Wang and Brown.

2.2 Use of Social Media for Prediction

The proliferation of social media services has prompted a surge of interest in using the associated data for various predictive purposes. For example, Twitter posts have been used to predict box office results [1], election results [2], and stock market trends [5]. Popular techniques in these studies include keyword volume analysis and sentiment analysis. These methods have proven useful for the tasks mentioned above; however, a deeper semantic understanding of tweets is required to predict discrete criminal incidents, which are not mentioned ahead of time (ruling out keyword volume and sentiment analysis).

3 Data Collection and Methods

Figure 1 shows the overall operation of our Twitter-based predictive model. We first collect a corpus of tweets from Twitter. We then extract events from the main textual content of each tweet using an NLP technique known as semantic role labeling (SRL). Next, we apply latent Dirichlet allocation (LDA) to identify salient topics within the extracted events. A predictive model is then built upon these latent topics. The following sections describe these steps in detail.

3.1 Data Collection

The user base of Twitter comprises a vast community of news agencies, journalists, and casual users who post tweets from their Internet-connected devices.

Each tweet is restricted to 140 characters and can be observed by those who subscribe to the poster's Twitter feed. As of March 11, 2011, Twitter was processing approximately 140 million tweets per day, with approximately 460,000 new accounts being created daily.[3] Traditional news stations and newspapers actively use Twitter to publish breaking news in real-time. For example, CBS19[4] in Charlottesville, Virginia published 3,659 tweets during the period of February 22, 2011 through October 21, 2011 (approximately 15 per day). We collected these tweets using the public interface provided by Twitter.

In addition to Twitter data, our investigation required ground-truth criminal incident data, which we used to estimate the parameters of our predictive model and evaluate its performance. We obtained these records from local law enforcement agencies, focusing on hit-and-run incidents during the same period covered by the Twitter data. In total, we collected records for 290 hit-and-run incidents (1.2 per day).[5]

3.2   Methods

Semantic Role Labeling. Our approach to Twitter-based crime prediction relies on a semantic understanding of tweets, one that goes beyond bag-of-words and sentiment representations. Such an understanding can be derived from a process known as semantic role labeling (SRL), which extracts the events mentioned in tweets, the entities involved in the events, and the roles of the entities with respect to the events. An example analysis (derived from Example 1, p. 232) is shown below:

(3) $[e_1{:}warning$ TRAFFIC] $[e_1$ ALERT]: $[e_2{:}entity$ Rt. 20] $[e_2$ closed] $[e_2{:}cause$ due to a wreck].

Two events were extracted from Example 3: (1) an *alert* event in which traffic is being brought to the reader's attention, and (2) a *close* event where a road is closed due to a wreck. Note that these events are signaled by a noun and verb, respectively. Gildea and Jurafsky documented the seminal investigation into SRL [9] and the NLP community has had a sustained interest in the technique since then (see [11] for a more recent survey and references). In our study, we used the system created by Punyakanok et al. [13] to analyze verb-based SRL structures and the system created by Gerber and Chai [8] to analyze noun-based SRL structures. In general, the SRL systems were well suited to the news tweets, since the systems were trained on news corpora. In the cited studies, the authors report F_1 scores of approximately 80% for verbal SRL and 72% for nominal SRL. The output from these systems forms the basis for event prediction, since it informs the model about current events, which (we hypothesized) might correlate with future criminal incidents.

---

[3] http://blog.twitter.com/2011/03/numbers.html (accessed November 1, 2011).

[4] http://www.newsplex.com

[5] http://www.charlottesville.org/index.aspx?page=257

Event-Based Topic Extraction via Latent Dirichlet Allocation. After processing the tweets with the SRL systems, we have multiple events e_i associated with each day. In topic modeling terms, each day d is associated with an abstract "document" doc_d that contains "words" $\{e_1, e_2, \ldots, e_{n_d}\}$, where n_d is the length of doc_d. These words describe what happened on day d.

As with topic modeling of actual textual documents, we hypothesized that a day's events would be related in a particular (though hidden) way. Thus, instead of using doc_d directly to predict future incidents, we further extracted topics $\{t_1, t_2, \ldots, t_k\}$ from doc_d using latent Dirichlet allocation (LDA) [4][3].[6] LDA is a probabilistic language model that can be used to explain how a collection of documents is generated from a set of hidden (or latent) topics. LDA efficiently discovers word-based topics and reduces the dimensionality of documents to lie within the k-dimensional space of topics. Given the number of topics k, LDA can estimate the topic-document distribution $\{T_{d,1}, T_{d,2}, \ldots, T_{d,k}\}$, where $T_{d,i}$ is the probability that document d is related to topic i.

We applied LDA to derive $\{T_{d,1}, T_{d,2}, \ldots, T_{d,k}\}$ for the events described in tweets on day d. Intuitively, this analysis tells us about the relationship between the k major (latent) events on day d and the observable events e_i that were reported by the news agencies. This reduces the dimensionality of doc_d and provides meaningful structured data for our predictive model, which is described next.

Predictive Model. doc_d contains the events that occurred on day d. Our goal is to use doc_d to make predictions about incidents in the future. Formally, we seek a function $y_{d+1} = f(doc_d)$, where y_{d+1} is a binary random variable indicating whether an incident will occur on day $d+1$. For example, we can use the following generalized linear regression model (GLM):

$$log\left(\frac{Pr[y_{d+1} = 1]}{1 - Pr[y_{d+1} = 1]}\right) = \beta_0 + \beta_1 T_{d,1} + \cdots + \beta_k T_{d,k} \tag{4}$$

Where each $T_{d,i}$ is derived via LDA. Parameters $\{\beta_0, \ldots, \beta_k\}$ can be estimated using the set of prior criminal incidents described in Section 3.1.

With both the estimated LDA model and GLM model, we can make a prediction using new tweets. To make a prediction, we first process tweets on day d' using the SRL systems described above. Then, the LDA model is used to infer the event-based topic distribution $\{T_{d',1}, T_{d',2}, \ldots, T_{d',k}\}$. Lastly, the predictive model (Equation 4) uses this distribution to predict the likelihood of an incident occurring on day $d' + 1$.

4   Evaluation and Results

We evaluated our predictive model using Twitter data and actual hit-and-run incidents that occurred in Charlottesville, Virginia. As described in Section 3.1,

---

[6] We used GibbsLDA++ in all of our experiments:
`http://gibbslda.sourceforge.net`

Table 1. Top 10 most likely words for each of the 10 topics

Topic 1	Topic 2	Topic 3	Topic 4	Topic 5	Topic 6	Topic 7	Topic 8	Topic 9	Topic 10
close	say	arrest	plan	report	expect	report	say	trial	come
fire	make	suspect	kill	say	remain	close	found	make	cbs
crash	hanchettjim	death	use	student	rsb	open	die	break	help
look	search	murder	ask	vote	cancel	confirm	crash	set	start
delay	confirm	shoot	plead	tell	follow	block	fall	traffic	lead
come	run	rsb	life	hear	close	wreck	want	begin	look
reopen	start	hear	sell	work	warn	follow	find	hope	check
stay	move	protest	convict	speak	make	accord	kill	bring	lawsuit
watch	begin	report	visit	head	price	move	shut	hit	arrest
driver	end	need	statement	call	list	check	come	stop	left

our data cover the period of February 22, 2011 through October 21, 2011. We studied the hit-and-run incidents per day using traditional time series methods, but discovered no trend, seasonality, or autocorrelation. Thus, without any additional information, a baseline system would assign a uniform probability of incidents to all future days. This approach constitutes our baseline model.

We used the data before September 17, 2011 to train the LDA and predictive models, setting k (the number of latent topics) to be 10. Table 1 presents the top 10 words for each topic. The nature of these topics is subjective; however, some structure is present. For example, topic 1 appears to be related to crashes, whereas topic 3 appears to be related to shootings and their associated criminal processes. We trained a GLM on these topics as described in Section 3.2, using stepwise selection to identify the most informative features. The resulting GLM is shown below:

$$\log\left(\frac{Pr[hit_{d+1} = 1]}{1 - Pr[hit_{d+1} = 1]}\right) = 0.4 + 0.71T_{d,1} + 0.88T_{d,4} + 0.72T_{d,6} + 0.61T_{d,8} \quad (5)$$

In Equation 5, $Pr[hit_{d+1} = 1]$ denotes the probability of at least one hit-and-run incident occurring on day $d+1$. $T_{d,\cdot}$ is the topic distribution on day d. As shown, the model emphasizes topics 1, 4, 6, and 8 in the prediction of future hit-and-run incidents.

We applied this model to predict hit-and-run incidents during the period of September 17, 2011 to October 21, 2011. Figure 2(a) shows the ROC curve of the prediction performance. Vertical bars are 95% confidence intervals derived with a bootstrap resampling procedure. The ideal ROC curve stretches toward the upper-left corner. A curve along the diagonal indicates no predictive power. As shown by Figure 2(a), the LDA/GLM model was able to predict future hit-and-run incidents; although, due to the limited amount of testing data, we observed fairly wide confidence intervals. The baseline system ROC curve lies on the diagonal, as it predicts hit-and-run incidents uniformly across all days.

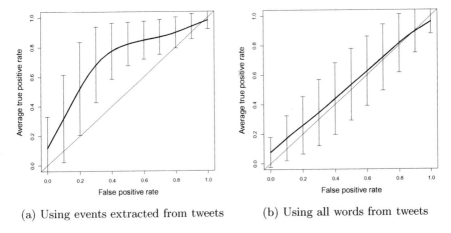

(a) Using events extracted from tweets   (b) Using all words from tweets

Fig. 2. ROC curves for predicting hit-and-run incidents

Our SRL systems have a runtime complexity of $O(n^3)$, where n is the number of words and punctuation marks in the sentence. In order to justify this added complexity, we re-trained the predictive LDA/GLM model on all words in the tweets instead of event words only. The remaining experimental conditions were held constant, resulting in the ROC curve shown in Figure 2(b). As shown in the figure, the system exhibits minimal predictive power, thus supporting the use of event extraction for incident prediction.

5   Conclusions and Future Work

This paper has presented a preliminary investigation into the use of social media for criminal incident prediction. Although our data source (Twitter) and the prediction domain (criminal incidents) are not novel, we are not aware of any prior work that brings these topics together. Our approach is based on the automatic semantic analysis and understanding of natural language tweets, combined with dimensionality reduction via latent Dirichlet allocation and prediction via linear modeling. Evaluation results demonstrate the model's ability to forecast hit-and-run crimes using only the information contained in the training set of tweets. Given the widespread use of social media services such as Twitter, our results indicate a fruitful line of future research.

There are many ways in which this work can be extended. In our semantic analysis step, we assumed that the events contained in a tweet occurred on the day that the tweet was posted; however, tweets often describe events that occurred days, weeks, or even years ago using overt linguistic expressions (e.g., "Last year's storm..."). This information needs to be taken into account when predicting future incidents, since events in the distant past might lose their influence. Regarding the GLM model, we used a simple stepwise selection method to choose features for prediction. A better alternative might be to apply the penalized GLM (PGLM) to select the most predictive features. Lastly, our approach does not leverage the massive amount of information produced by the Twitter service. Large-scale analysis of tweets might provide insights that are not apparent within a single feed.

Acknowledgments. We would like to thank the anonymous reviewers for their comments on a previous version of this paper. The work presented in this paper was funded by United States Army Grant W911NF-10-2-0051.

References

1. Asur, S., Huberman, B.: Predicting the future with social media. In: 2010 IEEE/WIC/ACM International Conference on Web Intelligence and Intelligent Agent Technology, pp. 492–499. IEEE (2010)
2. Bermingham, A., Smeaton, A.: On using twitter to monitor political sentiment and predict election results. In: Proceedings of the Workshop on Sentiment Analysis Where AI Meets Psychology (SAAIP 2011), Asian Federation of Natural Language Processing, Chiang Mai, Thailand, pp. 2–10 (November 2011)
3. Blei, D., Carin, L., Dunson, D.: Probabilistic topic models. IEEE Signal Processing Magazine 27(6), 55–65 (2010)
4. Blei, D.M., Ng, A.Y., Jordan, M.I.: Latent dirichlet allocation. J. Mach. Learn. Res. 3, 993–1022 (2003)
5. Bollen, J., Mao, H., Zeng, X.: Twitter mood predicts the stock market. Journal of Computational Science (2011)
6. Chainey, S., Tompson, L., Uhlig, S.: The utility of hotspot mapping for predicting spatial patterns of crime. Security Journal 21, 428 (2008)
7. Eck, J., Chainey, S., Cameron, J., Leitner, M., Wilson, R.: Mapping crime: Understanding hot spots (2005)
8. Gerber, M., Chai, J., Meyers, A.: The role of implicit argumentation in nominal SRL. In: Proceedings of Human Language Technologies: The 2009 Annual Conference of the North American Chapter of the Association for Computational Linguistics, pp. 146–154. Association for Computational Linguistics, Boulder (2009)
9. Gildea, D., Jurafsky, D.: Automatic labeling of semantic roles. Computational Linguistics 28, 245–288 (2002)
10. Howard, P.N., Duffy, A., Freelon, D., Hussain, M., Mari, W., Mazaid, M.: Opening closed regimes: What was the role of social media during the arab spring? Tech. rep., Project on Information Technology and Political Islam, University of Washington, Seattle (January 2011)
11. Màrquez, L., Carreras, X., Litkowski, K.C., Stevenson, S.: Semantic role labeling: an introduction to the special issue. Comput. Linguist. 34(2), 145–159 (2008)
12. Mohler, G.O., Short, M.B., Brantingham, P.J., Schoenberg, F.P., Tita, G.E.: Self-exciting point process modeling of crime. Journal of the American Statistical Association 106(493), 100–108 (2011)
13. Punyakanok, V., Roth, D., Yih, W.T.: The importance of syntactic parsing and inference in semantic role labeling. Comput. Linguist. 34(2), 257–287 (2008)
14. Wang, X., Brown, D.E.: The spatio-temporal generalized additive model for criminal incidents. In: ISI (2011)

Dynamic, Covert Network Simulation

Patrick O'Neil*

GeoEye Analytics, McLean, VA 22102
o'neil.patrick@geoeye.com
http://www.geoeye.com

Abstract. Covert networks are nearly impossible to fully describe due to inherent difficulties in obtaining data. To have complete information on a covert network, it is necessary to simulate such data. The simulation presented in this paper dynamically builds a covert network off of an existing network structure. Through parameter tuning, the user is able to generate a wide variety of networks which mimic real world covert networks. Due to the dynamic nature of the simulation, changes in network structure can be analyzed over time. Additionally, network response to intervention, such as the killing of members, can be modeled and compared with real world findings. As the simulation can generate a wide variety of networks, its use can help with testing intervention strategies in different situations.

Keywords: Network, simulation, terrorism, dynamic systems, covert, multi-agent.

1 Introduction

This paper presents a simulation tool, called the Dynamic Covert Network Simulator (DCNS), that builds a stochastic, multi-agent, dynamical system from an existing network structure. The network grows and evolves in response to internal characteristics as well as external intervention. While many network generation models [1, 2, 3] work with a preset number of nodes, DCNS adds new members to the network when they are needed. Then, by assigning each node a set of attributes and roles, we can define the node's behavior. The dynamics are modeled on the way real covert networks have acted in the past. For example, part of the network may act similar to the terrorists involved in the 2002 bombings in Bali. An advantage of this approach is that the model can be calibrated to mimic a wide range of real world networks. For this paper, the model was calibrated for generating terrorist networks. A sample network is generated and tested for validity.

2 Model Description

This simulation works by giving nodes the ability to communicate and work with each other to accomplish randomly generated objectives. Nodes operate

---

* Research funded by the Office of Naval Research, Code 30 Contract Number N00014-09-C-0263, Directed by Martin Kruger, Scott McGirr, and Joan Kaina.

S.J. Yang, A.M. Greenberg, and M. Endsley (Eds.): SBP 2012, LNCS 7227, pp. 239–247, 2012.

locally in a covert manner and from this local behavior, a globally secretive network is created [2]. To begin, an undirected network structure, $G(V, E)$, is loaded into the model. The network nodes, V, are leaders and subordinates and the network edges, E, are weighted edges which reflect when two nodes have the ability to communicate and work together. Large edge weights indicate the connected members work well together.

2.1   Member Attributes

Members are assigned a set of attributes and roles that define their behavior within the network. Attributes describe the node's vulnerability to intervention and the node's devotion to the network. Roles determine the member's abilities (such as recruitment and resource allocation) and were taken from the John Jay & Artis Transnational Terrorism Database (JJATT) [4].

Leadership. A node may also be a leader, which allows the node to build a cell and begin operations. This position is either inherited from the initial dataset or earned through involvement in successful operations. *Leaders* are in charge of individual cells, which are made up of *Subordinates*. For simplicity, every node is either a leader or a subordinate.

Chance of Discovery. A member's chance of discovery (*CoD*) indicates how exposed it is to external intervention. We denote the *CoD* of node $x \in V$ by x_d. High values of x_d indicate the node is vulnerable to being captured or killed. When involved in operations, x_d increases (opposite otherwise). Other factors affecting *CoD* include the capture of neighbors, the current location of the node, and the node's attributes (secrecy and roles). A parameter may be set to capture/kill any member whose *CoD* surpasses the set value.

Neighbors. When a node x is captured or killed, the *CoD* of all x's neighbors increases by the parameter ρ (i.e. $\forall y \in N_x, y_d(t+1) = y_d(t) + \rho$). This captures the fact that compromised nodes yield information about other members of the network [5].

Locations. At any given time, a member can be located in one of three types of locations: hostile, neutral, and friendly. Let L be the set of locations. For $l \in L$, we let $l_s \in [0, 3]$ indicate the location's stance. Then $l_s \in [0, 1) \to l$ is hostile; $l_s \in [1, 2) \to l$ is neutral; $l_s \in [2, 3] \to l$ is friendly. Neutral locations do not affect *CoD*, however hostile locations increase *CoD* and friendly locations decrease *CoD* by an amount proportional to l_s. Thus, being in a safe location lowers the chance a node will be discovered [6].

Secrecy. When a member is added to the network, it is assigned a *secrecy* score, x_s. This value reflects the node's ability to remain unexposed. Each round, x_s is subtracted from x_d (i.e. $x_d(t+1) = x_d(t) - x_s$). Thus, nodes with a high x_s are more likely to avoid intervention. This "survival of the fittest" approach reflects the reality that some covert agents are harder to detect than others.

Radical. A member's *Radical* score, x_r, indicates how committed it is to the success of the network [7]. A high score will lead to a promotion while a low score will lead to a demotion and eventually cause the member to leave the network. *Radical* is affected by the node's level of involvement. The score x_r will increase when engaged in an operation, receive a bonus for successful completion, and decrease when not involved.

Roles. Member roles were selected from the JJATT dataset (see Table 1), but may be customized to fit other types of networks. Some roles, such as *Foot Soldier* and *Bomber*, can be assigned when the member is recruited; other roles, like *Recruiter*, are acquired during the simulation or from the initial dataset. Many members will also have multiple roles. For example, there may be a member with the *Foot Soldier* and *Bomber* roles. The role abilities come into play during operations.

Table 1. Types of Roles

Role	Earned	Info
Bomb Maker	- -	Required for bombing operations
Bomber	- -	Attacks target locations
Financier	- -	Aquires resources
Logistician	- -	Required for co-Ops
Foot Soldier	- -	Attacks target locations
Weapons	- -	Aquires resources
Trainer	Yes	Gives members new roles
Recruiter	Yes	Recruits new members

2.2 Network Attributes

Network Security. Network security measures the secrecy (influenced by Lindelauf [8]) of the network by accounting for recent-past member attrition rates. During the simulation, highly exposed members will be captured or killed. If this begins to happen frequently, the network may try to protect itself from further intervention by lowering the degree to which members connect with each other [9]. Network security, $NS(c_m, k_m)$, is dependent upon the number, $c_m, k_m \in \mathbb{N}$, of members captured/killed in the past $m \in \mathbb{N}$ rounds. Higher values of NS yield lower connectivity within the network.

Operation Generation Rate. The probability of a new operation being generated, $P_O(c_m, k_m, r_c, r_k)$, depends on the number, c_m, k_m, of members captured/killed in the past m rounds [10]. The second set of arguments, $r_c, k_c \in [-1, 1]$, reflects the network's reaction to intervention. If $r_c > 0$, the network will generate more attacks in response to captures (defiance), and similarly if $r_c < 0$, the network generates fewer (deterence) [11]. Also, if $r_c = r_k = 0$, the network does not act differently because of intervention.

2.3   Node Update Process

The node update process (see Figure 2.3) updates members using a set of rules described here. To begin, if member $x \in V$ is not captured/killed, it is considered free and the member's chance of discovery, x_d, is decreased by the member's secrecy, x_s. If a member is involved in an operation, the node-level update process for x will be handled during the operation-level update process. Otherwise, it is evaluated for promotion or demotion. If the member's radical, x_r, is greater than the Promotion Level, $pL = \mu_r + w_p \sigma_r$ where μ_r is the average radical score in the network, σ_r is the standard deviation, and $w_p \in [0, 1]$ is a weight, it is promoted. Being promoted allows the member to create a new cell as well as begin operations. Similarly, if x_r is less than the Demotion Level, $dL = \mu_r - w_p \sigma_r$, then the member is demoted. If x is demoted, all of the members in x's cell search for a leader within a geodesic distance of 2 (neighbors of neighbors) and attach themselves to that leader's cell. If there is no such leader, the members leave the network. After this, the member's *Radical*, x_r is updated by subtracting the value associated with not being involved in an operation (τ_r). Then x_d is updated by subtracting the value associated with not being involved in an operation (τ_d).

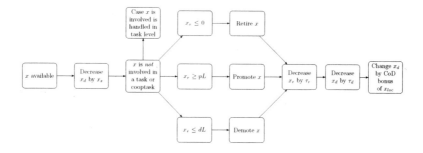

Fig. 1. Node Update Process

2.4   Connections

Connections reflect when two nodes can communicate with each other. If nodes i and j are connected, then $e_{ij} \in E$. The strength of connections is determined by the function $w : E \to [-1, 1]$. When an edge, e_{ij}, is first created, $w(e_{ij}) = 0.1$. For all subsequent times, $w_{t+1}(e_{ij}) = w_t(e_{ij}) + qO_c + sL_c$ where $q = 1$ when i, j are involved in the same operation (0 otherwise), $s = 1$ when i, j are in the same location (0 otherwise), and $O_c, L_c \in [0, .1]$. Negative values of s and q may be used instead of 0 to make edge weights decay. If an edge is removed from the network, the edge weight between the nodes is recorded so that if the nodes connect again in the future, the recorded edge weight can be used for the restored edge. There are three key parameters which determine the connection rates. The rate when two members have a mutual neighbor (C_n), are involved in the same task (C_t), or are involved in the same co-Op (C_c)

2.5   Operations

Leaders have the ability to begin operations, which are the driving force behind the model. Operations involve recruitment, resource acquisition, and communication. The two types of operations are *Tasks* (one cell) and *Cooperative Tasks* (multiple cells).

Operations: Tasks

Task Generation. Tasks consist of one leader and a number of subordinates working to accomplish a goal. The idea for the structure of operations is influenced by [12]. When a task, $T \in \mathbb{T}$, is generated, it is given a recruitment quota, T_{rec}, which is the required number of members to successfully complete the task. Additionally, the task is given a resource quota, T_{res}, indicating the number of resources required to complete the task. While aquiring enough members and resources does not guarantee successful completion of the task, not aquiring them guarantees failure. *Recruiters* can only recruit one new member each round, and *Resourcers* (*Weapons* and *Financier* roles) can only aquire one resource per round. As such, the amount of time allocated to complete the task, $T_{\Delta t}$, is subject to $T_{\Delta t} \geq \max(T_{rec}, T_{res}) + opTime$. The constant, $opTime$, is the amount of time after the preparatory stage during which the operation actually takes place. Other attributes attached to the task at the time of creation include the task's risk, T_d, and the task's radical bonus, T_r. These values are used to update the *CoD* and *Radical* scores, respectively, of the members involved. Finally, a target location is picked from the hostile locations.

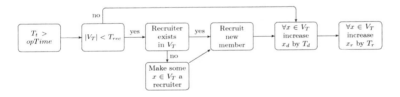

Fig. 2. Preparation Time for Tasks

Preparation Time. Recruitment takes place at the beginning of the task. If the recruitment quota has not been reached, an existing cell member may be attached to the task or a task member with a *Recruiter* role can recruit a new member to join the network. When recruited, the new member is randomly assigned a role (using a preset probability distribution for roles). If there are no task members with the *Recruiter* role, a task member is chosen at random and given the role. When a member is recruited, they are attached to the task leader's cell and their attributes are randomly generated. Before the round ends, all task members have their *CoD* and *Radical* scores updated.

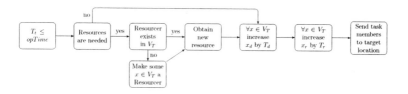

Fig. 3. opTime for Tasks

opTime. When the task enters *opTime*, i.e. $T_t \leq opTime$, resources are aquired and members travel to the target location. To aquire resources, a task member must have a *Financier* or *Weapons* role. If no task member possesses one of these roles, a task member is chosen at random and given one of the two roles (collectively referred to as Resourcer). Recruitment and resource aquisition are both probabilistic activites. The probability of successfully recruiting a new member is set prior to running the simulation and will play heavily into the growth rate of the network.

At the end of the task, if the number of task members meets the recruitment quota and the correct number of resources have been aquired, the task has a chance of being successful. This chance is determined by a utility function run on the subgraph, $G_T(V_T, E_T)$, induced by the task T. The utility function, $U(T)$, averages the edge weights of all involved members using weights of 0 when an edge does not exist. Since $w(e_{ij}) \in [-1, 1]$ for all $i, j \in V_T$, it is clear that $U(T) \in [-1, 1]$. Note that $U(T)$ indirectly captures the experience of the members involved since edge weights are accumulated over time and inexperienced members would have small edge weights. A base probability for success, p_s, is introduced and, assuming the recruitment and resource quotas have been reached, the chance of a task, T, being successful is $P_{suc}(T) = p_s \cdot U(T)$ (if $P_{suc}(T) < 0$, we set $P_{suc}(T) = 0$). If the task succeeds, then the edge weights of all the connected task members are increased by T_{socInc}. On the other hand, if the task fails, then the connections are decreased by T_{socDec}. Thus, members who have worked well together in the past will have a better chance of succeeding again in the future.

Operations: Cooperative Tasks (co-Op). While tasks are simple recruitment and resource aquisition activities run by a single cell, *Cooperative Tasks* (co-Ops) are based on real world covert operations and involve up to three different cells. In this calibration of the DCNS, there are three forms of co-Ops based on terrorist attacks in Bali (2002), Mombasa (2002), and Mumbai (2008). In this paper, the Mombasa attacks are discussed. Information about these attacks was obtained from [13]. The goal behind simulating a real attack is to capture the high level behavior of nodes. As previously discussed, recruiters are needed for recruitment and financiers/weapons are needed to aquire resources.

Fig. 4. Mombasa co-Op Attack Sequence

The Mombasa style co-Op is broken down into three stages in addition to the *opTime* stage (see Figure 2.5). During the first stage, the leaders meet at the co-Op's safehouse. They then proceed to travel to various hostile locations searching for a target. During this time, recruitment is taking place. At the end of stage one, all members travel to the safehouse and the target location is chosen. In stage two, recruitment continues, footsoldiers and bombmakers are trained, resources are aquired, and bombmakers travel to the safehouse. During stage three, the attack team (consisting of footsoldiers and bombers) travels to the target location. When *opTime* arrives, attack team members are captured or killed and the surviving members return to the safehouse. When the task is over, the remaining task members flee to friendly locations. The chance of the co-Op being successful is calculated in the same way as tasks (recruitment/resource quotas and the utility function).

3 Results

In this section, we present an example demonstrating the ability of the DCNS to mimic real world networks. The results presented were aquired by running the simulation using the 1995 Jemaah Islamiyah (JI) network, available in the JJATT database [4], as the initial dataset. The network parameters were chosen to mimic the JI network and it is shown that the resulting network structure resembles that of the actual JI network in 2005 [4]. For brevity, the parameter values have been withheld.

The actual and simulated networks are shown in Figure 5 and some network structure metrics are shown in Table 2. The similarity between the actual and simulated networks shows that creating a realistic structure is possible if the correct parameter values are chosen. Since network structures at individual times (in this example JI 2005) can be mimicked using the DCNS, this process can be repeated and parameter values can be changed over time so that DCNS can mimic any sequence of network structures.

Table 2. Simulation Results

Network	Nodes	Edges	Leaders	Avg Degree	Density	Avg Path Length
Real Start	22	83	14	7.55	0.36	1.60
DCNS Sim	44	236	15	10.72	0.25	2.02
Real Finish	45	233	14	10.36	0.24	2.07

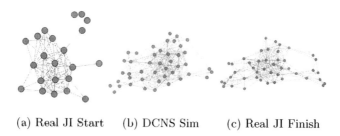

(a) Real JI Start   (b) DCNS Sim   (c) Real JI Finish

Fig. 5. Network Dynamics

4 Conclusion

A highlight of this model is that it can be tuned to generate a wide variety of
networks. It is possible to make highly connected, efficient networks (high connectivity values) as well as very secretive, cellular networks (low connectivity
values). While our example demonstrated its ability to simulate real networks
such as Jemaah Islamiyah (JI), the model's true use is in generating a range of
dynamic networks, taking subsets of those networks, running network analysis
algorithms and tools on the subnetworks, and then comparing the subnetwork
results to the full network results. Following this approach, the user can determine the effectiveness of some proposed network analysis tool by investigating
which types of networks can be accurately analyzed using the tool. It will also
allow the user to determine how sensitive the tool is to which subnetwork is chosen. Alternatively, since DCNS acts covertly, the simulations can be used to test
an intervention strategy on different types of networks. While one intervention
strategy might work well for a particular type of network, it might prove rather
ineffective at dealing with another.

References

1. Erdos, P., Renyi, A.: On the Evolution of Random Graphs. Publication of Mathematics Institute of Hungian Academy of Sciences 5(1761) (1960)
2. Tsvetovat, M., Carley, K.M.: Generation of Realistic Social Network Datasets for
   Testing of Analysis and Simulation Tools. CMU-ISRI-05-130 (2005)
3. Albert, R., Barabasi, A.-L.: Statistical Mechanics of Complex Networks. Reviews
   of Modern Physics 74 (2002)
4. Atran, S.: John Jay & ARTIS Transnational Terrorism Database. Technical report,
   John Jay College of Criminal Justice (2009)

5. Enders, W., Jindapon, P.: Network Externalities and Terrorist Network Structure. Journal of Conflict Resolution 54(2), 262–280 (2009)
6. Takeyh, R., Gvosdev, N.: Do Terrorist Networks Need a Home. The Washington Quarterly 25(3), 97–108 (2002)
7. Henke, M.G.A.: How Terrorist Groups Survive: A Dynamic Network Analysis Approach to the Resilience of Terrorist Organizations. School of Advanced Military Studies Monographs (2009)
8. Borm, P., Lindelauf, R., Hamers, H.: The Influence of Secrecy on the Communication Structure of Covert Networks. CentER Discussion Paper No. 2008-23 (2008)
9. Enders, W., Su, X.: Rational Terrorists and Optimal Network Structure. Journal of Conflict Resolution 51(1), 33–57 (2006)
10. Hussain, S.E.: Impact of Terrorist Arrest on Terrorism: Defiance, Deterrence, or Irrelevance. PhD thesis, University of Pennsylvania (2010)
11. Sherman, L.W.: Defiance, Deterrence, and Irrelevance: A Theory of the Criminal Sanction. Journal of Research in Crime and Delinquency 30, 445–473 (1993)
12. Don, B.W., et al.: Network Technologies for Networked Terrorists. Technical report, RAND Corperation (2007)
13. Fighel, J.: Al Qaeda - Mombassa Attacks November 28, 2002. Technical report, The Meir Amit Intelligence and Terrorism Information Center (2011)

Computing the Value of a Crowd

Manas S. Hardas[1,2] and Lisa Purvis[1]

[1] Spigit, Inc. 311 Ray Street, Pleasanton, CA 94566
[2] Computer Science Department, Kent State University, Kent, OH 44240
{mhardas,lpurvis}@spigit.com

Abstract. This paper proposes that the "value" of a crowd can be defined in terms of the overall engagement of the individuals within the crowd and that engagement is a function of certain characteristics of the crowds such as small world-ness, sparsity and connectedness. Engagement is hypothesized as messages being exchanged over the complex network which represents the crowd and the "value" is calculated from the entropy of message probability distributions. An initial random network is passed through a process of entropy maximization and the values of some structural properties are recorded with the changing topology to study the corresponding behavior. We show that as the small world-ness and connectedness of a crowd increases and the sparsity decreases, the engagement in the crowd increases.

1 Introduction

In the recent past it is increasingly being observed that social media plays an important role in putting a dormant crowd in action. The most recent examples of social media fueled crowd activism are the Egyptian revolution, the Spanish May 15th movement [5] and the most recent Occupy wall street protests [2] spread out across many cities in the United States. The common underlying theme in all these examples of activism is the intelligent use of social media like Facebook, Twitter, etc. to bring a crowd into action. Beneath any crowd is a network of people which have to be engaged towards a common goal for a crowd to perform a purposeful action. The engagement of a crowd can be measured in terms of the activity between the individual nodes. If the "activity" between the nodes is high, it means that the crowd is better engaged. The "activity" between any two nodes can then be measured as the number of messages exchanged between the two nodes.

We propose that the value of a crowd comes from its engagement and a very well engaged crowd has more value than a crowd which is not engaged. Engagement has certain properties like working towards a common goal, working in parallel on different tasks to achieve a common higher task, etc. The value of a crowd changes from time to time. Some movements start off great but die out soon, some movements start out small but pick up momentum and turn into a full-fledged revolutions. Of course there are many reasons for this happening and in this paper we briefly try to focus on the structural aspects of the crowd network.

Crowd networks are very different from typical static and structured communication networks. Although underlying the crowd is a social network of

S.J. Yang, A.M. Greenberg, and M. Endsley (Eds.): SBP 2012, LNCS 7227, pp. 248–255, 2012.

connected individuals, the phenomenon of a crowd happens only in the presence of certain characteristic conditions. One of the central conditions is the messaging activity which happens between the individuals which form the crowd. The so-called value of a crowd comes from the ability of a crowd to interact with each other and communicate with each other by sending and receiving messages – in essence the engagement of the crowd with one another. It is because of this activity on the links of crowd networks that creates the power and capability to achieve valuable actions and results.

Crowds can be represented as complex networks with different topologies and internal structural properties like small worlds, community organization, component structure, sparseness, etc. Therefore crowds can also be analyzed for their complex network properties like the clustering in a network, shortest path lengths, size of the giant component, etc. This paper presents a hypothesis that the engagement (and the resultant value generated from the engagement)of a crowd can be measured in terms of the entropy of the network. The entropy is calculated as a function of the probability distributions of the incoming and outgoing messages, which represents the entropy or uncertainty in the activity over the network. In this way we can translate the activity occurring over a network in terms of message exchange as a measure of the value of a network. An evolutionary algorithm is presented to optimize the entropy of a network by successively changing the network topology. Results indicate that the value of the crowd is very closely related to its small world-ness, sparsity and connectedness.

2   The Significance of Engagement in Crowd Networks

One of the principles articulating the social value of a network is the Metcalfe's law ($value \propto n^2$) which states "the value of a communications network is proportional to the square of the number of connected nodes which grows geometrically as the number of nodes grow." In other words, as a network gets bigger it increases in value than it would by just adding more nodes. For instance, a network of 5 nodes would have a value of 25 while a 10-node network would have a value of 100 (double the number of nodes, but 4 times the value).

While Metcalfe's law provides a useful view of a network's value, it focused on more static, communication networks as its basis and makes a variety of assumptions on the type of structures in the network. Most real complex networks are not homogenously linked by similar type of edges. Most real world complex networks display certain characteristic properties like small world phenomenon and scale free distributions [1, 7]. Today's hyper connected networks of people, information, and devices pose an entirely different challenge – in order to extract the value of today's crowd networks, it is not enough to understand simply the connectedness of the network. We must also understand the activity over those connections, representing the engagement of the crowd, and how it grows or shrinks over time and under what conditions. Only then can we begin to extract the true value of the crowd network. A crowd network has structural and behavioral complexities not present in a more static communication network. As such, crowd networks have significantly different dynamics and models of information and opinion spreading than a more predictable device communication network. Key differences are the social clustering that happens in a crowd network – effectively creating clusters of large strongly connected components, as well as a typically small amount of weak links that connect the entire network into one. Furthermore, the activity of communications across the connecting links as well as the triggers for growth or decline of that activity creates emergent network effects whose drivers today are largely unknown and undefined. Recent examples of real-life crowd network effects taking place are the May 15th movement in Spain, the self-organizing response network after the World Trade Center attack on September 11th 2001, and the Haiti earthquake response network. In each of these cases, we saw network effects at work, where the crowd drove results and action from the broad spreading of information, from a decentralized coordination and synchronization, and from the rapid information dissemination triggered in these instances. The value of the crowd response in each of these instances hinged on communication activity throughout the network, i.e. engagement and growth of that engagement. We next examine what characteristics of the crowd network structure could lead to this value and engagement over the network connections.

3   The Value of a Crowd

A crowd is defined as a complex network of people which can be represented by a graph consisting of vertices/nodes and edges/links. A vertex represents a person and an edge represents the connectivity between two people.

Let the crowd 'C' be represented by a graph $G = (V, E)$ where $V = \{v_1, v_2, \dots, v_n\}$ and $E = \{e_1, e_2, \dots e_m\}$ where n is the number of nodes and $m \le \frac{n(n-1)}{2}$ is the number of edges. The graph is represented by an adjacency matrix A in which each element $a_{ij} = [0, 1]$ denotes the existence of a link between nodes i and j. If no link exists then $a_{ij} = 0$. Each node is characterized by two probabilities. The first is the probability of a node to receive a message from any other node in the network and the second is the probability that the node sends a message to any other node. These probabilities are represented in a matrix of probability distributions in which each vector is associated with each node. The incoming probability distribution is represented by random variable $X = \{x_{ij} \mid i, j \le n\}$ and the outgoing probability distribution is represented by a random variable $Y = \{y_{ij} \mid i, j \le n\}$. The probabilities are stored in matrices $X_{n \times n}$ and $Y_{n \times n}$ respectively. Thus, the incoming probability of node v_i to receive a message from v_j is given by $x_{ij} \in X$ and the outgoing probability of node v_i to send a message to node v_j is given by $y_{ij} \in Y$.

3.1   Entropy of a Network

The entropy of a network is the measure of the uncertainty in a network. It measures the network's heterogeneity in terms of the diversity in the incoming and outgoing message distributions. The value of a node is calculated in terms of the incoming and outgoing message entropy, which is the entropy of the incoming and outgoing probability distributions associated with a particular node. Note that the entropy here does not try to measure the actual information content of the messages being exchanged but only the randomness in the incoming and outgoing message probability distributions. The cumulative incoming and outgoing message entropies of a network are calculated as the summation of all the individual incoming and outgoing node entropies. Thus,

$$H^{in} = \sum_{i=1}^{n} \sum_{j=1}^{n} a_{ij} * x_{ij} \log\left(\frac{1}{x_{ij}}\right) \tag{1}$$

$$H^{out} = \sum_{i=1}^{n} \sum_{j=1}^{n} a_{ij} * y_{ij} \log\left(\frac{1}{y_{ij}}\right) \tag{2}$$

Finally, the total value of a network (crowd C) is calculated as a weighted measure of the incoming and outgoing entropies of the network. Thus, the value V_C is represented as a function of weighing variable α as in Eq.3. The variable α allows for weighing in the incoming and the outgoing network entropies.

$$V_C(\alpha) = \alpha H^{in} + (1 - \alpha) H^{out} \tag{3}$$

3.2 Entropy Maximization Algorithm

The main objective of the algorithm is to find a network topology which maximizes the entropy of the network (value of the crowd) in terms of the incoming and outgoing messages calculated as above. We start with a randomly generated adjacency matrix with a random number of connections according to a predefined probability p. The algorithm is as follows,

1. Calculate the value of the objective function (Eq. 3) for the network represented by the adjacency matrix A.
2. If the value is greater than the previous value of network entropy then accept the new network. Otherwise randomly flip each bit in the current adjacency matrix with a probability p to obtain a new adjacency matrix representing a new network topology.
3. Repeat step 1 and 2 until the entropy of the newly generated network does not exceed the previous value for at least n^2 times.

The limit of n^2 times to check for the best candidate is selected arbitrarily. Consequently an upper limit can be set for the acceptable number of links in a network as the stopping criteria. In experience, if allowed to proceed then the algorithm keeps on looking for better network topologies in terms of the given incoming and outgoing distributions. The probability distributions are generated uniformly at random. In theory, the maximum entropy is achieved when all the nodes in the network are. However, a fully connected network generated by this process has no meaning from the point of view of the structural regularities which we are trying to analyze. Therefore a threshold is set to stop the optimization process after a certain number of time steps. Other more sophisticated measures of the network entropy can be found here [3, 6, 8, 9].

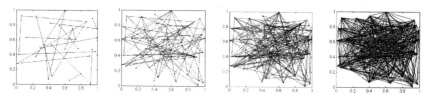

Cumulative entropy →

Fig. 1. Depicts the evolution of the graphical structure through the process of entropy maximization. As the cumulative entropy increases, so do the number of connections.

3.3   Analysis of the Structural Properties

In the above given model, the entropy of a network is a function of the topological features of the network. The algorithm tries to maximize the network entropy such that more nodes communicate with each other [Fig. 1]. We observe the behavior of four parameters with respect to the cumulative entropy of the network. Fig. 2 shows the behavior of the structural parameter versus the network entropy.

Total Number of Active Links (m). As the cumulative entropy increases so do the number of links in a network. This means that the value of a crowd increases with the number of people communicating with each other which seems to be self-evident.

Clustering Coefficient (C). Clustering coefficient is defined as the fraction of nodes which form triangles in a network [7]. The formation of a triangle is indicative of how closely related neighbors of neighbors are inside a network. Again it is self-evident that to the value of a crowd increases with the increase in the clustering in a network. As more nodes begin forming clusters (of basic size 3 in this measure), this results in high local clustering and therefore results in better engagement.

Average Shortest Path Length (l). This parameter measures the average distance in terms of number of hops between any two vertices in a network. It is observed that as the cumulative entropy increases the average shortest path length decreases from over 2 to 1.5. The value of a network increases if more nodes are accessible in shorter paths from each other.

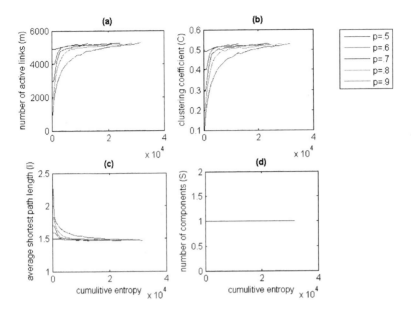

Fig. 2. Shows the behavior of structural parameters plotted against the increasing cumulative entropy of the network. For all the networks the number of nodes n=100 and the weighing parameter is constant at $\alpha = 0.5$. The link density probability (p) is varied from 0.5 to 0.9. For all values of $p < 0.5$, the parameters display more or less linear behavior. (a) The increase in the entropy is brought about by increase in the number of active links (m) in the network (b) shows as the number of links increases so does the clustering coefficient along with the network entropy, (c) shows the decrease in the average shortest path length along with the network entropy (d) shows number of disconnected components very quickly decreases to 1 forming a single giant component with increasing network entropy. In all cases after an initial sudden increase/decrease the values quickly normalize to the theoretical limits.

Number of Components (S). Connected components are those nodes in a network which are accessible from each other directly or through a different path. A network may contain many components in which the nodes within a component are accessible to each other but one component is not connected to another. In this case the nodes in different components are assumed to be at infinite distance from each other. A network with a single giant component is obviously better from the point of view of crowd engagement since all the nodes in the graph are accessible through a path.

4   Discussion

Small World Properties. Small worlds are characterized by low average short path lengths and high clustering. This phenomenon is observed when the average distance between two nodes grows logarithmically with the size of network. Small worlds have been observed in many real world networks. Watts and Strogatz [10] showed that small worlds existed in real world networks like the power grid of United States, collaboration of film actors, and the neural network of the worm C. elegans. Other real networks which show small world effect (reviewed in [10]) are, World Wide Web [1], collaborations in physics and mathematics literature [7], metabolic networks in biology [7], protein interaction networks [7], etc. From the above figures it is clearly seen that the entropy optimization process leads to networks which are small worlds with increasing clustering and decreasing shortest path lengths. Thus, the value of a crowd in term of network engagement is inherently associated with its small world-ness.

Sparsity. As the network gets denser in terms of the number of links the engagement increases. This means that the average number of nodes an individual node connects to increases with the entropy. This may seem self-evident as the number of links increase through the optimization process. The value of the crowd then is a function of the average number of nodes a node connects to (average degree).

Connectedness. The connectedness of a network is usually measured in terms of the number of components. As it is seen from the figures about, during the process of optimization the networks very quickly forms a single giant component consisting of all the nodes from a few disconnected components in the beginning as the entropy increases. This means that to achieve a higher value for a crowd, it is essential that most of the nodes be connected to each other through some path if not directly connected. Despite the sparsity of the network during the initial stages of optimization, it is observed the giant component is formed very quickly in proportion to the sparsity. Thus a connected crowd is a better engaged crowd.

5   Conclusion

It is hypothesized that certain statistical regularities seen in complex networks contribute to the "value" of a crowd as measured in terms of the messages being exchanged over the network. We show that indeed crowds which are better engaged do seem to be small worlds with high clustering and low average shortest path. They also generally seem to be densely connected into a single giant connected component. Understanding more deeply the aspects of crowd network structures that contribute to the overall collective value will enable new uses and applications of crowds and effective leverage of crowd behaviors for problem solving. In the future we would like to explore how certain influentials in the network contribute to the value of the crowd and how network phenomenon like percolation affect the value.

References

1. Barabási, A.L., Albert, R.: Emergence of scaling in random networks. Science 286, 509–512 (1999)
2. Caren, N., Gaby, S.: Occupy Online: Facebook and the Spread of Occupy Wall Street (October 24, 2011), SSRN: http://ssrn.com/abstract=1943168

3. Demetrius, L.: Robustness and network evolution–an entropic principle. Physica E 346 (3-4), 682 (2005)
4. Surowiecki, J.: The Wisdom of Crowds: Why the Many Are Smarter Than the Few and How Collective Wisdom Shapes Business, Economies, Societies and Nations. Brown Publishing (2004)
5. Borge-Holthoefer, J., Rivero, A., García, I.: Structural and Dynamical Patterns on Online Social Networks: The Spanish May 15th Movement as a Case Study. PLOSOne (August 19, 2011)
6. Anand, K., Bianconi, G.: Phys. Rev. E 80, 45102 (2009)
7. Newman, M.E.J.: The structure of scientific collaboration networks. Proceedings of the National Academy of Sciences 98, 404–409 (2001)
8. Cancho, R.F.I., Sole, R.: Optimization in Complex Networks. Lect. Notes Phys., vol. 625, pp. 114–125 (2003)
9. Sole, R.V., Valverde, S.: Information Theory of Complex Networks: On Evolution and Architectural Constraints. Lect. Notes Phys., vol. 650, pp. 189–207 (2004)
10. Watts, D.J., Strogatz, S.H.: Collective dynamics of 'small-world' networks. Nature 393, 440–442 (1998)

Mnemonic Convergence: From Empirical Data to Large-Scale Dynamics

Alin Coman[1,*], Andreas Kolling[1,*], Michael Lewis[1], and William Hirst[2]

[1] University of Pittsburgh, School of Information Sciences, Pittsburgh, PA 15260
[2] The New School for Social Research, New York, New York 10011

Abstract. This study builds on the assumption that large-scale social phenomena emerge out of the interaction between individual cognitive mechanisms and social dynamics. Within this framework, we empirically investigated the propagation of memory effects (retrieval induced forgetting and practice effects) through sequences of social interactions. We found that the influence a public figure has on an individual's memories propagates in conversations between attitudinally similar, but not attitudinally dissimilar interactants, further affecting their subsequent memories [3]. The implementation of this transitivity principle in agent based simulations revealed the impact of community size, number of conversations and network structure on the dynamics of collective memory.

1 Introduction

Community identity and both intergroup hostility and cooperation often rest, at least in part, on the collective memories communities form of their past [9, 14]. We are interested here in using psychologically informed agent-based simulations to understand the dynamics underlying the formation of collective memory. We treat collective memory, following Hirst and Manier [7], as a representation of the past shared across a community that bears meaningfully on the identity of it's members. We want to understand the factors that affect the spread of a memory across a network of individuals and the convergence of the community onto a shared representation of the past. In this paper, we integrate the empirical data emerging out of the distributed cognition literature into an agent-based modeling framework [5, 8]. The assumption underlying this work is that macro-level social phenomena could emerge in predictable ways out of micro-level local dynamics.

In using the recent psychological work on social aspects of memory as our starting point for modeling, we are not only avoiding what Sun [13] has referred to as a shallow conceptualization of cognition in agent-based modeling, we are also extending the psychological work. This work mainly focuses on interactions in small groups, often consisting of just two or three people. The use of agent-based simulations that involve a large number of psychologically plausible agents allows us to overcome this limitation. In this paper, we focus on two psychological processes: practice effects and induced forgetting effects.

---

* The first two authors contributed equally to the paper.

S.J. Yang, A.M. Greenberg, and M. Endsley (Eds.): SBP 2012, LNCS 7227, pp. 256–265, 2012.
© Springer-Verlag Berlin Heidelberg 2012

We envision a community as a network of connected nodes. Each node possesses a memory, which can be transmitted to connected nodes while affecting memories of both the recipient and sender. We chose the two processes because they have received a great deal of attention to date [1, 4]. Future work can incorporate other psychological processes governing social influences on memory.

Although the model could be used to address a large number of substantive issues, we address the following questions here:

(1) Is the mnemonic convergence on a shared representation dependent on community size and number of conversations among individual agents? Inasmuch as groups differ radically in size, it is important to understand the consequences of these different group sizes on the ability of a group to form a collective memory. In recent years, some researchers have argued that, from an evolutionary perspective, core group configurations afford the evolution of specialized functions [2] and have highlighted group size as a pertinent variable [12]. They hypothesized, for instance, that bands or demes, usually with a group size of around 30, are adapted for the shared construction of reality. Following this lead, we expect that smaller networks would provide an appropriate context by which psychological principles of memory might promote swift formation of collective memory, while larger networks would not.

(2) How does the rate of convergence vary across different network parameters (e.g. network density)? There are good reasons to believe that network structure plays a role in the formation of collective memories. In particular, researchers have highlighted how network structure affects information transmission, and through it, collective problem-solving [11].

1.1   The Empirical Work Underlying the Two Processes

We mean by practice effects (PE) the finding that, when one is exposed to an event, and recalls it subsequently, information that is recalled is much more likely to be remembered later on than information that was not recalled [10]. Because this effect is well-established, we focus in this introduction on the other psychological process we plan to incorporate into our model: retrieval-induced forgetting (RIF). RIF has been extensively studied in situations in which a person initially studies material, then selectively practices some of the studied material, and then finally is asked to recall the initially studied material again. RIF occurs when people find it more difficult to remember memories unpracticed in the selective practice phase that were related to the practiced ones than those that were unrelated to the practiced memories [1]. Cuc et al.[4] refers to this as within-individual retrieval-induced forgetting (WI-RIF). Recent work established that the same pattern of induced forgetting occurs for those listening to other selectively practice. For instance, it occurs for listeners who attend to the selective recollections of speakers in a conversation. Such socially shared retrieval-induced forgetting (SS-RIF) is thought to occur because, on many now well-specified occasions, listeners concurrently, albeit covertly, recall selectively along with the speaker. This concurrent retrieval triggers the same processes in listeners as in speakers [4, 6].

In a recent study, Coman & Hirst [3] explored the propagation of practice effects and RIF through short sequences of social interactions, with the goal of understanding the formation of mnemonic convergence in small groups. They looked at how listening to a lecture on the legalization of euthanasia reshapes memories of learned material and whether the influence of the lecture propagates into a conversation and then through the conversation to a final recall test (see Figure 1). In the first phase, participants were exposed to arguments in favor and against the legalization of euthanasia, grouped into categories; each category contained two arguments in favor and two arguments against euthanasia. In a practice phase, similar to a lecture, participants were exposed to half of the arguments (only pro-euthanasia) from half of the categories. We call this phase Person-Pro exposure. After their individual memories were assessed for practice and induced forgetting effects triggered by Person-Pro, two participants were paired and were asked to jointly remember the arguments that they were exposed to initially. Subsequent to the conversation, in a final recall test, participants were asked to remember individually the arguments.

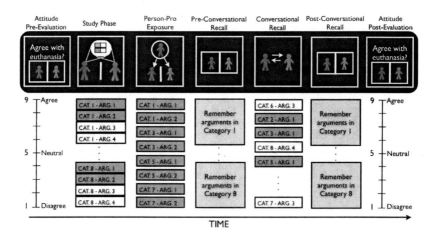

Fig. 1. The different phases of the experimental procedure in [3]. In the Study, Person-Pro Exposure and the Conversational Recall phases, darker shades indicate arguments in favor of the legalization of euthanasia.

Exposure to Person-Pro created three types of items: items mentioned by Person-Pro, which are always in favor of legalization (Rp+), unmentioned items related to the ones mentioned by Person-Pro, which are always against the legalization (Rp-), and items unrelated with the ones mentioned by Person-Pro, which are either in favor or against euthanasia (Nrp-pro and Nrp-anti). A practice effect due to Person-Pro emerges if the recall proportion of Rp+ items is greater than the recall proportion for Nrp-pro items. Similarly, an induced forgetting effect due to Person-Pro emerges if the recall proportion of Rp- items is less than the recall proportion of Nrp-anti items. Coman & Hirst [3] examined whether the practice and induced forgetting effects triggered by the lecture propagated into the conversations between participants, and through the conversation, into the final recall. They found that, when conversations were between like-minded individuals (both participants in the pair in favor of euthanasia or both against euthanasia), Person-Pro shaped what was remembered in the conversation. This influence further propagated into a final recall test, suggesting a lasting influence. Moreover, as practice effects and RIF effects propagate through the conversation to the final recall, they increased in size, thus strengthening with transmission.

The results indicate that mnemonic influences exhibit a principle of transitivity as they propagate through a sequence of social interactions. Importantly, mnemonic convergence around the information provided in the lecture emerged following conversations between like-minded individuals, even in situations where like-minded people disagreed with Person-Pro (see [3] for detailed analyses). Politicians can have a profound influence on what people remember, even when their listeners turn to each other to discuss the issue in an effort to remember the original material as best as they can. Coman & Hirst [3] explored how micro-level processing in a single social interaction shapes the emergent memories in a small sequence of social interactions. Their work suggests that macro-level phenomena specifically, the formation of a collective memory – could emerge in predictable ways out of micro-level dynamics. In what follows we want to extend their work to large networks, using agent-based simulations.

2 Agent Model

We designed an agent-based model that incorporates the sequence of interaction in [3]. In the study phase, we set the level of initial activation for the agents' memory. Agents were then exposed to Person-Pro. Subsequent to the exposure, agents were allowed to interact with each other. As a consequence of an interaction, activation was increased for items recalled during the interaction, and decreased for other items, with a greater decrease for those more closely related to recalled items. This differential decrease captured the induced forgetting effect. In the simulations presented here, activation updates were based on values obtained from empirical data collected by [3]. We first sought to determine whether our model produced results similar to those found by these researchers without introducing additional artifacts. Once satisfactory fitting was achieved, we ran simulations to explore whether mnemonic consensus is influenced by: 1) community size, 2) number of conversations among agents, and 3) the conversational network structure.

2.1 Model Details and Formalism

Let $\mathcal{A} = \{a_1, \ldots, a_n\}$ be a set of n agents. Agents are connected in a communication graph $G^c = (\mathcal{A}, E^c)$ with agents receiving edges whenever they are capable of exchanging messages. An agent a_i has a unique identifier and a memory model, i.e. $a_i = \{i, M_i\}$. The memory model $M_i = (X, f_i, G^m)$ consists of m_{size} items written $X = \{1, \ldots, m_{size}\}$, an associated activation function $f_i : X \to [0, 1]$, and a weighted memory graph $G^m = (X, E^m, w^m)$ that describes the strength of connections between items in X with a weight function $w^m : E^m \to \mathbb{R}$. Notice that only f_i depends on the agent and X, G^m, w^m are identical across all agents. Patterns of activation given by f_i are the distinctive feature of an agent's memory. This simplifies the memory model and forces to relate all changes in memory across agents to changes in f_i. Two effects on activation levels are modeled whenever an agent is exposed to an item $x \in X$, namely a practice effect and an induced forgetting effect. To describe these we interpret $f_i(x)$ as a function that changes with regard to the number of practice and induced forgetting instances. As a shorthand, we write $f_x(n)$ for $f_i(x)(n)$ where n denotes the number of practice or forgetting instances. Upon exposure to a item $x \in X$ its activation f_x changed according to: $f'_x = \sigma(f_x)$, with f'_x denoting first order derivative of f_x and $\sigma : [0, 1] \to [0, 1]$ modulating the increase in activation. We require that $\sigma(a) \le 1 - a$ and $\sigma(a) > 0, \forall a \in [0, 1]$ to ensure that f_x remains within $[0, 1]$. The choice of σ determines whether low activation items and high activation items are affected differently during exposure. To model induced forgetting, all neighbors of x in G^m, written $\bar{x} \in N(x)$ with $[x, x'] = e' \in E^m$, are subject to: $f'_{\bar{x}} = -w^m(e') \cdot \bar{\sigma}(f_{\bar{x}})$, where $\bar{\delta} : [0, 1] \to [0, 1]$, analogous to σ, modulates the decrease in activation. Notice that the magnitude of the induced forgetting effect also depends on the $w^m(e')$.

A conversation is now defined as the repeated recall and exposure of m_{conv} items from X for two agents a_1 and a_2. At first we have $X' = X$ and the probability of an item x being recalled by agent $a_i, i = 1, 2$, is:

$$P(x, a_i) = \frac{f_i(x)}{\sum_{j=1,2} \sum_{x \in X'} (f_j(i))}. \tag{1}$$

After an item is recalled both agents are exposed to it and it is removed from X'. Notice that an agent with a higher overall activation level will get to recall more items during a conversation. In addition to the simple conversational dynamics above we also consider biased conversations in which agents have an attitude bias given by $b_i \in \mathbb{R}^+$. This bias models the differences in attitude towards the recall of items in memory X. More precisely, two agents a_1 and a_2 in a conversation are said to have a different bias if $b_1 \ge \delta$ and $b_2 < \delta$ or vice-versa for some threshold δ. In this case, the biased activation level $f_b(x)$ for agent a_1 for a subset of items $x \in X_1 \subset X$ becomes: $f_b(x) = f(x) + f_{bias}$. Similarly, for agent a_2 and a subset of items $x \in X_2$ the biased activation is given by: $f_b(x) = f(x) - f_{bias}, \forall x \in X_2$. Agent a_1 therefore mentions X_1 more frequently and a_2 mentions X_2 less frequently. This biased activation models changes in recall probabilities observed in experimental data as we shall see later.

3   Application of the Model

In this section we determine the parameters for our model to fit the experiment data from [3]. The memory graph G^m is set to represent the relationships between the items in each category. A total of eight categories with four items each leads to $m_{size} = 32$. Every four consecutive items are in the same category, i.e. $\{4i - 3, 4i - 2, 4i - 1, 4i\} = C_i \subset X, i = 1, \ldots, 8$. Furthermore, every two consecutive items are Pro and Anti items, $X_{pro} = \{1, 2, 5, 6, \ldots, 29, 30\}$ and $X_{anti} = \{3, 4, 7, 8, \ldots, 31, 32\}$. With slight abuse of notation we define G^m as a weighted adjacency matrix:

$$G^m(i,j) = \begin{cases} w_1 + w_2 & \text{if } j \in C_{\lfloor i/4 \rfloor} \text{ and } i, j \in X_{pro} \text{ or } i, j \in X_{anti} \\ w_1 & \text{otherwise if } i, j \in C_{\lfloor i/4 \rfloor} \\ w_2 & \text{otherwise if } i, j \in X_{pro} \text{ or } i, j \in X_{anti} \\ 0 & \text{otherwise} \end{cases}$$

In essence, we interpret X_{pro} and X_{anti} as broader categories in addition to the connections within the smaller categories $C_i, i = 1, \ldots, 8$. To specify σ and $\bar{\sigma}$ we make two additional assumptions, namely that the practice effect decays exponentially and induced forgetting is linear, leading to: $\sigma(x) := \delta_l \cdot (1 - x)$ and $\bar{\sigma}(x) := \delta_f$, which, enforcing the boundary condition $f_x(0) = 0$, is solved by $f_x(n) = -e^{-\delta_l \cdot x} + 1$, with e denoting Euler's constant. Finally, we determine values for $\delta_l, \delta_f, w_1,$ and w_2 from experimental data, specifically the recall of RP+,RP-,NRP-pro, and NRP-anti items. After a study phase, all participants were exposed to Person-Pro discourse. Person-Pro recalls items from a specified list $\{4 \cdot i - 3, 4 \cdot i - 2\}, \forall i = 1, 3, 5, 7$. After this conversation, items of type RP+, RP-, NRP-pro, and NRP-anti, are recalled with probabilities $0.4537, 0.3199, 0.3067,$ and 0.4366, respectively. In our model the recall for NRP-anti is unaffected by Person-Pro and hence the initial activation for all items, assuming uniform study effects, is $p_{init} = 0.4366$. The difference to p_{init} for RP+, RP-, and NRP-pro, allows us to determine the values of our parameters. After setting $w_1 = 1$ the others are completely specified by fitting the above recall probabilities with $w_2 = 0.2781, \delta_l = 0.3357$ and $\delta_f = 0.0584$. Using these parameters, an initial activation level of 0.4366 for all items and Person-Pro's influence, we reproduce the following recall probabilities: $0.4826, 0.3199, 0.3067,$ and 0.4366 for RP+,RP-,NRP-pro, and NRP-anti respectively. When comparing these values with the ones obtained in [3], only RP+ items have a small error of 0.0299 suggesting that the initial recall probabilities are in fact close to uniform and that this assumption is not too strict. After exposure to Person-Pro the conversation phase starts. Two agents are paired into a heterogenous pair with an attitude bias or a homogenous pair without bias. Agents each have an attitude sampled from $[1, 9] \subset \mathbb{N}$ with $\delta := 5$ and are paired with another random agent. Items affected by the bias are RP- items, as X_1, and RP+ items, as X_2. This models the observed increase in recall of RP- items by participants with a pro bias and the reduced recall of RP+ by participants with an anti bias. The conversation length m_{conv} is eleven, the mean conversation length in [3].

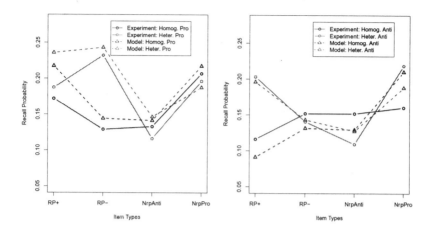

Fig. 2. Model and experimental data for pro (left) and anti (right) participants and agents in the conversation

To determine f_{bias} we tested values between 0.0 and 0.5 in intervals of 0.1 and after obtaining 0.15 as the best fit in terms of the mean squared error (MSE) for values compared in 2 we tested intervals of 0.01 from 0.1 to 0.3, leading to $f_{bias} = 0.23$ with a MSE of 0.0083. The recall probabilities, measured by the frequencies with which pro and anti agents recall items of a certain type, are seen in Fig. 2 and compared to the experimental data on which the MSE was computed. The main effects from [3] are modeled with the exception of reduced RP+ recall for pro participants (left on Fig. 2).

4   Results

To measure the level of convergence of memory in a network we define mnemonic distance between any two agents by $\|f_i - f_j\| = \sum_{x \in X} |f_i(x) - f_j(x)| \in \mathbb{R}^+$. Across the network we consider the mean mnemonic distance of all agents to all other agents. In the following simulations we keep the parameters as determined for the experiments with the exception of the G^c which is now a *small-world* network with rewiring probability (the proportion of links that cut across the network) $p = 0.20$, and $k = 3$ neighbors for every agent. This network configuration corresponds to the configuration of a plausible social network [15]. Every agent had C conversations in the network, choosing a random neighbor in G^c for each, and thus making homogeneous and heterogeneous conversations equally likely. Each model was run for ten trials.

Is mnemonic consensus influenced by network size and by the number of conversations among agents? In order to answer this question, we varied $C = 0, 1, 2, \ldots, 9, 10, 20, \ldots, 60$ and $n = 10, 50, 100$. Results in Fig. 3 indicate that smaller communities arrive at a mnemonic consensus with fewer conversations than larger communities and also converge to a different asymptote. Convergence occurs after an initial increase in mnemonic distance due to the conversational dynamics following exposure to Person-Pro. A possible explanation for this effect is that most memory items are unlikely to be mentioned in consecutive conversations in smaller communities. In larger communities the likelihood for consecutive mentions, at least for some local neighborhoods, is larger and once mentioned sufficiently often these memory items can persist locally and hence lead to a larger mnemonic distance in the network.

Does Person-Pro have a differential impact on mnemonic consensus as a function of network density? To explore this dynamic, we manipulated network density by varying $k = 1, 2, 3, 4, 5$, and we ran simulations for networks with and without the influence of Person-Pro where $n = 100$, $p = 0.10$, and $C = 10$. As seen in Fig. 3, the number of neighbors k influences consensus with denser networks reaching consensus faster. In addition, Person-Pro's discourse decreases the average mnemonic consensus, most likely because it differentially affects homogeneous and heterogeneous pairs. We expect that introducing homophily in the network would result in a reversed pattern [15]. Finally, these results also suggest that mnemonic convergence is possible even for large networks in which agents have relatively few conversations ($C = 10$), but only if the agents are richly connected.

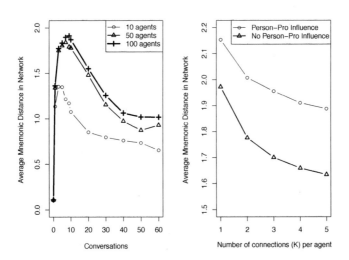

Fig. 3. Consensus dynamics (lower score=larger consensus) as a function (left) of community size and number of conversations per agent; and (right) of network structure (k neighbors) and presence/absence of Person-Pro's influence. Each data point represents the average of 10 runs.

5   General Discussion

In this paper we constructed an agent-based model with psychologically realistic assumptions about memory and conversational dynamics. It builds upon solid empirical work on the propagation of memories in conversations, focusing on two ways memories can be altered: through practice effects and through retrieval-induced forgetting. By extending this empirical work into the domain of agent-based modeling, the current work has underscored how: (1) psychological principles can guide the construction of simulations, and (2) how agent-based modeling can allow for the study of memory propagation and convergence across large network structures. Even at this early stage of model development, the results bear on critical issues concerning the formation of collective memory that should be of interest to both social scientists and policy makers. Specifically, the results supply some insight into why researchers might insist that bands or demes are well-suited for the construction of social reality [15]. It also underscores the fact that certain network structures can facilitate the construction of shared reality even in larger communities. In particular, the results indicate that mnemonic convergence depends on group size and number of conversational exchanges, but also hint at the fact that connectedness of agents in a network, as well as the nature of these connections are critical factors to be considered. With technological advances in human subjects experimentation (e.g. Mechanical Turk), we're planning on validating our results by using what Mason and Watts [11] call virtual experiments; that is, experiments in which groups of up to 20 participants interact in experimenter controlled tasks. Finally, although more detailed models charting the interaction between network structure, network size, memory, and conversational exchange still need to be formulated, the research presented here provides a solid framework for undertaking this endeavor.

References

[1] Anderson, M.C., Bjork, R.A., Bjork, E.L.: Retrieval induced forgetting: Evidence for recall specific mechanism. Psychonomic Bulletin & Review 7, 522–530 (2000)
[2] Caporael, L.R.: The evolution of truly social cognition: The core configuration model. Personality and Social Psychology Review 1(4), 276–298 (1997)
[3] Coman, A., Hirst, W.: Cognition through social networks: the propagation of practice and induced forgetting effects. Journal of Experimental Psychology: General (in press, 2012)
[4] Cuc, A., Koppel, J., Hirst, W.: Silence is not golden: A case for socially-shared retrieval-induced forgetting. Psychological Science 18, 727–733 (2007)
[5] Epstein, J.: Generative social science: Studies in agent-based computational modeling. Princeton University Press, Princeton (2006)
[6] Hirst, W., Echterhoff, G.: Remembering in conversations: the social sharing and reshaping of memories. Annual Review of Psychology (in Press)
[7] Hirst, W., Manier, D.: Towards a psychology of collective memory. Memory 16, 183–200 (2008)
[8] Hutchins, E.: Cognition in the Wild. MIT Press, Cambridge (1995)
[9] Kaplan, R.D.: Balkan Ghosts: A Journey Through History. Picador (1993)

[10] Karpicke, J.D., Roediger, H.L.: Repeated retrieval during learning is the key to long-term retention. Journal of Memory and Language 57, 151–162 (2007)

[11] Mason, W., Watts, D.: Collective problem solving in networks. Proceedings of the National Academy of Sciences (in press)

[12] Power, C., Dunbar, R., Knight, C. (eds.): The Evolution of Culture. Edinburgh University Press (1999)

[13] Sun, R.: Cognition and Multi-Agent Interaction: From Cognitive Modeling to Social Simulation. Cambridge University Press, New York (2006)

[14] Zerubavel, Y.: Recovered Roots: Collective memory and the Making of Israeli National Tradition. University of Chicago Press (1997)

[15] Watts, D.: Six degrees: The science of a connected age. Norton, New York (2004)

Sizing Strategies in Scarce Environments

Michael D. Mitchell[1], Walter E. Beyeler[1], Robert E. Glass[1],
Matthew Antognoli[2], and Thomas W. Moore[1]

[1] Complex Adaptive System of Systems (CASoS) Engineering
Sandia National Laboratories, Albuquerque, New Mexico, USA
{micmitc,webeyel,rjglass,tmoore}@sandia.gov
[2] School of Engineering, Unicersity of New Mexico
Albuquerque, New Mexico, USA
mantogn@sandia.gov

Abstract. Competition is fierce and often the first to act has an advantage,
especially in environments where there are excess resources. However,
expanding quickly to absorb excess resources creates requirements that might
be unmet in future conditions of scarcity. Different patterns of scarcity call for
different strategies. We define a model of interacting specialists (entities) to
analyze which sizing strategies are most successful in environments subjected
to frequent periods of scarcity. We require entities to compete for a common
resource whose scarcity changes periodically, then study the viability of entities
following three different strategies through scarcity episodes of varying
duration and intensity. The three sizing strategies are: aggressive, moderate, and
conservative. Aggressive strategies are most effective when the episodes of
scarcity are shorter and moderate; conversely, conservative strategies are most
effective in cases of longer or more severe scarcity.

Keywords: Scarcity, Interacting Specialists, Agent-Based Model, Sizing
Strategies, Complex Adaptive Systems, First-Mover Advantage.

1   Introduction

We conducted a study to determine the selection of sizing strategies in environments
subjected to dissimilar patterns of scarcity. The study utilized a configuration of the
Exchange Model developed at Sandia National Laboratories to investigate complex
adaptive systems (CAS) [Beyeler et. al. 2011]. A system is defined as a set of
interacting entities that together serve a common objective; an adaptive system is one
in which behavior changes over time due to interactions or environmental conditions;
in a complex adaptive system interactions among elements additionally produce
emergent, non-linear behavior. Complex adaptive systems (CAS) may share many of
the same underlying processes and characteristics. The agent-based Exchange model
provides a framework in which a collection of interacting specialists, or entities,
which produce and consume resources, may be described. This model can be
configured to represent various biological or non-biological systems. The
environment determines the availability of resources that entities require for survival.

S.J. Yang, A.M. Greenberg, and M. Endsley (Eds.): SBP 2012, LNCS 7227, pp. 266–273, 2012.

Entities use environmental signals to determine the amount of resources to consume and produce. First movers, entities that consume/produce aggressively, are strategically the most successful in environments with excess resources, as follows the preemption of assets advantage identified by Lieberman and Montgomery [Lieberman, Montgomery 1988]. We study environments in which resource scarcity increases in both frequency and intensity to determine if there is a point at which an aggressive strategy is no longer the most advantageous strategy.

In this paper we focus on how entities utilize rate of growth and sizing to adapt to diverse, variably scarce environments (periods of scarce resource availability followed by periods of recovery). The frequency and duration of the episodes of scarcity are differentiated to investigate how entities with different sizing strategies select their adaptation for the various environments. The growth strategies are analogous to the risk tolerance of a firm. Our goal in this research is to determine which growth strategies are most effective for different levels and durations of scarcity.

2   Model Formation

For the purposes of this study, the Exchange model is configured as a system containing six entity types which produce four resources required for survival. In the Four-By-Six model configuration (detailed in Table 1), the four resources are labeled A, B, C and D. A fifth resource required by the entities, M, is not produced by any of the six entity types; the M resource is discussed in Section 2.1. This configuration sets up a symbiotic relationship among producers and consumers within the system. Each entity type consumes two resources and produces two other resources. Each resource is produced and consumed by more than one entity making the system robust while increasing competition among the entity types. An entity's health is expressed by means of a homeostatic process involving the consumption of resources.

Table 1. Four By Six Model Configuration

Entity Type	Produced Resources	Consumed Resources
CD Maker	C,D	A,B,M
BD Maker	B,D	A,C,M
BC Maker	B,C	A,D,M
AD Maker	A,D	B,C,M
AC Maker	A,C	B,D,M
AB Maker	A,B	C,D,M

2.1   Scarcity

In order to force competition among entities, a global resource M is added to the configuration. M represents a resource needed by all entities but which the entities cannot produce. Instead M is produced by an outside process at a fixed rate. Examples of such a global resource are entrepreneurship, labor, and capital investment. Initially the system produces an amount of M required by the entities. Episodes of scarcity are introduced by reducing the rate of M production for a period of time.

M is a resource that entities must qualify to receive. We create competition for the scarce resource M, which is distributed based on a measure of merit that reflects the entity's performance. The resource M is made available and exchanged in a market. Merit is determined by an entity's ability to purchase M in an open market. Entities bid for the amount of M they want to consume. The market sets the price of M based on the availability and on entity bids for M. We use an entity's bid price for M to determine the merit of entities to receive M. Entities with less M in their stores, and with more money, will bid a higher price.

2.2   Growth

Entities have the capacity to change their size over time as a function of their environment. Size is a mechanism employed by an entity to exploit or adapt to the availability of resources in an environment. Entities use size to adjust their production and consumption rates. The size of an entity is determined by the following state equation.

$$\frac{dS}{dt} = S * \frac{\left(\frac{h(t)}{h_0}\right)}{tGrowth} \tag{1}$$

Where S is the current size of an entity, $\frac{h(t)}{h_0}$ is a normalized value of health, and tGrowth is a time constant which governs the rate at which an entity can change its size. A smaller tGrowth value indicates a more aggressive sizing strategy; conversely a larger tGrowth value indicates a more conservative sizing strategy.

Initially, a size of 1 is assigned to entities. Entities can increase or decrease their size based on current health relative to nominal health. Favorable conditions cause an entity to exploit its environment and increase its size. An entity's size increases when the normalized value of health is greater than 1 causing an increase in production and consumption rates which systematically push the normalized level of health down to 1. An entity can use the same mechanism to decrease its size when the normalized level of health is less than 1. An entity strives to find equilibrium of health $\frac{h(t)}{h_0} = 1$ via size manipulations limited by tGrowth.

3   Model Configuration

The Four-By-Six configuration is suitable for three sizing strategies due to a natural pairing of entities which collectively produce all four consumed resources. Each of the symbiotic entity pairs is assigned a sizing strategy. These parings allow some entities to survive in the event that other entities pairs fail. This allows us to compare the success or failure of strategies in the model. The success or failure of entities relative to one another is reported as an aggregated statistic for each strategy. Table 2 lists the entities and their strategies. Subsequently, we will discuss strategies rather than individual pairings.

Table 2. Entity Types Sizing Strategy

Entity Type	Sizing Strategy	tGrowth
CD Maker	Aggressive	1.E+04
BD Maker	Moderate	5.E+04
BC Maker	Conservative	5.E+05
AD Maker	Conservative	5.E+05
AC Maker	Moderate	5.E+04
AB Maker	Aggressive	1.E+04

We ran each simulation for total time of 5.E+05. Five entities were realized for each entity type. The three tGrowth constants represent the three sizing strategies we are modeling. The tGrowth parameter specifies how quickly an entity can change its size by a factor of 2. The aggressive strategy has a tGrowth of 1.E+04 allowing it to change its size 500 times during the simulation. The moderate strategy has a tGrowth of 5.E+04 allowing it to change size 50 times during the simulation. Finally, the conservative strategy has a tGrowth of 5.E+05 allowing it to change size 1 time during the simulation.

We configured nine simulations to study different environments of scarcity (detailed in Table 3). Frequency describes the percent of the time entities are subjected to episodes of scarcity. Intensity is the percent reduction in the availability of the global resource M during episodes of scarcity. The model is stochastic; each simulation was run 10 times to capture the difference in outcome associated with random behavior in the model.

Table 3. Environments Configured

Simulation ID	Frequency	Intensity
1	50%	10%
2	50%	20%
3	50%	30%
4	75%	10%
5	75%	20%
6	75%	30%
7	90%	10%
8	90%	20%
9	90%	30%

3.1   Measuring Market Share

The primary output of the model is a time series of state variables describing the entities $\{e\}$ in the model [Beyeler, et. al. 2011]. There are state variables describing the health and size of entities. For comparison, we define a single value ϕ_j to measure the market share for each strategy j for a simulation. The following equation describes how we derive the valuation of market share for a strategy.

$$\phi_j \equiv \frac{\sum\limits_{e \in T_j} \int\limits_{t_f - p}^{t_f} h_{t,e} s_{t,e} dt}{\sum\limits_{e} \int\limits_{t_f - p}^{t_f} h_{t,e} s_{t,e} dt} \tag{2}$$

Where h_t is an internal representation of health, s_t is an internal representation of size, and T_j is the set of entities of type j. Multiplying health and size allows us to compare entities which have different growth strategies. For instance, an entity using an aggressive strategy may acquire a size of 5 and health of 1 whereas an entity using a conservative strategy may acquire a size of 1 and health of 5. In this case, the two entities would be considered equal. Summing $h_t s_t$ across entities using the same strategy allows for comparisons of strategies which have different population compositions due to size. Simulation results are reported as the average market share for each strategy in a simulation over the final environmental cycle, running from time $t_f - p$ to time t_f.

4   Model Results

First we discuss general observations of the mechanics for when entities experience alternating periods of scarcity and recovery. Second, we present the detailed results of two simulations demonstrating how the strategies selected differently for each environment.

Model results are stated in terms of the average market share gained by each strategy (detailed in Table 4). The strategy with the highest percent of market share for each simulation is highlighted. Intensity is the percent reduction in the availability of resource M in periods of scarcity. Frequency is the percent of simulation time subject to periods of scarcity.

Table 4. Aggregated Simulation Results

Intensity	10%			20%			30%		
Frequency	A	M	C	A	M	C	A	M	C
50%	59%	25%	15%	52%	35%	11%	31%	25%	23%
75%	58%	24%	17%	63%	21%	15%	26%	19%	34%
90%	31%	30%	38%	33%	33%	32%	28%	40%	31%

4.1   Entities' Response to Scarcity and Recovery

Each simulation begins with entities reaching the same equilibrium for health and size. Sufficient resources exist such that no strategy has an advantage. When the first period of scarcity is initiated, entities are not able to acquire the quantity of resources they need and their health begins to decline. Entities using an aggressive strategy respond by rapidly decreasing their size in an effort to decrease their consumption rate to a point at which health is no longer declining. Entities utilizing the moderate strategy reduce their size at a much slower rate than the aggressive entities. Entities in conservative strategy group are not able to shrink their size very much and may need to survive with unmet consumption needs. Entities' reducing their size to minimize their exposure to the resource scarcity cause consumption rates to decline, consequently lessening the intensity of scarcity.

A period of recovery begins with the restoration of the amount of the global resource to pre-scarcity conditions. Nominally this would result in entities being able to converge on the same initial equilibrium for health and size. However, entities that reduced their consumption rates via a reduction in size created a reduction in demand for the global resource at the beginning of the recovery period, leading to an environment in which there is a temporary excess of the global resource. At this point, entities using the aggressive strategy begin to absorb excess resources and rapidly expand in size. Concurrently, those using the moderate strategy are growing, but at a slower rate, and entities using the conservative strategy, although not growing, are trying to consume slightly more than normal resources to restore the health lost during the episode of scarcity.

4.2   An Environment 50% Scarce

The first simulation we present is configured to have 10% fewer global resources over 50% of the time (see Fig. 1).In this simulation, users of the aggressive strategy

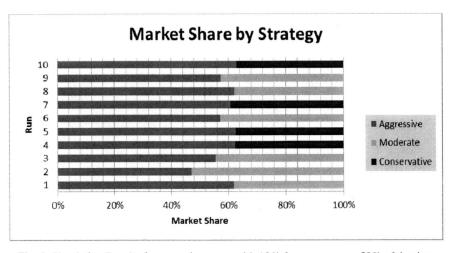

Fig. 1. Simulation Results for an environment with 10% fewer resources 50% of the time

dominate their competition, capturing an average of 59% of the market share. In each simulation, aggressive strategy entities compete against only one other strategy, due to the fact that their growth during a period of recovery forced users of one competing strategy out of the market. The market share gained by users of the aggressive strategy is gained from the failure of a competing strategy.

Users of the aggressive strategy are able to take over market share by expanding during the recovery period causing the competing strategy to fail. The failure of the competing strategy occurs before the next period of scarcity begins. Once a competing strategy fails, there is more global resource availability. The increased availability of the global resource lessens the impacts of future periods of scarcity.

4.3   An Environment 90% Scarce

The second simulation we present is configured to have 10% reduction in the availability of the global resource over 90% of the time (see Fig 2) In this environment, the conservative strategy is the most successful, holding an average of 38% market share. This strategy is successful for users due to its slow growth. Conservative-strategy entities also benefit from the fast growth of those using the aggressive strategy due to two factors.

- The difference in how strategies respond to a decline in resources. A decline in resources triggers a decline in health because the resources necessary for consumption are limited. Strategies resulting in a faster growth rate trade a reduction in health for a reduction in size, thereby decreasing their consumption rates. Strategies resulting in a slower growth rate experience a more rapid rate of health decline due to not being able to reduce their consumption rate. A steep decline in health triggers a desire in entities to consume more resources in an effort to raise their health level. Scarcity causes aggressive entities to desire fewer resources and conservative entities to desire more resources.
- Users of the aggressive strategy free up some of the scarce resource by rapid size reduction. In the 50% scarce environment, entities using the aggressive strategy were able to dominate by growing quickly during the periods of recovery and forcing one of the competing strategies to fail. In the 90% scarce environment, the period of recovery is not long enough for aggressive entities to grow and thus they remain smaller. This frees up resources needed by the users of the conservative strategy causing them to be more successful.

The margin by which the conservative strategy succeeds is small; users control just over one third share of the total market. The captured market share was won from entities using the aggressive and moderate strategies, reducing their size and thus somewhat lessening the scarcity of the global resource. In order to gain a larger percentage of market share, a competing strategy would have to fail. The conservative strategy does not enable users to gain enough market share to force entities using a competing strategy to fail.

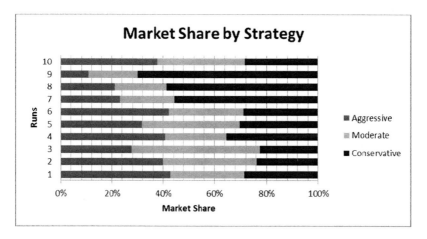

Fig. 2. Simulation Results for an environment with 10% fewer resources 90% of the time

5   Conclusion

The results of the study indicate that the strategy to best select during periods of scarcity depends more on the duration of the scarcity or, more aptly, the duration of the recovery than the intensity of the scarcity. Longer periods of recovery offer entities using an aggressive strategy the opportunity to benefit from a first-mover advantage, the preemption of assets [Lieberman, Montgomery 1988]. By absorbing excess resources, users of the aggressive strategy can force entities employing a competing strategy out of the market. Once a competing strategy has failed (causing the death of those entities using it), there is more of the global resource available, enabling users of the aggressive strategy to survive future periods of scarcity more easily. Longer periods of scarcity are more advantageous for entities using a conservative strategy. The reduction in size of the aggressive entities, and subsequent decline in consumption rates, benefits entities using the conservative strategy by decreasing demand for the scarce resource. This allows users of the conservative strategy to survive the period of scarcity with a larger portion of the market share than it would otherwise have.

References

1. Beyeler, W.E., Glass, R.J., Finley, P.D., Brown, T.J., Norton, M.D., Bauer, M., Mitchell, M., Hobbs, J.A.: Modeling Systems of Interacting Specialists. In: 8th International Conference on Complex Systems (June 2011)
2. Lieberman, M.B., Montgomery, D.B.: First-Mover Advantages. Strategic Management Journal, Special Issue: Strategy Content Research 9, 41–58 (Summer 1988)

Applying System Dynamics to Military Operations

Corey Lofdahl

Charles River Analytics, 625 Mount Auburn Street,
Cambridge, Massachusetts 02138, USA
clofdahl@cra.com

Abstract. This paper addresses the application of computer-based tools to aid senior decision makers in a modern military environment. It is organized in three sections. First, the key features of successful computer-based solutions are presented. Past database centric efforts are reviewed and found insufficient. Second, simulation generally and System Dynamics (SD) specifically are discussed, developed, and contrasted with more database centric solutions. Third, an example SD model is developed that addresses a classic COIN problem—counterintuitive insurgent subtraction—which explains why insurgent numbers can actually increase after military forces engage them. This paper concludes by discussing several ways in which computational models can be made more relevant to military operations.

Keywords: Counterinsurgency, COIN, decision support, system dynamics, simulation, operational modeling, insurgent subtraction, scenario analysis.

1 Introduction

"The commander must work in a medium which his eyes cannot see, which his best deductive powers cannot always fathom, and with which, because of constant changes, he can rarely become familiar." —**Carl von Clausewitz**

Warfare has always presented a challenging decision environment. Modern warfare in the form of counterinsurgency (COIN) has proven especially challenging as it blends both governance and development with more traditional security concerns. High levels of complexity result from the number of moving parts because these security, governance, and development lines of effort do not fall into separate, well defined lanes. Instead they combine and blend in ways that confound decision makers. Senior military decision makers and commanders are increasingly confronted with complex decisions that encompass a range of interconnected and dynamic social system features. General and flag level officers especially find themselves choreographing and synchronizing the full range of Diplomatic, Information, Military, and Economic (DIME) elements of national power even though they may only be expert in one or two. Recognizing that the modern military decision-making context is "complex" however is insufficient—some way must be found to analyze and make sense of this complexity. It is natural to use computers and computer-based tools to supplement the

S.J. Yang, A.M. Greenberg, and M. Endsley (Eds.): SBP 2012, LNCS 7227, pp. 274–281, 2012.

commander's cognition. The tool should provide clarity and grant the commander a sense of familiarity with this difficult and demanding operational environment.

Using computer-based tools to provide decision-support to senior decision makers remains an ongoing opportunity because getting such tools to fit and contribute within an operational environment is a hard problem. This paper addresses the application of computer-based tools to the problem of aiding senior decision makers in a modern military environment, and it does so in three sections. First, the key features of a successful computer-based solution are presented. Initial database centric efforts are reviewed and found lacking. Second, simulation generally and System Dynamics (SD) specifically are discussed, developed, and contrasted with more database centric solutions. Third, an example SD model is developed that addresses a classic COIN problem—counterintuitive insurgent subtraction—which explains why when military forces engage insurgents they can end up with more of them. The conclusion discusses several ways in which computational models can be made more relevant to military operations.

2   Complexity, Computers, and Decision Support

The insurgencies that modern military forces seek to counter can be thought of as a complex social system [1], which provides several strategic and operational benefits. First, in a long war—and counter insurgencies do tend to be extended with operations often lasting more than a decade—success requires campaign continuity. Treating an insurgency as a complex social system helps to support, achieve, and maintain the long-term perspective necessary for COIN success. Second, treating an insurgency like a complex social system requires developing metrics that are tracked over time to help provide that long-term perspective and measure progress against key objectives. The metrics need to be specified by identifying clearly what is being measured and why. Metrics also need to be quantified by giving values, units, and expected ranges. Third, metrics need to be related to each other and connected to form a system. In this manner, the interactions among security, governance, and development lines of effort can be better identified, monitored, and understood.

Modern computers provide a natural way to try to handle system complexity given their prodigious memory and speed. While much experimentation has been undertaken to address military and COIN complexity with computation, successfully applying the resulting intuitions and insights in an actual operational theatre remains an open problem. Part of the problem is due to the computer technologies employed, a combination of databases, computers, and networks. For example, the Palantir system connects multiple databases together and allows analysts to search and display the extracted information [2]. While the tool provides greater access to data, it does not help to select what data is most critical to a particular problem. Furthermore, Palantir does not "boil down" data into products that are more understandable and thus usable by senior decision makers. While the analyst can decide what data is most pertinent and can condense information for senior decision makers, it is the analyst rather than the computer that is doing the hard work. The computer and database are instead providing easier access to ever more data.

The DARPA Nexus 7 project suffers from similar shortcomings in that considerable effort was put into gaining access to and connecting multiple databases, but the work did not result in the envisioned analytic insights [3]. Much of the problem was due to the team's inability to establish insightful connections among the data. This situation has been encountered frequently when policy makers reach out to engineers and have them develop computer-based products to deal with the complexity that confounds their decision-making. However, reaching out to engineers for help with decision support problems places them into the position of analysts, and while engineers know much about building and programming computers, they know comparatively little about analysis. The gap usually comes in having the appropriate mental models that allow them to weigh, fit, and prioritize the data in their computers for products that inform decision making. The Nexus 7 engineers were trained in the latest pattern matching technology, but it was not clear to them what patterns needed to be matched. Models help construct and identify such patterns and their absence can limit success. So computers play an important role, but by themselves they cannot support decision making.

The military's experience with Effects Based Operations (EBO) tells a similar story. Early works articulated the reasons why a new generation of tools was necessary to help decision makers [4,5]. Proponents argued that a new breed of tool was necessary to understand direct, indirect, and cascading effects. But while the need was clear, the way forward to address that need was less so. Once again the military reached out to engineers to help them build tools for decision makers, and the engineers again provided systems built around computers, databases, and networks. The systems were operated by System of System Analysts (SoSA) who filled the databases with data that they thought would be useful to senior decision makers. However, senior decision makers did not find the data helpful. Moreover, the databases did not address the fundamental problem of EBO, to provide insight into direct, indirect, and cascading effects of complex social systems. Databases store data. They do not "go" or progress; they do not generate scenarios or address causality, which is necessary to address the effects questions. EBO's overreliance on databases caused much frustration, and eventually the effort was discontinued [6].

These examples reveal that while the need to address complex military problems to support senior decision makers is clear, the way to provide that support is less so. Moreover, reaching out to engineers for computer-based tools to support decision making has been shown to result in database-centric solutions that senior decision makers, their intended beneficiaries, have found inadequate. Political scientist and Presidential Special Advisor Richard Beal studied these problems and reached the following conclusions [7]: 1) too much time and effort is spent on data collection, 2) not enough attention is given to synthesis at the macro level for senior level decision making, and 3) there is not enough systems thinking by virtue of education and training. So Beal specifically addresses data collection and finds it insufficient. Instead, those building tools to support senior decision makers need to focus on, "synthesis at the macro level." Information synthesis is a hard computer science problem, which explains why workable solutions have yet to be found. Information synthesis will be addressed in more detail in the following section.

3   Modeling, Simulation, and System Dynamics

One of the key gaps identified in the previous section concerned models—constellations of ideas, variables, and relationship—that help people understand a complex world. Models help order and narrow possible variables and relationships, thereby making variable selection and testing more intentional. Physicist Richard Feynman, in a 1964 lecture at Cornell, describes science as a combination of the following three activities [8]: 1) make a guess, 2) create an experiment to test the guess, and 3) if the guess does not agree with the experiment, then it is wrong. Guesses are always based on some model of the way the world works: the better the model, the more accurate the guess. Experimentation and data collection without an accurate model becomes mere trial and error that ultimately becomes unproductive because too much effort is expended for too little benefit.

Theories and models are inherently used to make predictions about the future, what Feynman calls a "guess." Simulations can be used in a complementary fashion to generate and analyze scenarios that tease out over time. In fact simulations are a type of model and can be used to develop theories through a process of specification and quantification [9]. Instead of providing a prose-based description, simulation requires providing a logical or mathematical description. This entails defining a suite of connected metrics that form a system. Traditional mathematical proofs are limited in the scale and scope of variables that can be considered and tend towards closed-form solutions. With simulation supported by powerful modern computers, a much larger set of variables can be combined to generate a much richer set of scenarios.

System Dynamics (SD) is used here as a tangible and mature methodology to create simulations [10, 11], which provides several benefits. First, SD provides a way to develop, fit, filter, and organize metrics. Too often metrics are developed to acquire data that are unconnected to other relevant metrics. The SD methodology provides a well developed way to create systems of causally connected metrics. Second, SD provides a working definition of complexity, a combination of nonlinear, feedback, and stock-flow (i.e., integration) causal relationships. Each of these confuses human cognition, and together they overwhelm it. SD simulation provides a way for analysts and decision makers to handle this complexity, understand it, and become familiar with it. Third, while SD simulation can be used to develop theory [12], it can also be used to synthesize data and show how multiple data sources together impact key output metrics. To the extent that SD can synthesize data, it begins to address the key concerns articulated by Beal. The following section works though an example that clarifies these descriptions.

4   A Complex Social Systems Example

This paper has discussed both social systems and complexity. A complex social system example is developed in this section to provide specificity. The example is taken from MG Michael Flynn who talked about the calculus of counter insurgency, in which direct action against insurgents can counter-intuitively result in even more insurgents [13]. This example provides an example of how the complexity of social systems can result in counterintuitive behavior. The crux of the problem is this: if you have one hundred terrorists and neutralize ten of them, then how can you end up with even more?

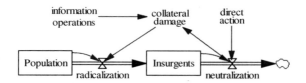

Fig. 1. Insurgent Subtraction Model

Figure 1 shows the "Insurgent Subtraction Model," which consists of seven variables. Starting at the beginning, military units undertake "direct_action," which results in "neutralization" of "Insurgents." This reduces the square stock of Insurgents, which starts at 100. However there is a secondary consequence to direct action, "collateral_damage." The application of military force can result in noncombatants getting hurt. When direct action events occur, collateral damage—whether real or imagined—can be exploited through "information_operations." These operations can take the form of television, radio, internet, leaflets, or even word-of-mouth, but they can achieve "radicalization" of a certain percentage of the "Population," who in turn become "Insurgents."

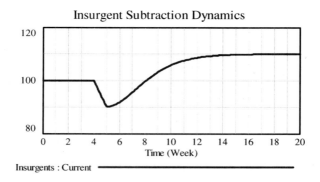

Fig. 2. Insurgent Subtraction Dynamics

Figure 2 provides an Insurgent Subtraction Dynamics scenario that demonstrates several simulation features. First, the scenario takes place over time, here 20 weeks. Second, it charts the dynamics for a single variable in the simulation, "Insurgents." The graph shows the number of insurgents over those 20 weeks, which starts at 100, dips to 90 as a result of direct action, and then grows back to 110 as a result of collateral damage, information operations, and radicalization. Each of the seven variables have values that can be measured and displayed, which means that simulated social system can be changed, observed, and measured much more easily than real social systems. Finally, the variables are causally connected, which allows for the system's direct, indirect, and cascading consequences to be analyzed.

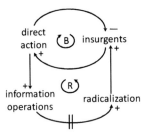

Fig. 3. Insurgent Subtraction Causal Loop Diagram

Figure 3 shows a Causal Loop Diagram of the Fig 1 model. The arrows indicate causality and begin with the cause and point to the effect, so cause and effect are defined. The arrows each have polarity, with a positive sign indicating change in the same direction. So the greater the cause, the greater the effect. Arrows with a negative sign indicate change in the opposite direction. So the greater the cause, the smaller the effect. These causal arrows combine to form systems of circular causality or *feedback*, which also takes two forms. Feedbacks with an odd number of negative connections result in a *Balancing* loop, which tends to maintain system equilibrium. Feedbacks with zero or an even number negative connections result in a *Reinforcing* loop, which causes growth until a balancing loop intervenes to re-establish system equilibrium. Figure 3 features two feedbacks: a balancing loop between `direct_action` and `insurgents`. The causal arrows represent that the more direct action there is, then the fewer insurgents. The positive arrow in that loop indicates the greater the number of insurgents, the more direct action is required to neutralize them. The positive loop explains the counterintuitive behavior of this simple social system. Direct action indeed decreases insurgents, but it also provides the subject matter for `information_operations` that leads to `radicalization` and more insurgents. Figure 3 also features two parallel lines on the positive connection between `information_operations` and `radicalization`, which indicates a time delay. The delay represents the real-world process of a person hearing a media message, becoming radicalized, and then taking the steps to become an insurgent.

It is not enough to simply articulate the problem with a simulation. Some policy recommendations should result. For this model, scenario analysis shows that if information operations are reduced, then so are the number of generated insurgents. Conversely, if information operations are more effective, then more insurgents are generated. Thus, Figure 4 shows that information engagement is as important as the security, governance, and development lines of effort. Looking to Figure 1, information operations works in conjunction with the reality of collateral damage, so reducing collateral damage also provides a way to reduce insurgent creation. Finally,

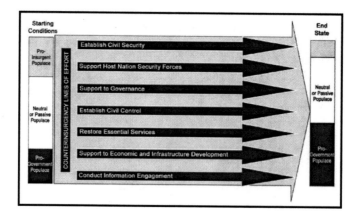

Fig. 4. COIN Lines of Effort [14]

the model can be extended because insurgent activities are subject to the same information operations. Too much collateral damage from insurgent operations can result in reduced popular support, which diminishes the insurgents' capabilities.

5   Conclusion

This paper draws from the complex social system and COIN literatures and discusses ways that computer technology aids senior decision makers. Systems focused on database technology have not helped decision makers because they push information rather than synthesize it. Moreover, they don't tease out scenarios over time in a way that addresses the direct, indirect, and cascading consequences of complex social systems. Simulation generally and SD specifically is offered as an analytic alternative because SD naturally synthesizes information, represents complex causal connections, and calculates their behavior over time. These claims are developed in a simple simulation that depicts the calculus of insurgent subtraction, which shows how neutralizing insurgents can counter-intuitively result in more insurgents.

SD is not offered as a panacea, but as a way to operationalize certain characteristics of computer-based tools that must be present to support decision making in an operational rather than a research environment. In so doing, gaps can be identified that then become requirements for the next generation of tools. Three are discussed here. First, better visualizations are needed for senior military decision makers. The system depictions of Figures 1, 2, and 3 are comfortable and meaningful for an engineer but they are opaque and off-putting to a flag or general officer. The behavior over time graphs can yield good results, though senior decision makers would rather see real rather than simulated data. Second, decision makers want to see disaggregated analyses that depict regions. The insurgent subtraction model is aggregated in that only a single region is addressed. However, multiple variables can be defined to represent different regions. This data can be used to drive a map-based display, which would be more comfortable for a senior decision maker. Third, decision support systems in operational commands must be able to support non-expert users as opposed to research communities that feature a small group of skilled developers and users. Making complex social systems models and simulations more modifiable, maintainable, and interpretable by non-expert users continues to present a significant research challenge [15].

References

1. Kilcullen, D.: Counterinsurgency, ch. 6. Oxford, New York (2010)
2. Sankar, S.: Intelligence Infrastructure. PowerPoint presentation downloaded from Palantir Technologies (September 13, 2011),
   `http://www.palantir.com/infrastructure`
3. Shachtman, N.: Inside DARPA's Secret Afghan Spy Machine. Wired (July 21, 2011)
4. Davis, P.K.: Effects-Based Operations (EBO): A grand challenge for the analytical community. RAND, Santa Monica (2002)
5. USJFCOM. Operational Net Assessment (ONA): Version 2.0. Concept Paper for the United States Joint Forces Command (May 2004)
6. Mattis, J.: Assessment of Effects Based Operations. US Joint Forces Command Memo (August 14, 2008)
7. Beal, R.: Decision Making, Crisis Management, Information, and Technology. Program on Information, Center for Information Policy Research. Harvard University, Cambridge (1985)
8. NOVA. The Best Mind since Einstein. Television documentary on Richard Feynman by PBS (1993)
9. Davis, J.P., Eisenhardt, K., Bingham, C.B.: Developing Theory through Simulation Methods. Academy of Management Review 32(2), 480–499 (2007)
10. Forrester, J.W.: Industrial Dynamics. Productivity Press, Cambridge (1961)
11. Sterman, J.D.: Business Dynamics: Systems thinking and modeling for a complex world. McGraw-Hill, Boston (2000)
12. Davis, Eisenhardt, Bingham. Ibid (2007)
13. Flynn, M.T.: Operational Need. In: Keynote Talk given at HSCB Focus 2011: Integrating Social Science Theory and Analytic Methods for Operational use Conference, Chantilly, VA, February 8 (2011)
14. US Army. Tactics in Counterinsurgency: Field Manual (FM) 3-24.2, pp. 3–8. Headquarters, Department of the Army, Washington, DC (2009)
15. Davis. Ibid (2002)

Lessons Learned in Using Social Media for Disaster Relief - ASU Crisis Response Game

Mohammad-Ali Abbasi[1], Shamanth Kumar[1],
Jose Augusto Andrade Filho[2], and Huan Liu[1]

[1] Computer Science and Engineering, Arizona State University
[2] Department of Computer Science-ICMC, University of Sao Paulo
{Ali.Abbasi,Skumar34,Huan.Liu}@asu.edu, Augustoa@icmc.usp.br

Abstract. In disasters such as the earthquake in Haiti and the tsunami in Japan, people used social media to ask for help or report injuries. The popularity, efficiency, and ease of use of social media has led to its pervasive use during the disaster. This creates a pool of timely reports about the disaster, injuries, and help requests. This offers an alternative opportunity for first responders and disaster relief organizations to collect information about the disaster, victims, and their needs. It also presents a challenge for these organizations to aggregate and process the requests from different social media. Given the sheer volume of requests, it is necessary to filter reports and select those of high priority for decision making. Little is known about how the two phases should be smoothly integrated. In this paper we report the use of social media during a simulated crisis and crisis response process, the *ASU Crisis Response Game*. Its main objective is to creat a training capability to understand how to use social media in crisis. We report lessons learned from this exercise that may benefit first responders and NGOs who use social media to manage relief efforts during the disaster.

1 Introduction

Social Media is a term ascribed to the current generation of Internet-based social information sharing and social interaction platforms. Some popular examples of social media are Facebook, Twitter, Youtube, and Flickr. Advances in mobile devices, have allowed social media to become available and accessible to anyone who is connected to the Internet. Social media such as Twitter has even let users share microblogging messages via SMS.

In August, 2010, Red Cross published a report[1] that for the first time surveyed 1,054 respondents from the United States population and reported on the expectations from and usage of social media during disasters. The study discovered that Facebook was the most popular social media website with more than 58% of users maintaining a Facebook account and it was also the most preferred channel for posting eyewitness information or sending information about one's safety. The study also discovered that population between 18-35 was more

---

[1] http://www.redcross.org/www-files/Documents/pdf/SocialMediainDisasters.pdf

S.J. Yang, A.M. Greenberg, and M. Endsley (Eds.): SBP 2012, LNCS 7227, pp. 282–289, 2012.

likely to use social media to send and receive information during emergencies. The number of users on social media sites has been constantly increasing. Now Facebook has more than 800 million users and over 75% of these users are from outside the United States[2]. Twitter, a popular microblogging site has more than 200 million users. Several million messages are posted on these two sites every-day and it is clear that they can play a key role as an information source during disasters. In the light of these facts, the results from the ASU Crisis Response Game show that people do have expectations from social media that must be addressed by agencies responding to a disaster. Moreover, it shows that social media can be a valuable means to reach out to people and offer assistance.

In the past, social media has been used to publish eyewitness accounts after a disaster. Twitter was one of the first sources of eyewitness information during the Mumbai terror attacks in 2008[3]. After the earthquake in Japan in early 2011, a Japanese Twitter user reached out to the American Ambassador in Japan, John Roos, who was heading the American rescue operations after the earthquake. With the following tweet: "Kameda hospital in Chiba needs to transfer 80 patients from Kyoritsu hospital in Iwaki city, just outside of 30km(sic) range"[4].Social media can be a valuable source of information to obtain situational awareness during and after a disaster. More recently, the disaster relief agencies have recognized the potential of social media as an information outlet. Hurricane Irene was the first natural disaster where the official agencies used social media to spread information about disaster awareness and preparation[5].A recent Congressional research report [6] on social media outlines how social media was used by agencies during disasters. The focus of agencies is now shifting from passively observing on social media to actively communicating with people. Social media has been used previously for disseminating disaster preparation information and at times for community outreach. Obtaining situational awareness through monitoring of information flow is another utility of social media sources as outlined above. The report also suggests several ways in which an agency like FEMA can use social media in disaster recovery efforts.

Existing HA/DR systems that aid agencies need to adapt to the changing focus of the agencies. For this change to occur, we think it is essential to understand how information from a disaster can be effectively harnessed. A suitable way to start this process is by testing the systems in a real environment. However, due to the nature of the domain this is not advisable. Therefore, we need a simulated controlled environment where we can conduct these tests. Hence, we create a disaster simulation game that can be played with real people to simulate a disaster and test the systems for crisis and disaster response.

---

[2] http://www.facebook.com/press/info.php?statistics

[3] http://www.guardian.co.uk/technology/2008/nov/27/mumbai-terror-attacks-twitter-flickr

[4] http://www.usatoday.com/tech/news/2011-04-12-1Ajapansocialmedia12_CV_N.htm

[5] http://www.huffingtonpost.com/2011/08/28/hurricane-irene-fema-response_n_939545.html

2   Related Work

Disaster simulation games have been played prior to our game. Exercise 24 was one of the first humanitarian assistance and disaster relief exercise to test the usage of social media and communication tools in response to a simulated disaster. First conducted in September 24, 2010, an earthquake and a tsunami were simulated off the coast of California. People were encouraged to create social media accounts and participate in the testing of social media sources. In the second installment of this event conducted on March 28, 2011[6], the USEUCOM and the San Diego State University's Viz Lab, simulated an earthquake in the Adriatic Sea off the coast of Montenegro. Volunteers on site participated using social media and programmed messages were posted at each of the three stages of the disaster. Several volunteers assisted in mapping and geolocating responses which were collected and visualized using a Ushahidi[7] based crowdmap.

Another disaster simulation, the Great California Shake Out[8], was conducted in October 20,2011. Nearly 9 million people participated in this exercise where people were encouraged to imagine that an earthquake had hit California and to perform the "Drop, Cover, and Hold On" procedure. Although users were not required to publish any information on social media in this simulation, the size of the population that participated is of interest.

The exercises mentioned above focus on testing the ability to collect information from social media and preparing people for a disaster. In order to effectively use social media for disaster response, we need to be able to collect reports from the various social media sources and act upon the ones that require a response. Trustworthiness of such data has been of some concern lately as social media is a free medium. This issue was investigated in the study performed by Mendoza et.al. [7] on tweets generated during the 2010 Chilean earthquake. The authors discovered that immediately after the earthquake several rumors were posted on Twitter which increased the already existing chaos after the disaster. In a subsequent study [1] they propose a model based on features constructed using the qualities of the user, tweet, topic, and propagation that can be used to predict the credibility of a message. A system to detect rumors and misuses of Twitter has also been built by researchers at the University of Indiana, called Truthy[9].

Crowdsourcing is one of the approaches that demonstrate the power of social media. Crisis maps are one type of social media based system which are being used more prominently for disaster response during crises. Authors Gao et.al. discuss some existing systems using examples from crisis maps deployed for real disasters [2]. One such platform is Ushahidi which can collect, organize, and visualize SMS reports on a map. Challenges in using this system include a manual verification and preprocessing step necessary to ensure that the received information is accurate and actionable. At NSCWDD[10], the Quicknets team has built

---

[6] https://sites.google.com/a/inrelief.org/24/media-report
[7] http://www.ushahidi.com/
[8] http://www.shakeout.org/
[9] http://truthy.edu
[10] http://www.navsea.navy.mil/nswc/dahlgren/default.aspx

a plugin for the Ushahidi system which facilitates identification of actionable information by trained volunteers from organizations, such as Humanity Road Inc. It is also capable of replying back to a user via SMS to request additional information or to send an update on the response. Quicknets focuses on SMS but can be potentially used with other social media for disaster relief. In [4], the author outlines some of the problems with existing crisis map platforms and suggests guidelines that will help in the future development of such systems. Recently, Twitter has attracted a lot of attention for its role in various disasters. It represents one popular way to produce timely and instant data. At ASU, we have built a Twitter monitoring and analysis system - TweetTracker [5] that can be easily customized for tracking and retrieving disaster-related information to assist first responders to make critical decision and effective response. These systems concentrate on using the information from social media to respond to a disaster. Such a response has not been tested previously.

3   ASU Crisis Response Game

The game is a live-action role-playing exercise in which volunteers take part as either victims or first-responders involved in a disaster. In this game, there is intensive use of Twitter and SMS. In the game victims use social media to ask for help. These requests are collected and processed by a relief system (TweetTracker or QuickNets). Based on these reports, missions will be generated upon which the first-responders can act. Figure 1, shows the architecture of the game.

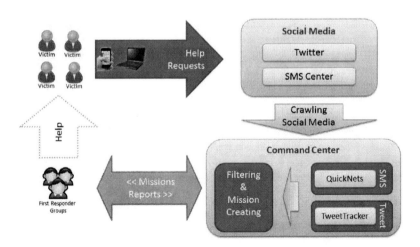

Fig. 1. ASU Crisis Response Game, Architecture

3.1   Game Components

There are three main components, Victims, First-responders, and Filtering team.

Teams Victim (TV). These teams perform the role of victims in a disaster scenario. They use social media (Twitter in this case) or send SMS to ask for help or report injuries. In each message they should include their location, a specific game related hashtag, and their request. They can report any kind of problem, such as fire, rioting, injuries, among others. Individuals or a group of two or more people make a team. Each team should have a device for communication (e.g., smart phone, laptop, or tablet) with ability of connecting to the Internet to send tweets or use their phones to send SMS. At the beginning of the game Victims scatter themselves around the pre-defined game locations. They only can send requests when they are settled in the game locations. Then they should stay till first-responders arrive to resolve their problem. Using social media or sending SMS to a center is the only way that victims can communicate with relief organizations. Teams should activate geo-location related features on their device when they use it to ask for help. This feature helps first-responder teams to locate them easily.

First-Responder Teams (FR). These teams perform the role of NGOs, government agencies, and any other organization able to help. They use a relief coordination system (QuickNets, ACT or TweetTracker in the ASU Game) to find requests for which they can take responsibility. Each team is composed of two or more volunteers. Each team member has one or more capabilities such as Security, Fire and Rescue, Medical, and Wildcard. Teams can perform missions according to their capabilities. In addition some teams can have Wildcard which enables teams to perform all kind of missions. At the beginning of the game, first-responders are at their designated homebase. After Victims send requests for help, FRs select missions based on team members' capabilities and go to the TV's location. Team-Victim members have to stay in place until first-responders arrive and fulfill their request. If the FR team does not have all of the necessary capabilities they should collaborate with other FR teams to accomplish the mission. When a problem has been solved they report back the new situation to the center. When they run out of resource cards to spend, they must go back to their head quarter and get resources renewed.

Filtering Team. During a disaster people send many tweets about the disaster. The filtering team is responsible for reviewing the tweets and selecting ones that are related to the mission of first responders. After the selection process they generate missions. These missions includes details about the problem and the location. Each mission can be generated by analyzing one or more tweets. It is possible that they generate more than one mission from a single tweet. This team is using specific software that is able to collect data from social media or receive SMS from victims. Then they use their own system to publish missions that will be used by FR. This team is responsible for generating missions. To do this task they need to perform the following sub-tasks:

- Selection (Deciding whether a message is actionable)
- Categorization (for the game, what capabilities are necessary to do the task)
- Geolocation (Finding the exact locating of the victims)

These tasks can be done manually or by using software systems. After this step, first-responders will be able to see and select the missions.

4   Game Exercise and Lessons Learned

We ran our game during the last week of the August 2011, at ASU main campus. Over 75 volunteer students participated in the real game and played for more than 4 hours. We assigned them to 25 teams of victims, and 8 teams of First-Responders. In addition around 20 more people participated in the game as the Filtering team and Line-Judges[11] Victims had the opportunity to scatter into 7 different buildings in the campus as shown in Figure 2. Finally, 17 (68%) Victim teams and 4 (50%) First-Responder teams visited at least 1 out of 7 game location which is labeled on the ASU map. All of the seven buildings were visited by both teams. We collected 212 Short Messages (SMS) and 230 tweets from 13 distinct tweeters. Victims used #ASUGAME as part of their tweets to make it possible to be found by the First-Responder's twitter crawler software.

4.1   Social Media Data Collection

TweetTracker was used to collect tweets from Twitter. In addition, we used ACT and QuickNets to collect SMS. *TweetTracker* is a tool for collecting and analyzing tweets to obtain situational awareness. *ACT (ASU Coordination Tracker)* is a tool for crisis event visualization, communication, monitoring, and coordination [3]. *QuickNets* is a plugin of Ushahidi for NGOs to collect and respond to SMS and email requests that have been validated by domain experts.

4.2   Lessons Learned

Many valuable lessons are learned. We summarize the major ones below.

Collecting Tweets. Each day people use Twitter to send more than 200 million tweets. Employing a proper method to find related tweets is very important. In this game, we used two different systems to collect tweets. One uses Twitter streaming API to collect and the other uses the jTwitter API to increase coverage. Twitter steaming API does not have limitation but jTwitter has the limitation of 150 or 20,000 requests per hour for regular and white-list users, respectively. In both methods we searched for tweets that mentioned the hashtag #ASUGAME. The most reliable way to find tweets related to a crisis is by searching for hashtags that people create and use during the crisis.

Creating FR missions. The filtering team was responsible for reviewing all of the tweets and generating missions based on the tweets. Even during the ASU-Game in which we only had 220 tweets our filtering team was overloaded. In a

---

[11] Line-Judges act as moderators, with free will to decide if a particular capability can be applied to a particular mission or not. They also help volunteers with any doubts.

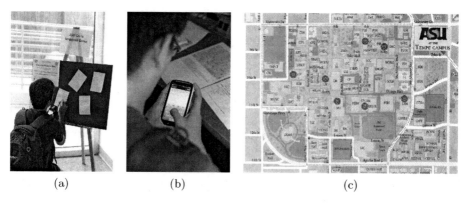

(a)   (b)   (c)

Fig. 2. (a) and (b) show victims that tweeting and posting a help request during the game. (c) Shows the game map. We used 7 locations (which are highlighted by a circle) in ASU main campus to simulate the real disaster relief.

real crisis, the number of tweets and SMSs would be significantly higher, thus we must use automated systems. Finding related and actionable tweets is the most complicated task in this process. During the disaster a small fraction of tweets are related and can be used to create missions. To find that tiny percent, all of the tweets should be processed. In addition, there are many conversational tweets about the crisis that using the same hashtags and keywords used by victims. Spammers are another problem that generate lots of tweets that should be found and ignored. A practical way to handle this process is using a system that rank tweets according to their importance for a specific crisis. One practical way to help first responders is asking victims to include as much information as they can in their tweets. For example, we asked each team to include location and team identification. This task consumed resources and quickly became a bottleneck. Two solutions for this problem are (1) to develop an intelligent analytic system that can prioritize the tweets or (2) to recruit a sufficient number of experienced people to work for the Filtering team.

GeoLocation. People are able to attach their location information to new Tweets through the web and mobile clients. The knowledge of the exact location of victims and areas that require assistance is invaluable during disasters. In the past, we have observed that less than 5% of users provide location information with their tweets due to privacy concerns and lack of awareness about this feature. In the game, even after being given explicit instruction to add location information to their tweets only few of the tweets had geotag.

NOT REAL THIS IS A GAME!. During the gameplay we were concerned that our tweets might be observed by a real agency and they might be mistaken to have been generated in response to a real disaster in ASU. We coordinated with the campus police department and instructed the participants to begin their messages with "NOT REAL THIS IS A GAME!"

Languages. In crises such as Haiti's earthquake people used different languages to tweet. The crawler should be able to search among tweets with popular languages in the area that disaster happens. In the game, we encouraged victims to use languages other than English, and tried to have one non-English speaker in each team if possible. Translation is critical for information gathering and integration and could be time-consuming and error-prone.

5   Conclusions

ASU Crisis Response Game simulates a disaster scenario (e.g., earthquakes) and tests how social media based disaster relief systems would work in a crisis simulation on ASU campus. Around 75 students voluntarily played different roles (victims and first responders) in this disaster scenario. In the simulation, victims had sent out their requests using Twitter or SMS. First-responders used different software to collect the requests from victims and manage actions, and responses. The process helped us evaluate how these systems work on real situations. The experience of using social media during a controlled disaster relief process had valuable lessons for our team.

Acknowledgments. This research was sponsored in part by the Office of Naval Research (ONR) grant: N000141010091. We would also like to thank members of DMML Lab at ASU and our collaborators Catherine Graham from Humanity Road Inc., Mark Bradshaw and his team from NSWCDD, Office of Joint Staff, Dr. Rebecca Goolsby from ONR, and Professor Kathleen Carley and her team from CMU.

References

1. Castillo, C., Mendoza, M., Poblete, B.: Information credibility on twitter. In: Proceedings of the 20th International Conference on World Wide Web, pp. 675–684. ACM (2011)
2. Gao, H., Barbier, G., Goolsby, R.: Harnessing the crowdsourcing power of social media for disaster relief. IEEE Intelligent Systems 26(3), 10–14 (2011)
3. Gao, H., Wang, X., Barbier, G., Liu, H.: Promoting Coordination for Disaster Relief – From Crowdsourcing to Coordination. In: Salerno, J., Yang, S.J., Nau, D., Chai, S.-K. (eds.) SBP 2011. LNCS, vol. 6589, pp. 197–204. Springer, Heidelberg (2011)
4. Goolsby, R.: Social media as crisis platform: The future of community maps/crisis maps. ACM Transactions on Intelligent Systems and Technology (TIST) 1(1), 7 (2010)
5. Kumar, S., Barbier, G., Abbasi, M.A., Liu, H.: TweetTracker: An Analysis Tool for Humanitarian and Disaster Relief. In: Fifth International AAAI Conference on Weblogs and Social Media, ICWSM (2011)
6. Lindsay, B.R.: Social Media and Disasters: Current Uses, Future Options, and Policy Considerations. In: CRS Report for Congress (2011)
7. Mendoza, M., Poblete, B., Castillo, C.: Twitter under crisis: Can we trust what we rt? In: Proceedings of the First Workshop on Social Media Analytics, pp. 71–79. ACM (2010)

Intergroup Prisoner's Dilemma with Intragroup Power Dynamics and Individual Power Drive

Ion Juvina[1,*], Christian Lebiere[1], Cleotilde Gonzalez[2], and Muniba Saleem[2]

[1] Department of Psychology, Carnegie Mellon University, 5000 Forbes Ave.,
Pittsburgh, PA 15213
[2] Department of Social and Decision Sciences, Carnegie Mellon University,
5000 Forbes Ave., Pittsburgh, PA 15213
{ijuvina,cl,coty,muniba}@cmu.edu

Abstract. This paper introduces a game paradigm to be used in behavioral experiments studying learning and evolution of cooperation. The goals for such a paradigm are both practical and theoretical. The design of the game emphasizes features that are advantageous for experimental purposes (e.g., binary choice, matrix format, and tractability) and also features that increase the ecological validity of the game (e.g., multiple players, social structure, asymmetry, conflicting motives, and stochastic behavior). A simulation of the game based on human data from a previous study is used to predict the impact of different levels of power drive on payoff and power, to be corroborated in future studies.

Keywords: Prisoner's Dilemma, Nested repeated games, Power, Cooperation.

1 Introduction

The US President Barak Obama and the House Speaker John Boehner played four hours of golf before returning to the negotiation table. It was everyone's hope that playing golf would help the two leaders to build a relationship and a sense of camaraderie that would transfer to their negotiation situation and ultimately would increase the chances of bipartisan cooperation in Congress.

Learning from games and transfer of learning across games have been vastly documented. For example, Haruvy and Stahl [1] showed that players were able to learn sophisticated beliefs about others and use those beliefs in subsequent games. In a related project, we have observed that players' ability to learn to cooperate in a game increases their chance to cooperate in a subsequent unrelated game. More generally, human studies show ensemble effects, that is, spillovers of strategies and/or beliefs across games suggesting that games are not treated as independent of each other [2].

Real-world cooperation and negotiation situations are rarely (if ever) independent of each other. For example, domestic and international politics are usually entangled:

---

* Corresponding author.

S.J. Yang, A.M. Greenberg, and M. Endsley (Eds.): SBP 2012, LNCS 7227, pp. 290–297, 2012.
© Springer-Verlag Berlin Heidelberg 2012

international pressure leads to domestic policy shifts and domestic politics impact the success of international negotiations [3]. The paradigm of nested repeated games has been suggested to represent real-world situations better than specific games. For instance, a two-level conflict game extensively studied is the Intergroup Prisoner's Dilemma [4]. In this game, two levels of conflict (intragroup and intergroup) are considered simultaneously. The intragroup level consists of an n-person Prisoner's Dilemma (PD) game while the intergroup level is a regular PD game.

Previous research in our labs shows that players' awareness of interdependence influences their chances to establish mutual cooperation. Awareness of interdependence was manipulated by gradually displaying information about the existence of another player, the other's choices, and their payoffs. The condition hypothesized to afford the maximum level of awareness of interdependence was that in which the game matrix was presented at each round, because the game matrix shows all conceivable combinations of strategies and their associated payoffs. The last condition was indeed the one that increased mutual cooperation the most.

In this paper we introduce a three-level nested game called Intergroup Prisoner's Dilemma with Intragroup Power Dynamics and Individual Power Drive (IPD^3, pronounced IPD-cube). We intend to use this game in a series of behavioral studies to investigate the role of power in interpersonal and intergroup conflicts. We also intend to use it as a learning tool to develop strategic interaction skills. For these reasons, we developed a game interface and a learning protocol that are intended to allow human participants to: (1) learn to play the game, (2) play it in an engaging fashion, and (3) ultimately develop skills of strategic interaction that can be transferred to real-world settings.

1.1   Background

IPD^3 is a direct descendent of Intergroup Prisoner's Dilemma with Intragroup Power Dynamics (IPD^2, pronounced IPD-square). IPD^2 was extensively described elsewhere [5] and it is only briefly summarized here.

IPD^2 introduces the concept of power which is relevant for many real-world situations involving cooperation and conflict. For example, power imbalances are known to be involved in radicalization and escalation of political conflicts. In IPD^2, two groups play a Repeated Prisoner's Dilemma game. Each group is composed of two players. Within a group, each player chooses individually whether to cooperate or defect, but only the choice of the player with the greatest power within the group counts as the group's choice. This is equivalent to saying that the two players simultaneously vote for the choice of the group and the vote of the powerful player bears a heavier weight. On a given round, individual power and payoff increases or decreases depending on the group payoff, the power status, and whether or not there is consensus of choice between the two players on a group. The key feature is that in the absence of consensus, positive group payoffs will result in an increase in power for the powerful player while negative group payoffs will result in an increase in power for the powerless player. The players make simultaneous decisions and they receive feedback after each round.

The main results from the previous study were as follow. The level of mutual cooperation gradually increased as the game unfolded. The intragroup power dynamics prevented the groups from settling in the mutual defection equilibrium: prolonged mutual defection would cause power shifts which would effectively reset the interaction and give cooperation a chance to start anew. The human participants were perfect reciprocators, that is, on average, their level of cooperation or defection at each trial matched that of their opponents. The human participants played a strategy that was most similar to a combination of the Tit-for-tat and Pavlov strategies (this result will be demonstrated in Section 3.2). The same study also revealed a number of limitations of IPD^2: The game was administered in a tabular format with brief instructions. Some participants reported that they did not fully understand the rules and the dynamics of the game. The intragroup power game did not include an important component of power – prestige. The motivation to acquire power was partly confounded with that of acquiring payoff. Only in a few cases, when the player's behavior was extreme, it was possible to disentangle the drive for power from the drive for payoff. Sometimes players found themselves in a situation where they wanted to increase their power but they did not know the right combination of moves to do so. In these cases their power drive did not materialize in their behavior.

2   Conceptual Design of IPD^3

IPD^3 adds a third level to IPD^2. Thus, IPD^3 has three levels: (1) intergroup, (2) group, and (3) individual.

2.1   The Intergroup Game

The intergroup game is Repeated Prisoner's Dilemma. In this game, two players, "Player1" and "Player2," each decide between two actions that can be referred to as "cooperate" (C) and "defect" (D). The players choose their actions simultaneously and repeatedly. The two players receive their payoffs after each round, which are calculated according to a payoff matrix setting up a conflict between short-term and long-term payoffs. If both players cooperate, they each get one point. If both defect, they each lose one point. If one defects while the other cooperates, the player who defects gets four points and the player who cooperates loses four points. Note that the Repeated Prisoner's Dilemma is a non-zero-sum game: one player's gain does not necessarily equal the other player's loss. In our case, each "player" is a group of two players that have to produce a group choice. This choice is the result of the intragroup game presented next.

2.2   The Intragroup Game

The intragroup game is an interpersonal power game. As in IPD^2, the power game is a zero-sum game because only one of the two players in a group can be in power at a given time. The choice of the powerful player gets counted as the group's choice.

The rules that govern changes in power correspond to important characteristics of the power concept. We included in IPD^3 three of the most relevant aspects of power: outcome power, dominance, and prestige. The outcome power or "power to" is a measure of a player's ability to bring about positive outcomes. Dominance or "power-over" is the ability of a player to impose its decisions over another player whether the latter accepts it or not. Prestige or "power-from" is the empowerment a player gets from other players willingly supporting or agreeing with the player's decisions.

Outcome power is implemented in IPD^3 by a rule that makes power updating dependent on the group payoff. Thus, if the powerful player makes a choice that results in positive (or negative) payoff for the group and the powerless player opposes that choice, the power of the powerful player increases (or decreases) with an amount that is proportional with the group payoff.

Dominance is reflected by the rule that makes the choice of the powerful player count as the group decision. The powerful player not only makes the decisions for the group but also takes a larger share of the group payoff, because individual payoff is a function of group payoff and individual power. The powerless (dominated) player cannot influence the outcome of the intergroup game. The choice of the powerless player is only consequential for the intragroup power game. Thus, the power of the powerless player increases (or decreases) when he or she opposes a bad (or good) choice (negative and positive outcome, respectively) of the powerful player.

Prestige conferral occurs when the powerless player supports (i.e., makes the same choice as) the powerful player. In such cases, the power of the powerful player increases (and consequently the power of the powerless player decreases because the power game is a zero-sum game).

2.3   The Intraindividual Game

Up to this level, players could only influence their intragroup power through their choices in the intergroup game (cooperate or defect) and in the intragroup game (support or oppose the other player's choices). Players' drive for power could not be expressed independently of their ability to make good decisions for themselves and for the group. Presumably, players could have a range of preferences related to power and payoff. Some players may be unable to make good (i.e., lucrative) decisions but could have as high (or higher) drive for power as all other players (see, for example, the case of some dictators). Other players might want to secure power because they are confident in their ability to make positive payoffs for the group. Also, some players might prefer to remain powerless as long as the powerful player makes lucrative decisions for the group.

At each round in the game, each player has the option to raise their payoff or their power with one unit. When a player raises his power, he forfeits the opportunity to raise payoff. It is as if the payer pays a cost for raising power (or buys power). If both players in a group raise their power, the actions cancel out and the power standings remain the same. The choice to raise power or payoff indicates players' preferences and motives. This choice also affects the dynamics of the intragroup and intergroup games. Thus, for example, if a powerless player A decides to raise her power and her group mate B decides to raise his payoff, the power standings might change: A might become the powerful and B the powerless. This, in turn, might change the dynamics of the intergroup game, as A (the new leader) might decide to change the group's strategy.

3   Design of IPD^3 as an Experimental Paradigm

We developed an interface and a learning protocol that allows human participants to learn to play the game and play it in an engaging way.

3.1   The IPD^3 Interface

Previous research in our labs [6] shows that displaying the available choice and payoff in the game matrix at each round increases the chances of mutual cooperation. We decided to use the game matrix not only to display all the options and payoffs but also as a game interface. The cells of the game matrix showing the players' available options are also used as buttons that can be used to select those options. The payoffs resulting from the selected options are displayed and highlighted on the same matrix. We extended the classical matrix format to show interdependencies between the nested games, running totals for payoff and power, and visualization of the players power standings (Figure 1).

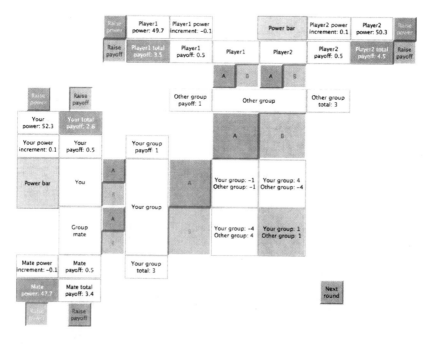

Fig. 1. The IPD^3 interface. The sunken buttons show individual and group selections. The power bars indicate the powerful players and their power levels.

3.2   The Other Player's Strategies

Although the game can be played between four human players, it is useful sometimes for experimental and learning purposes to have a set of preprogrammed strategies. This makes the game more tractable and allows for implementing various manipulations. We designed a set of three strategies to be used as players together with a human player. Tit-for-tat and Pavlov are two strategies that are frequently used in behavioral experiments in which human participants are paired with computer strategies. Tit-for-tat repeats the last move of the opponent while Pavlov repeats its winning move and switches its losing move (win-stay-lose-switch). We added a stochastic component (5%) to make them less predictable and more human-like. We also added a simple strategy to use when they lack power. This strategy was "stochastic always cooperate" because it was determined to be the most effective strategy to gain power when the game settles in the mutual defection equilibrium.

However, neither Tit-for-tat nor Pavlov matches the human data on repetition propensities from our previous study [5]. Repetition propensities refer to players' probabilities to repeat their last choice contingent upon the outcome of that choice. Rapoport, Guyer and Gordon [7] defined four repetition propensities: alpha (D after DD), beta (C after CD), gamma (D after DC), and delta (C after CC). The correlation of the human repetition propensity data with Tit-for-tat is 0.85 and with Pavlov is 0.51. We noticed that a combination of Pavlov and Tit-for-tat that simply averages their repetition propensities is better at matching the human data (correlation 0.96). We called this strategy Pavlov-tit-for-tat and used it as another preprogrammed strategy in our simulations. Although Pavlov-tit-for-tat is the best at matching the human data among the preprogrammed strategies, there are important differences (see Figure 2): humans are more retaliating (higher alpha), more forgiving (higher beta) and more exploiting (higher gamma) than Pavlov-tit-for-tat.

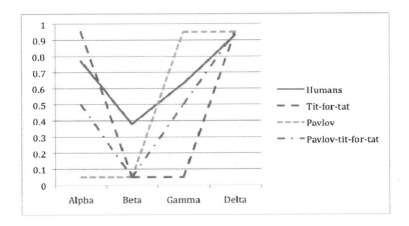

Fig. 2. Comparison between humans and preprogrammed strategies with regard to repetition propensities

3.3   The Learning Protocol

We learned from the previous study on IPD^2 that the game is fairly complex and thus challenging for the human participants. For this reason we designed a protocol to facilitate learning and improve players' experience with the game. The participants are invited to play a sequence of progressively more complex games: IPD (Iterated Prisoner's Dilemma), IPD^2, and IPD^3. Detailed instructions and quizzes are administered before each level. At each level, the human participants play three games of 100 rounds where the human is paired with each of the three preprogrammed strategies described above. We are in the process of collecting pilot data with this paradigm, and the preliminary results are encouraging: the pilot human participants report not only complete learning of the game but also a pleasant experience.

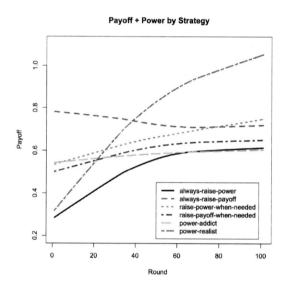

Fig. 3. Predicted time course of the sum of payoff and power for different strategies

4   Simulation and Prediction

In preparation for running a human study with IPD^3, we ran a number of simulations aiming to predict the human behavior in this game. We simulated a human player based on the data from our previous study on IPD^2, that is, we set the probabilities to cooperate or defect in various conditions according to the average frequencies observed in the previous study. In addition, the simulated human was given six different strategies to play the intraindividual game (raise power vs. payoff). These strategies were as follow: (1) Always-raise-power, (2) Always-raise-payoff, (3) Raise-power-when-needed (the probability to raise power is inversely proportional to the value of power), (4) Raise-payoff-when-needed (raise payoff when payoff is negative), (5) Power-addict (the probability to raise power is directly proportional to the value of power), and (6) Power-realist (the probability to raise power is inversely proportional to the difference in power between the two players in a group). The preprogrammed strategies (stochastically and dynamically) matched the simulated human with regard to these strategies. Under the assumption that humans are motivated by both power and payoff, IPD^3 predicts that they will adopt the power-realist strategy because this strategy maximizes both power and payoff on a long run (see Figure 3). However, we expect to find significant individual differences in the drive for power. Based on our previous study we hypothesize that extreme drive for power will be associated with extreme aggressive behavior.

5   Conclusion

This paper introduced IPD^3 – a game paradigm to be used in behavioral experiments and learning of strategic interaction skills. We have designed the game so that it is solidly grounded in state-of-the-art game theoretic and socio-cognitive research. This design keeps features that are advantageous for experimental purposes (e.g., binary choice, matrix format, computational tractability) while adding features that increase ecological validity (e.g., multiple players, social structure, asymmetries, conflicting motives, and stochastic behavior). We will run a series of experiments with IPD^3 to test its conceptual and ecological validity.

Acknowledgments. This research is supported by the Defense Threat Reduction Agency (DTRA) grant number: HDTRA1-09-1-0053 to Cleotilde Gonzalez and Christian Lebiere.

References

1. Haruvy, E., Stahl, D.O.: Between-game Rule Learning in Dissimilar Symmetric Normal-form Games. Games and Economic Behavior (in press)
2. Bednar, J., Chen, Y., Liu, T.X., Page, S.: Behavioral Spillovers and Cognitive Load in Multiple Games: An Experimental Study. Games and Economic Behavior (in press)
3. Putnam, R.D.: Diplomacy and Domestic Politics: The Logic of Two-level Games. International Organization 42, 427–460 (1988)
4. Bornstein, G.: Intergroup Conflict: Individual, Group and Collective Interests. Personality and Social Psychology Review 7, 129–145 (2003)
5. Juvina, I., Lebiere, C., Martin, J.M., Gonzalez, C.: Intergroup Prisoner's Dilemma with Intragroup Power Dynamics. Games 2, 21–51 (2011)
6. Martin, J.M., Gonzalez, C., Juvina, I., Lebiere, C.: Interdependence information and its effects of cooperation (2011) (submitted)
7. Rapoport, A., Guyer, M.J., Gordon, D.G.: The 2X2 Game. The University of Michigan Press, Ann Arbor (1976)

Modeling South African Service Protests Using the National Operational Environment Model

Chris Thron[1], John Salerno[2], Adam Kwiat[3],
Philip Dexter[4], and Jason Smith[5]

[1] Texas A&M University – Central Texas Department of Mathematics,
1901 South Clear Creek Road, Killeen TX USA 76549
[2] Air Force Research Laboratory, Rome Research Site, 525 Brooks Road, Rome NY 13441
[3] CUBRC, 725 Daedalian Drive, Rome NY 13441
[4] SUNY Binghamton Department of Computer Science, 4400 Vestal Parkway East
Binghamton, NY 13902
[5] ITT Corporation, Advanced Engineering & Sciences, 474 Phoenix Drive
Rome, NY 13441
thron@ct.tamus.edu, john.salerno@rl.af.mil,
adam.kwiat@cubrc.org, pdexter1@binghamton.edu,
jason.e.smith@itt.com

Abstract. The Air Force Research Laboratory's National Operational Environment Model (NOEM) is a strategic analysis/assessment tool that provides insight into the complex state space that depicts today's modern nation-state environment. A key component of the NOEM is its Populace Behavior Module, an agent-based model that describes activist behavior in terms of populace agents' perceptions of hardship and government legitimacy. Although based on Epstein's grievance vs. net risk model [1], enhancements have been made that make the model more practically applicable. We applied the model to an actual scenario, namely service protests in Gauteng Province, South Africa from 2004-2010. When model parameters were fit based on the data, the model successfully duplicated both qualitative and quantitative characteristics of the historical data.

Keywords: Agent-based modeling, behavior modeling, population simulation.

1   Introduction

The Air Force Research Laboratory's National Operational Environment Model (NOEM) is a large-scale stochastic model representing the environment of a nation-state or region, with a focus on population behavior. The NOEM enables the user to identify potential problem regions within the environment, test a wide variety of policy options on a national or regional basis, determine suitable courses of action given a specified set of initial conditions, and investigate resource allocation levels that will best improve overall country or regional stability. The different policy options or actions can be simulated, revealing potential unforeseen effects and general trends. An overview of the NOEM may be found in [2].

S.J. Yang, A.M. Greenberg, and M. Endsley (Eds.): SBP 2012, LNCS 7227, pp. 298–305, 2012.

At the heart of the NOEM is the Populace Behavior Module, which receives inputs from other modules and generates alternative possibilities for the population's response to regional conditions. Unlike other modules it relies on an agent-based model, where each agent represents a certain fraction of the population (this fraction is called the scale factor). As is typical with such models, randomness plays an important part – the same set of inputs can yield considerably different outcomes.

A key difficulty in using the Behavior Module for practical scenarios is furnishing suitable input parameters. We were able to develop a methodology for estimating model parameters from historical data from a particular scenario (Gauteng Province, South Africa from 2004-2010). Furthermore, we obtained model outputs that were consistent with observed levels of activism.

2   Background

2.1   Model Specification

The NOEM behavior model is based on Epstein's agent-based model [1], in which a region's population is modeled as a set of agents located on a two-dimensional grid. Each agent corresponds to a subset of individuals in the population, and the agent's status represents in some sense the cumulative effect of those individuals' behavior. At each time step, each populace agent is either active, inactive, or in "jail". Each agent that is not in jail makes an individual decision every time step whether or not to be active at that time step. This decision is based on the agent's perceived "grievance" (G) and "net risk" (N) according to the criterion: $G–N > T$, where T is a threshold that is the same for all agents for all time steps.

The quantities G and N may be further broken down. The grievance G depends on "hardship" (H) and "legitimacy" (L) according to the relation: $G = H(1–L)$, where both H and L are assumed to take values from 0 to 1. On the other hand, net risk N depends on the agent's estimated arrest probability P and risk aversion R according to the formula $N = P \cdot R$

Besides populace agents, the model also includes "cop" agents that represent the government's efforts at law enforcement. These patrol the grid and "arrest" active agents they come across. The location of cop agents is a major determining factor in each populace agent's estimated arrest probability P: specifically, the more cops that are "nearby" a given agent, the higher that agent's estimated arrest probability. On the other hand, the more active agents that are "nearby" a given agent, the lower that agent's arrest probability. For more specific details on the mathematical specification of P and the dynamics of arrest and imprisonment, see reference [1]. Further details of the specific NOEM implementation may be found in [3].

2.2   Service Protests in Gauteng Province, South Africa 2004-2010

In order to evaluate the behavior model's applicability to real data, we chose to look at service delivery protests in Gauteng Province, South Africa from 2004-2010: one major reason for this choice was the availability of fairly complete data over an extended period of time.

Since 2004, an increasing number of protests and outbreaks of violence have occurred in South Africa due to the government's inability to provide basic services (such as electricity, running water, housing and sanitation). These protests are normally restricted to the country's townships, the squatter camps/shanty-towns often located within the boundaries of larger cities. Townships, occupied by rural poor or migrant workers who moved to urban areas in hopes of finding employment, are rife with poverty and crime and are generally in squalid condition. Many demonstrations turn violent, with looting, destruction of property, stone-throwing, the burning of tires, and arson not uncommon. South African police are often on-site, and at times have had to enact crowd-control measures such as tear gas to quell protests.

Gauteng, the province with the highest number of protests, is South Africa's most urbanized and heavily populated province. It is the location of Johannesburg, the largest city and home to several townships. The province was formerly divided into six districts, including Johannesburg: two of these districts merged in May 2011.

Table 1 shows the increasing frequency of service protests in Gauteng Province over the period from 2004 to 2010.

Table 1. Protests in Gauteng Province by year

Year	2004-06	2007	2008	2009	2010
Number of Protests	10	9	9	25	44

Ironically, data obtained from the Gauteng provincial government[4][5] indicates that over this seven-year period basic services were considerably improved, which would seem to indicate that the populace's hardship was significantly decreased. In terms of our mathematical model, this can be explained as the result of a decrease in legitimacy that more than offset any decreases in hardship.

Data from Municipal IQ, (a South African intelligence service which tracks service delivery protests) indicates that 97 service protests took place from 2004 to 2010[6]. From South African media archives we found dates and locations for 67 of these protests. Figure 2 shows a cumulative plot of protest occurrence times by district.

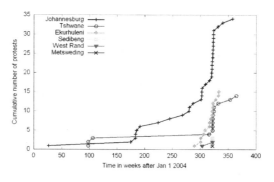

Fig. 1. Cumulative plot of protest occurrence times in Gauteng Province districts

The curves in Figure 1 exhibit a large number of "humps," where each hump indicates multiple protests occurring in a relatively short period of time. Apparently protests tended to occur in bursts, with one "superburst" occurring around week 320.

Not surprisingly, the more populous districts had the most protests. But we also found that the most densely populated districts had more protests per population than those more sparsely populated: a weighted least-squares regression of number of protests per million population versus population density (in thousands per square km) gave a regression coefficient of 0.37 ± 0.1. Surprisingly, service levels actually tended to be higher in more urbanized districts. Some observers[6] suggest that this counterintuitive situation is due to the urban poor's greater "relative deprivation" resulting from their exposure to nearby wealth which they themselves cannot attain.

Interestingly, during the "superburst" of March 2010 protests were spread evenly across all districts without regard to their urbanization. There is no apparent event that touched off the superburst, and there is no evidence of coordination between the protests: rather, this superburst has the characteristics of a completely spontaneous "flash mob"-type phenomenon that caught up the entire province.   This sudden increase in activism is inconsistent with Epstein's model: Epstein found that gradual increases in grievance never led to sudden increases in activism[1].

3   Modeling Considerations

3.1   Basic Grid Configuration

In order to model the Gauteng Province data, we had to decide how to map the South African populace onto a grid populated by agents. Since the protests were reported by township, in one sense it seemed appropriate to use townships as the effective agents in the model, and to model each township protest as one incident of agent activism. On the other hand, townships consist of tens of thousands of individuals, and we wanted to include effects due to the statistical behavior of the township populations. We did this by introducing modifications to the model, as described in Section 3.2.

Using data drawn from a World Bank study[7], we determined that there were 37 significantly underserved townships in Johannesburg. For the other districts, for want of detailed data we assumed that the number of underserved townships per population was the same for all districts. Altogether, our model used 100 township-agents: one example model configuration is shown in Figure 2 (configurations were re-randomized for each simulation run).

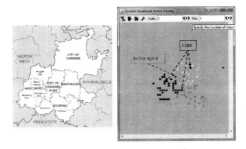

Fig. 2. Gauteng province (left); Example agent configuration (right)

3.2 Modifications to the Model

Due to the characteristics of our model, we found it necessary to make some key modifications to Epstein's model.

First, as mentioned above we addressed the statistical effect of the large populations within each township-agent. For this purpose, we considered each township itself as an agent-based system similar to Epstein's. Such systems can be mathematically described as finite Markov chains, since the system state change from time step to time step via fixed transition probabilities. Protests comprise a subset O of the possible states of the Markov chain. Aldous[8] has shown that if the chain is rapidly-mixing and the set O is rarely-visited, then the hitting time distribution is approximately memoryless, so that the distribution of inter-protest times is roughly geometric. As a result, we assumed that intra-township populations gave rise to a geometric distribution of township protest times. To obtain this, we added a mean-zero normal random variable to the grievance G for each agent for each time step. The standard deviation of this random factor was a fixed parameter input to the simulation. Under constant neighboring conditions, for each township-agent this produces a constant protest probability at each time step: while changing neighbor conditions changes the protest probability by changing the net risk level.

An additional modification to Epstein's model was made to reflect the influence of media on legitimacy. Current world events show that episodes of activism within a nation that are reported in the media (either mass media or social media) can stir up additional activism at long distances. One could model this by increasing the township-agents' "vision", that is the distance used in computing an agent's estimated arrest probability. However, enlarging the vision also increases the number of visible cops, which is both unrealistic (long-range cops cannot readily arrest an agent) and effectively counterbalances any increase in visible activism. So we took another approach, and modified the expression for legitimacy L as follows:

$$L = L_{\text{basic model}} - K(\#\text{TotalActivists} / \#\text{TotalPopulace}), \tag{1}$$

where the positive constant K reflects the long-range effects of activism. Equation (1) expresses the fact that an increase in overall average activism, as reported by the media, should increase agents' inclination to protest regardless of their location.

3.3 Estimation of Model Parameters

For NOEM to be a practical tool for describing the behavior of populations, it is necessary to develop a systematic methodology for estimating model parameters from real data. In the case of service protests in Gauteng Province, we used heuristic methods based on statistical analysis to the data to come up with plausible values for the different model parameters: these are summarized in Table 2.

Table 2. Model parameters used in NOEM simulation

Parameter	Value
Number of agents (townships)	100
Number of cop agents	3
Size of grid (in km & cells)	270 & 50
Populace vision & cop vision (in km)	10 & 25
Hardship H (all districts)	0.5
Legitimacy L (by district)	0.8(Joh),0.83(Tsh),0.85(Ek), 0.87(other)
Legitimacy decrease per time step δ	000037
Activism-determining threshold T	0.366
Agent grievance standard deviation σ	0.0
Jail term J (in time steps)	25
Long-range activism constant K	6.3

Temporal parameters were based on the choice of one week as time step: this choice was made because the time step in the model functions corresponds to agent "memory," and analysis of the experimental data suggested that agents were much more strongly influenced by events less than one week prior. The jail term parameter was chosen so that the median jail term was equal to the median time between protests at the same location. Agent vision was determined by examining protests less than one week prior to a given protest: a disproportionate fraction occurred within 10 km. Cop vision was set as the radius of a typical district. The threshold T, legitimacies, and legitimacy decrease were found via a system of equations based on pre-outburst per-province protest frequency data. The long-range activism parameter was chosen so as to account for the discrepancy between protest frequency during the superburst and the expected protest frequency without long-range activism.

4 Results

Figure 3 shows a typical simulation run: details varied from run to run.

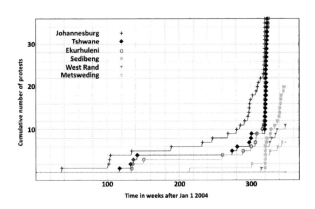

Fig. 3. Example NOEM simulation output for cumulative protest occurrence times, by district (compare Figure 1)

Figure 3 exhibits striking qualitative similarities with Figure 1, showing both isolated protests and small bursts of protests leading up to the superburst, Once the superburst breaks out all districts become involved, even those that have had no previous protests. In the NOEM model, once the superburst occurs the model breaks down and activism overtakes virtually all agents that are not imprisoned. Of course in the real world, widespread violence introduces new dynamics into the picture, such as the government calling in additional police and/or military resources. Hence, the model should not be expected to remain valid.

In order to make quantitative comparisons between simulation and observation, we ran ten simulations while recording the following statistics:

- Number of isolated protests preceding the superburst, where an isolated protest is one for which no other protest occurs during the same, preceding, and subsequent time steps;
- Superburst time, defined as the time step during which the maximum number of protests was observed;
- Precursor interval, defined as the number of time steps prior to the superburst time during which at least one protest occurred per time step;
- Number of pre-superburst protests, defined as the total number of protests up until the beginning of the precursor interval.

Table 3. Comparison of simulation results with Gauteng Province data

	# isolated protests	Superburst time (weeks)	Precursor interval (weeks)	#pre-superburst protests
Simulation	9.4±6.6	269±70	6.2±5	20.7±19
Observed	10	320	5	26

The error bars in Table 3 correspond to two standard deviations. The table shows that there is a wide variation in results from simulation to simulation: this suggests that unpredictable random factors play a large role in determining the space-time distribution of protests. The observed data was within error bars for all values.

We also investigated the sensitivity of the model to parameter changes. Table 4 shows averages of five simulation runs for various parameter changes, together with 2-standard deviation error bars.

Table 4. Effect of parameter changes on simulation averages

Parameter change	# isolated protests	Superburst time (weeks)	Precursor interval (weeks)	#pre-superburst protests
Baseline	9.4±2.2	269±23	6.2±1.7	20.7±6.3
J:25→30	8.4±2.2	281±49	4.2±0.8	26.2±13
K: 6.3→7	7.8±2.9	268±36	4.6±1.5	30±16.7
T:0.366→0.4	11.8±0.8	471±47	5.6±3.2	33±15
δ:.00037→.0004	8.4±5.5	272±47	5±0.7	29±13
σ: 0.085→0.09	9±1.9	200±40	4.8±1.3	49±9.9

Table 4 shows that the 10-20% changes for most parameters did not have a large effect on the simulation averages. One exception is the activism threshold T: a 10% increase in T (from 0.366 to 0.4) led to large increases in superburst time and non-isolated pre-superburst protests. Increasing the agent grievance standard deviation by 5% also had a large effect, reducing the superburst time by roughly 30% while more than doubling the average number of non-isolated pre-superburst protests.

5   Conclusion

We have modified Epstein's seminal model to obtain a model for which parameters can be estimated from actual data, and which produced simulation results that agreed with real data in many qualitative and quantitative respects. Widespread, non-localized outbursts of activism arise naturally from our model, in contrast to Epstein's findings with his model that gradual decreases in legitimacy never lead to bursts in activism. Our model provides a plausible quantitative mechanism to explain how media reporting of protests can lead to widespread outbreaks of activism.

Acknowledgements. This work was partially supported by the Air Force Research Laboratory (AFRL), Air Force Office of Science and Research (AFOSR) under a Laboratory Research Initiation Request (LRIR) and a grant from the Air Force Summer Faculty Fellowship Program (SFFP).

References

1. Epstein, J.M.: Modeling civil violence: An agent-based computational approach. PNAS 99(suppl. 3), 7243–7250 (2002)
2. Salerno, J.J., Romano, B., Geiler, W.: The National Operational Environment Model (NOEM). In: Proceedings of Information Systems and Networks: Processing, Fusion, and Knowledge Generation, Defense & Security, pp. 25–29 (April 2011)
3. Salerno, J., Geiler, W., Hudson, B., Romano, B., Smith, J., Thron, C.: The National Operational Environment Model: A Focus on Understanding the Populace. In: 2011 MODSIM World Conference and Expo, NASA, Virgina Beach (2011)
4. Gauteng Provincial Government, Socio-Economic Review and Outlook (2010), http://www.treasury.gpg.gov.za/docs/sero_2010.pdf (retrieved January 5, 2012)
5. Gauteng Provincial Government, Socio-Economic Review and Outlook (2009), http://www.treasury.gpg.gov.za/docs/SocioEconomicReview2009.pdf (retrieved January 5, 2012)
6. Karamoko, J.: Community Protests in South Africa: Trends, Analysis, and Explanations (July 2011), http://www.ldphs.org.za/publications/publications-by-theme/local-government-in-south-africa/community-protests/Community_Protests_SA.pdf (retrieved January 5, 2012)
7. Chandra, V., et. al.: South Africa: Monitoring Service Delivery in Johannesburg (April 2002), http://siteresources.worldbank.org/INTURBANPOVERTY/Resources/SoAfricaMonitoringServiceDeliveryJoburgApri2002.pdf
8. Aldous, D.: Markov Chains with Almost Exponential Hitting Times. Stochastic Processes and their Applications 13, 305–310 (1982)

Modeling and Estimating Individual and Population Obesity Dynamics

Hazhir Rahmandad and Nasim S. Sabounchi

Grado Department of Industrial and Systems Engineering,
Virginia Tech, Northern Virginia Center, Falls Church, VA, 22043, USA
{Hazhir,Sabounchi}@vt.edu

Abstract. The obesity trend in the U.S. and many other countries has increased the need for models that can assess the potential impact of alternative interventions to reverse this trend. In this paper we report on building a generic dynamic model that can be used for obesity policy analysis at multiple levels. We build an individual level model for both childhood and adulthood to capture the energy balance and weight change throughout the life of individuals, and aggregate individual level models to create population level trends of obesity. Simulated method of moments is used to estimate uncertain parameters of this model from NHANES data. The resulting model enables community, state, or national policy analysis building on a calibrated model.

Keywords: Obesity, simulation, simulated method of moments, energy balance, system dynamics.

1   Motivation and Background

The obesity trends in the U.S. and many other countries are alarming. The percentage of Americans who are obese has doubled to near 30% during the past four decades [1, 2]. Multiple levels of factors, from biological to environmental, are involved in creating the obesity problem and thus a systems approach to analyze the problem and assess interventions is called for [3]. Models that can assess the potential impact of alternative interventions are much needed in turning the obesity trend. Such models can facilitate policy analysis by expanding the boundaries of our mental models and enhancing learning from evidence [4]. However due to ethical and practical considerations in data collection available dynamic models for obesity rely on short-term time series data and small sample sizes [5-9] which reduces their direct applicability for policy analysis at the population level. While this literature provides a great starting point for modeling individual level body weight dynamics, none of the current models include both childhood and adulthood dynamics. Moreover, the current models focus on modeling a single "average" individual. Extending them to capture variations across individuals is critical for population health policy analysis. Simulating population level weight gain and loss dynamics, and assessing alternative interventions in a new population group, requires dynamic models that 1) Capture the

S.J. Yang, A.M. Greenberg, and M. Endsley (Eds.): SBP 2012, LNCS 7227, pp. 306–313, 2012.
© Springer-Verlag Berlin Heidelberg 2012

individual-level body weight dynamics realistically, building on biological processes that regulate energy balance in body. 2) Connect individual level and population level dynamics in a robust and generalizable fashion. 3) Express the impact of interventions on energy intake and physical activity for different individuals.

2   Analysis and Results

The purpose of the research is to study the dynamics of obesity in the United States over time to build a generic system dynamics model that can be used for obesity policy analysis at multiple levels. We first introduce the individual level model of body weight dynamics used in this study. The population level model which consists of multiple replicas of individual model and their relationships will then be discussed. Finally we explain the calibration and parameter estimation processes and some of the results of the analysis.

2.1   Modeling Body Weight Dynamics

Several models of body weight dynamics have been discussed in the literature [5-14]. These models vary in their level of complexity and the feedback mechanisms they capture. Common across most these models are the state variables fat mass (FM) and fat free mass (FFM) which constitute the majority of body weight in a normal person. More detailed models may consider the stock of glycogen, protein, and extracellular fluid mass and adaptive thermogenesis among other stock variables [7, 15, 16]. While additional complexity could be important in evaluating dynamics that unfold in hours or days, results of comparative studies by Hall [8, 17] suggest that for longer term dynamics FM and FFM provide much explanatory power with very little complexity. We therefore rely on these two variables as the main state variables (stocks) in our individual model. Because we also model growth, a third stock variable captures individual's height. There is currently no unified model for childhood and adulthood body weight dynamics in the literature we therefore create one by combining insights from two of the models in literature by Hall [8] and Butte, Christiansen et al.[9] and developing a new framework that considers energy supply and demand at any point in time and allocates the supply to demand based on a set of priorities.

Energy supply comes from energy intake (EI) and consuming body mass. Total energy intake is the most important factor about food and beverage consumed by an individual Energy could also come from burning FM or FFM (either due to starvation, or if either mass is beyond what the body needs). These three sources create the total energy supply in our equations.

Factors influencing energy demand include demand for maintenance of body and energy demand for growth. The maintenance energy demand depends on resting metabolic rate (RMR) which contributes to 50-75% of energy expenditure, the physical activity energy needs, and the energy for digestion of food and nutrients consumed. RMR itself depends on the body composition (energy needs for maintaining FM and FFM are different). Energy expenditure attributed to physical activity (PA) is largely proportional to the total weight (BW=FM+FFM) and the intensity of PA. Following Hall [8] we represent the total maintenance energy demand (TME) by equation (1). Also following Butte Christiansen et al. [9] we note that γ_F and γ_L are a function of age (Tanner stage) in children, before they stabilize in adulthood.

$$TME = K + \gamma_L.FFM + \gamma_F.FM + PA_{Total}.BW + \eta_L.\frac{dFFM}{dt} + \eta_F.\frac{dFM}{dt} + \lambda.EI \quad (1)$$

We model energy demand for growth based on comparing current weight with the desired weight of individual in near future (e.g. one year ahead). The weight change rate indicated in this period is used to calculate energy needs for growth. Desired weight is determined based on the desired height and desired body mass index (BMI) for the individual, both coming from CDC growth charts, and adjusted based on variations in individual's potential height and current BMI (desired BMI is a weighted average of CDC median for that age and current individual BMI). This demand is then partitioned into energy demand for FFM (Fat Free Energy Demand: FFED) and FM (Fat Mass Energy Demand: FMED) based on regression equations that predict FFM as a function of age, gender, ethnicity, BMI, and height. The resulting gaps (or excesses) in FM and FFM determine the energy demand (supply) that comes from body's growth and BMI and body composition homeostasis processes. Included in these calculations are the energy costs for the implied tissue deposition and disintegration. These parameters are all taken from the literature [8].

Our model can take energy intake (EI) as an exogenous input. However, in practice reliable EI data is not available for large samples and over long time horizons, so we use an alternative set of expressions to endogenously generate EI values for simulated individuals:

$$EI_i = (TME_i + FFED_i + FMED_i)(1 + f(Time) + g(Age)$$
$$+ \; ae.Ethnicity + ae * Gender)(1 + \varepsilon_i) \quad (2)$$

Energy intake for individual i at time t depends on the energy demand for that individual (TME+FFED+FMED), adjusted based on time, age, and demographic effects (to be estimated from data) and an individual variation factor ($\varepsilon_i = VF(Age_i)*S_i$) that captures individual differences in energy intake. The magnitude of individual variation is age-dependent (function VF). Because an individual's eating habits will change only slowly the S_i term is modeled to be serially correlated, with its value at time t+dt being a linear combination of its value at time t and a new normal random number. Moreover, the variations in this term over time depend on its level: people with S_i values far from zero are more likely to change their eating habits (because otherwise they will grow exceedingly obese or starve), while those close to zero change their ε more slowly. These feedback effects are captured in modeling the auto-correlated individual variation term, S_i.

Finally, the three sources for energy demand (TME, FFED, FMED) are compared with three sources for energy supply (EI, energy from reserve FM (if any), and energy from reserve FFM (if any)). Once energy demand and supply are determined, the model uses a priority-based allocation scheme to allocate the supplied energy among demand items. If total supply exceeds demand, then additional energy will be deposited as FM and FFM, based on a partitioning function. The partitioning function we use is the empirical equation $1/(1+0.502*FM)$ [17, 18] as the fraction of energy surplus contributing to changes in FFM. If energy demand exceeds supply, then shortage is covered by not supplying some energy needs or burning essential FM or FFM, depending on priorities that are summarized in Table 1. In this table, a row with YY indicates that the demand source has higher priority than supply source, therefore getting allocated from supply source as much as needed. For example body will use supply of EI to satisfy TME with no reservation. Y means the priorities are close and therefore some supply will be provided, while some of the demand may remain unmet, e.g. in case of using essential FFM to support TME. Finally N means the priority of supply source is higher than demand, and therefore no supply will be provided (e.g. essential FFM will not be used to generate reserve FM).

The model also captures the changes in individual height over time. The height equation assigns each individual a potential that modifies the desired future height for that individual around CDC growth charts median. Height will grow based on the desired values, unless energy supply falls short of demand, in which case height growth is slowed. The logic of equations is discussed above and full equations are available from the first author upon request.

Table 1. Priorities of different energy supply and demand sources in energy allocation process

Supply\demand	TME	FFED	FMED	Reserve FFM	Reserve FM
EI	YY	YY	YY	YY	YY
Essential FFM	Y	N	N	N	N
Essential FM	Y	N	N	N	N
Reserve FFM	YY	NA	Y	N	N
Reserve FM	YY	YY	NA	N	N

Multiple replications of the individual level model can be simulated together to generate population level characteristics of interest such as percentage of the population that is overweight or obese. However creation of the population model requires more than just replicating the individual level model. Specifically, parameters that specify individual EI variation (ε) should be estimated to match the observed behavior in data.

2.2   Model Calibration and Parameter Estimation

An innovative feature of this study is its methodological contribution towards estimating dynamic models based on cross-sectional individual level data from a population. No current database offers the large scale time series data typically used for estimating dynamic models similar to the one we are working with here. However, we note that the data in National Health and Nutrition Examination Survey (NHANES) provides relevant information in terms of the distribution of weight, height, and body composition for different subgroups in the population. A good population level model should be able to match those distributions closely, and the quality of that match can inform parameter estimation and hypothesis testing. If the three statistics (sample mean, variance, and skewness) for the real population match those in the simulated population, we can have some confidence about the ability of the model to recreate the historical results. Further, we can also change model parameters to reduce the discrepancy between these moments of data and model, i.e. to calibrate the model and estimate the unknown parameters. Such comparison can be scaled to many more sub-population moments to increase the precision of the comparisons, and to find better parameter estimates. This is the core idea in the Simulated Method of Moments (SMM), one of the most versatile econometric estimation methods available [19-21]. In this study we use NHANES data for 10 demographic groups (Male and Female of Mexican-American, Other Hispanic, White, Black, Other), over 5 rounds of NHANES (2000, 2002, 2004, 2006, 2008). For each demographic group we divide the population into 26 different age groups (20 groups for ages 0-20, 6 groups for ages 20 and higher, with 10 year periods, i.e. 20-30, 30-40, etc.). Finally, for each age group we use data on 10 moments (mean, variance, and skewness of body weight and BMI, mean and variance of fat mass fraction and height) to estimate the model parameters. We start the model from year 2000, using data from that year to initialize the model, and compare the moments for simulated individuals in different population groups against the next four rounds of NHANES data. This gives us 4*10*26*10 moments to compare against simulated moments[1]. Denoting x as the vector of observed moments for a population group, $x^s(\theta)$ the simulated vector for the same moments using parameter vector θ in the model, the following optimization problem is numerically solved to estimate the parameter vector θ:

$$\hat{\theta} = \arg\min_{\theta}[x - x^s(\theta)].W.[x - x^s(\theta)]' \tag{3}$$

For the weighting matrix W we use the diagonal matrix populated with reciprocal of variances for different moments. We calculate these variances by comparing, for each age group, the variation in each moment across different NHANES rounds and demographic groups. This weighting mechanism provides relatively efficient parameter estimates (for asymptotic efficiency the reciprocal of covariance matrix should be used for W; but that is computationally too expensive). Numerical optimization to find ~ 20 parameters requires in the order of one million simulations and simultaneous estimation of all population groups requires simulating over 20000 agents per simulation, which becomes infeasible for our computational resources. We therefore estimate each demographic group separately.

2.3   Initial Results

Calibration is conducted using a population of 505 simulated individuals, distributed uniformly across different age groups in year 2000 and allowed to grow and change

---

[1] In practice fat mass fraction data is only available for a subset of NHANES rounds and age groups, reducing the total number of moments to slightly under 9,000.

for the next eight simulated years. Minimizing a total of 892 weighted (based on reciprocal of expected variance) moments' squared error from data across years 2002, 2004, 2006, and 2008 the calibration's error was 1045 after 582355 simulations using Powell's conjugate search method and 22 parameters to estimate. The model shows some systematic difference with data as the error value is higher than 942, the 95% confidence level for data and model being indistinguishable; however the difference is fairly limited (the difference could be attributed to only % of moments with significantly large errors) and suggests a generally good fit. The key relationships estimated are the time trends of energy intake, the age trend of variations in energy intake, and the structure of noise terms that drives energy intake (See Table 2). The calibrated model is used next to simulate the distributions of the weight, BMI, height, and fat fraction for a population of 500 white female children from year 2000, to 2008 when they are 13-15 years old.

Table 2. Overview of key estimated parameters

Energy intake for individual i :$EI[i]=Expected\ EI[i]*(1+VF(Age[i])*S[i])$
$VF(x)=0.18*(1-3.79*exp(-((x-2.34)/4.16)^2/2)/4.16)$ See Fig. 1-a.
$Expected\ EI[i]=DesiredEI[i]*(0.99+TF(Time))$ See Fig. 1-b
$DesiredEI[i]$: Calculated based on individual's current state to keep her on track for a normal growth trajectory, or in equilibrium if adult.
$S[i]$: Auto correlated noise, with half-life of $T=4.9/(1+|S[i]|)^{w[i]}$ See Fig. 1-c
$w[i]=1\ if\ S[i]>0,\ w[i]=6.04$ otherwise

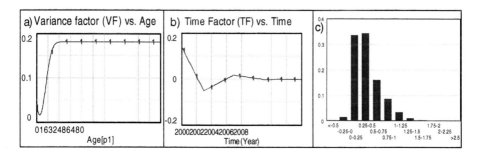

Fig. 1. Estimated relationships from calibration a) Individual variance factor (VF) as a function of age. b) Individual energy intake (TF) as a function of time and c)Distribution of individual variation in energy intake (S) resulting from estimated parameters.

The weight, BMI, height, and fat fraction distributions are compared with NHANES 2008 data for the same group (only 71 subjects available). The averages are largely consistent with the data while the standard deviation of height is more in data than in simulated results (Fig. 2-a). Note that the model has been calibrated to all 26 age groups simultaneously, should calibration been conducted against only these three age groups, even better fit could be expected.

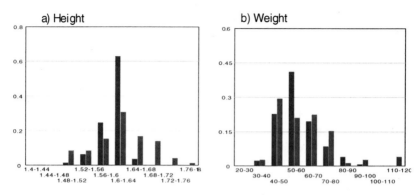

Fig. 2. Comparison of Height and Weight data (blue, right hand bars) (N=71) and simulated results (red, left hand side bars) (N=500) for NHANES 2008, 13-15 year old white female

3   Discussion and Limitations

The contributions of this work are twofold. First, for the first time in the public health literature we provide an integrated model of weight gain and loss that covers both childhood and adulthood and connect it to population level weight dynamics. The resulting model provides a flexible, validated, module to be integrated in any policy analysis project. The model is robust to extreme conditions, does not require parameter estimation, and can be plugged with any hypothetical interventions.

Second, we adopt the SMM for application to arbitrary system dynamics models to be estimated using not only panel or time series data but also cross sectional population statistics. This can open the door to much wider use of nonlinear feedback-rich models in data intensive domains traditionally dominated by simpler regression models. The results presented here provide a first cut at a complex modeling and estimation project, and future work should focus on improving the fit between the model and the data in some settings. Finally, the models main utility will be for policy analysis settings where energy intake or physical activity values are changed from their base levels; that work goes beyond the scope of current paper.

Acknowledgments. Financial support provided through National Collaborative on Childhood Obesity Research (NCCOR) Envision's Comparative Modeling Network (CompMod) program and NIH Office of Behavioral and Social Sciences Research. Dr. Alice Ammerman and Tosha Smith have made valuable contributions to the research project that has lead to this paper. We are also very thankful to Drs. Patty Mabry, Terry Huang, and the members of CompMod program for their continued support and their insightful comments and suggestions.

References

1. Bray, G.A., Bouchard, C.: Handbook of obesity: etiology and pathophysiology. Marcel Dekker, New York (2004)
2. Ogden, C.L., et al.: Prevalence of overweight and obesity in the United States, 1999-2004. JAMA: The Journal of The American Medical Association 295(13), 1549–1555 (2006)

3. Huang, T.T., et al.: A systems-oriented multilevel framework for addressing obesity in the 21st century. Preventing Chronic Disease 6(3), 1–10 (2009)
4. Sterman, J.D.: Learning from evidence in a complex world. Am. J. Public Health 96(3), 505–514 (2006)
5. Kozusko, F.: Body weight setpoint, metabolic adaption and human starvation. Bulletin of Mathematical Biology 63(2), 393–403 (2001)
6. Christiansen, E., Garby, L., Sørensen, T.I.A.: Quantitative analysis of the energy requirements for development of obesity. Journal of Theoretical Biology 234(1), 99–106 (2005)
7. Flatt, J.-P.: Carbohydrate-Fat Interactions and Obesity Examined by a Two-Compartment Computer Model. Obesity 12(12), 2013–2022 (2004)
8. Hall, K.D.: Mechanisms of metabolic fuel selection: modeling human metabolism and body-weight change. IEEE Engineering in Medicine And Biology Magazine 29(1), 36–41 (2010)
9. Butte, N.F., Christiansen, E., Sorensen, T.I.A.: Energy Imbalance Underlying the Development of Childhood Obesity. Obesity 15(12), 3056–3066 (2007)
10. Abdel-Hamid, T.K.: Modeling the dynamics of human energy regulation and its implications for obesity treatment. System Dynamics Review 18(4), 431–471 (2002)
11. Christiansen, E., Garby, L.: Prediction of body weight changes caused by changes in energy balance. European Journal of Clinical Investigation 32(11), 826–830 (2002)
12. Thomas, D.M., et al.: A mathematical model of weight change with adaptation. Mathematical Biosciences and Engineering: MBE 6(4), 873–887 (2009)
13. Song, B., Thomas, D.M.: Dynamics of starvation in humans. Journal of Mathematical Biology 54(1), 27–43 (2007)
14. Kozusko, F.P.: The effects of body composition on setpoint based weight loss. Mathematical and Computer Modelling 35(9-10), 973–982 (2002)
15. Hall, K.D.: Predicting metabolic adaptation, body weight change, and energy intake in humans. American Journal of Physiology. Endocrinology and Metabolism 298(3), E449–E466 (2010)
16. Hall, K.D.: Computational model of in vivo human energy metabolism during semistarvation and refeeding. American Journal of Physiology. Endocrinology and Metabolism 291(1), E23–E37 (2006)
17. Chow, C.C., Hall, K.D.: The Dynamics of Human Body Weight Change. PLoS Computational Biology 4(3), e1000045 (2008)
18. Forbes, G.B.: Body fat content influences the body composition response to nutrition and exercise. Ann. N.Y. Acad. Sci. 904, 359–365 (2000)
19. McFadden, D.: A Method of Simulated Moments for Estimation of Discrete Response Models without Numerical-Integration. Econometrica 57(5), 995–1026 (1989)
20. Lee, B.S., Ingram, B.: Simulation Estimation of Time-Series Models. Journal of Econometrics 47(2-3), 197–205 (1991)
21. Duffie, D., Singleton, K.J.: Simulated Moments Estimation of Markov Models of Asset Prices. Econometrica 61(4), 929–952 (1993)

Creating Interaction Environments: Defining a Two-Sided Market Model of the Development and Dominance of Platforms

Walter E. Beyeler, Andjelka Kelic, Patrick D. Finley, Munaf Aamir,
Alexander Outkin, Stephen Conrad, Michael D. Mitchell, and Vanessa Vargas

Complex Adaptive Systems of Systems (CASoS) Engineering, Sandia National Laboratories,
Albuquerque New Mexico, USA
{webeyel,akelic,pdfinle,msaamir,avoutki,shconra,
micmitc,vnvarga}@sandia.gov

Abstract. Interactions between individuals, both economic and social, are increasingly mediated by technological systems. Such *platforms* facilitate interactions by controlling and regularizing access, while extracting rent from users. The relatively recent idea of two-sided markets has given insights into the distinctive economic features of such arrangements, arising from network effects and the power of the platform operator. Simplifications required to obtain analytical results, while leading to basic understanding, prevent us from posing many important questions. For example we would like to understand how platforms can be secured when the costs and benefits of security differ greatly across users and operators, and when the vulnerabilities of particular designs may only be revealed after they are in wide use. We define an agent-based model that removes many constraints limiting existing analyses (such as uniformity of users, free and perfect information), allowing insights into a much larger class of real systems.

Keywords: Two-sided markets, platform economics, platform competition, agent simulation.

1 Introduction

A *platform* is a collection of equipment, facilities, and standards that facilitates a particular kind of interaction. Telecommunications systems, social networking sites, the internet as a whole, DVD players, and credit card networks are a few examples of the platforms that increasingly mediate interactions among people and institutions. Rochet and Tirole [1] and Evans [2,3] recognized the distinctive economic features of these systems, and initiated their formal study as two-sided markets. Many important results have been derived in the short time since, however almost all are derived for systems simple enough to be treated analytically. Some common assumptions include perfect information about demand functions, homogeneity of demand, and uniformity of fees across users of a given class. Most analyses obtain equilibrium results rather than exploring the dynamics of platform development and adoption. Because the

S.J. Yang, A.M. Greenberg, and M. Endsley (Eds.): SBP 2012, LNCS 7227, pp. 314–321, 2012.
© Springer-Verlag Berlin Heidelberg 2012

basic dynamics contain reinforcing feedbacks (for example platform attractiveness to prospective users increases with the number of current users) the equilibrium configuration is likely to be sensitive to small variations in development details.

Some of these analytical constraints are being removed by ongoing research. For example Alexandrova-Kabadjova et al. [4] use an agent-based model to study platform competition when geographical constraints influence interactions. The influence of platform security on users' adoption decisions has received little attention, despite the increasing use of platforms to carry personal and financial data. Creti and Verdier [5] have pioneered the study of fraud costs and liability allocation on platform selection using a staged optimization model amenable to analytical solution.

The agent-based model defined here removes constraints that analytical approaches impose. We focus on the interacting decisions of platform users and creators, including users' decisions to subscribe to or abandon a particular platform, creators' decisions to allocate tariffs and to invest in capacity and marketing. Because we are especially interested in platform security, the model includes intruders' decision to attack the platform in a way that imposes costs on users and creators. The prospect of such losses is a factor in users' adoption decisions and creators' investment decisions.

2   Platforms as Two-Sided Markets

In economic terms platforms create two-sided markets. They are used to interconnect two sets of users, which constitute the sides of the market. Sides typically play distinct roles, such as merchants and credit card customers, or application developers and application users, or musicians and audience. The two sides may use very different technology to connect to the platform, and may face different connection costs and fee structures. The platform operator creates and maintains the infrastructure, and gets revenue from one or both sides of the market.

The different costs faced by different kinds of users, and the platform operator's ability to determine prices and control access, can lead to surprising strategies for optimizing operators, such as subsidizing one side of the market at the expense of the other. The very recent recognition of two-sided markets as a distinctive category has produced important general insights of this kind; however almost all are derived for systems simple enough to be treated analytically.

The increasing variety of platforms through which economic and social interactions are conducted suggests that a model general enough to provide broad insights, yet rich enough to relax assumptions that constrain analytical approaches, would be a useful way of understanding, and setting policies for, important systems of this kind. We define such a model below by describing the essential dynamics of the system. Building from the basic interactions characteristic of two-sided markets, we demonstrate how operators' investments on performance and security can bring in new constraints on the adoption and use of platforms.

3   Model Components and Behavior

The model includes the basic classes of Platform, Operator, and User (Fig. 1). The systems of interest typically have two major subclasses of Users which define the sides of the two-sided market. One class is often a Producer of some good or service or content, while the other is a Consumer. Figure 1 shows the three principle classes and the flows of value considered in the model. These values create motivations for the actions of each class of decision maker. How the Platform creates these values depends on intrinsic features of the domain, performance properties of the Platform, and the number of users of the Platform. The model can be applied to specific cases by specifying parameter values, however the components of the model and their dynamics are meant to be generally applicable.

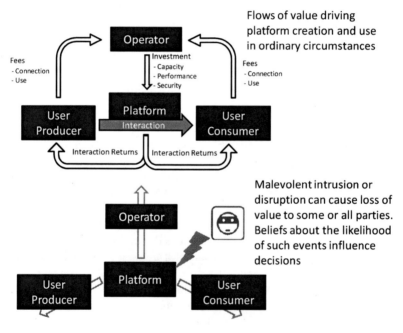

Fig. 1. Main classes in the platform model, and flows of value created by platform use and intentional disruption

Users derive some benefit from interacting on the Platform, and pay fees to the Operator, generally for both access and for usage. Often these fees are explicit; however they might be imposed indirectly, by means of advertising for example. There may be more than one Platform available to Users, so that Users can choose to subscribe to or use alternatives on the basis of their costs and returns. Operators can set the subscription and usage fees borne by each user, and these might vary across users.

Some platforms, especially those mediating financial transactions, compete on the basis of security. We include security as a consideration by means of random acute costs, which can be imposed on individual Users, on groups of Users, and on the Operator. These costs represent losses that would occur as a consequence of a security breach, such as theft of assets or expenses incurred as a result of identity theft. User's expectations about such costs will influence their platform choice and use. These expectations are based on prior beliefs, on the actual security history for the platform, on reports of trusted social contacts, and on marketing messages created by the Operator. Operators, in turn, may invest in security measures that reduce costs or probabilities of a breach as well as in marketing messages designed to shape Users' beliefs about the security of their Platform and other Platforms. Operators' interest in maintaining security comes both from costs they might incur as a direct consequence of a breach and costs of any loss of subscribers or usage arising from Users' changed perception of risk.

3.1   Model of Producer and Consumer Behavior

The success or failure of a particular platform, and the value produced for its users, are the result of interacting decisions by the Operator and by members of the two subclasses of User, Producers and Consumers. We assume Users derive some specified basic value from conducting a single transaction of the kind the platform supports. This basic value may be different for Producers and Consumers, but is the same for all platforms that compete for Users' business. Platforms differ in the number of transaction opportunities they provide, and in the costs they present to Users. Some of these costs can be directly controlled by Operators, while others are the indirect consequences of decisions Operators make, such as investments in capacity and security. These costs and decisions are the strategic variables that Operators use to compete for market share and profit.

The dynamical model of Users' decisions about their participation in a particular platform is shown as a causal loop diagram[1] in Figure 2. The defining dynamical feature of platforms is the reinforcing feedback that causes an increase in the number of producers using the platform to attract additional consumers, and vice-versa. Such two-sided network effects have been identified in many technological systems [6].

This reinforcing feedback can lead to exponential growth in platform users, as well as to exponential collapse, depending on how users' costs compare with the benefits they obtain from using the platform, and the costs presented by competing platforms. The basic value that a user obtains from a single transaction on the platform is exogenous. The model distinguishes subscription to the platform, which involves getting whatever equipment and authorization is required to use it, and usage, which involves conducting transactions on the platform. These two actions may have very

---

[1] Causal loop diagrams are qualitative model specifications showing the variables considered in the model and the causal relationships between pairs of variables. The effect of increasing the value of a variable on causally dependent variables is indicated by a +/- sign near the arrow to the dependent variable.

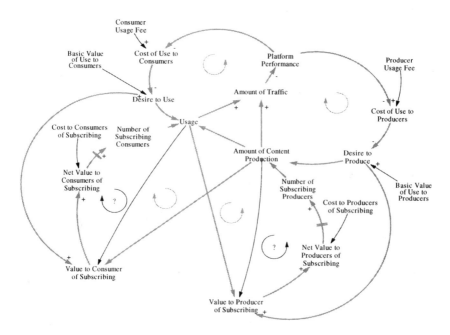

Fig. 2. Causal loop diagram showing the dynamics of platform subscription and usage by producers and consumers. The central reinforcing feedback can lead to exponential growth (or collapse) as additional users of one kind increase the value of using the platform to users of the other kind. Increased traffic on the platform may degrade users' perceived performance and limit growth.

different costs, which differ between producers and consumers. Platform operators can try to encourage growth by manipulating those aspects of costs that they can control. For example a new platform with few producers or consumers will present little value for either side to subscribe. Initial subsidies for subscription, reducing or inverting subscription costs, may attract enough initial users to allow the subsidy to be eliminated for later subscribers. The subscription costs and usage fees in Figure 1 are two important targets of operators' decision-making.

3.2   Model of Operator Behavior

Many existing analyses of two-sided markets derive pricing strategies for operators which maximize their profits, given differing price sensitivities of producers and consumers [1]. In these analyses Users' costs can be directly controlled by Operators. Many systems impose significant indirect costs on Users which may have considerable influence on their decisions, but which are not directly controllable by operators. We include the effects of platform performance and security as indirect costs. Figure 2 shows the potential for increased platform usage to degrade performance and so increase the usage costs of consumers or producers. This increased cost can place limits on prospective users' uptake of a platform. Investments in capacity can be used to improve performance and encourage further growth. The transactions hosted on many platforms have a financial component, and some platforms (such as credit cards) are specifically designed for financial purposes. The security of such platforms is a special concern to users. Security compromise might lead to loss of personal information, initiation of fraudulent transactions in the guise of legitimate users, corruption or blockage of transaction data, and many other undesirable consequences. Such events might lead to direct financial loss to users, or simply to inconvenience and delays which we represent as an indirect cost. Users do not need to experience such events directly in order to weigh such costs in their decisions to subscribe to and use the platform. The expectation of loss from lax security, which includes both the prospective cost of a breach and the users' probability that a breach will occur, is included as a component of cost. Managing platform security, and users' perceptions of security, is a second means by which operators can indirectly control costs.

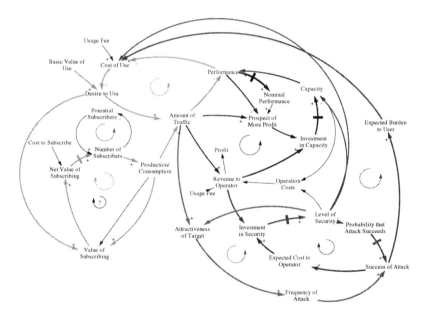

Fig. 3. Causal loop diagram showing the dynamics of platform operator's behavior. Platform growth can be driven by investments leading to improved performance and greater security. Perceptions of improved security may lag investments in security, and reductions in security may take time to manifest as attacks. These delays can create oscillations in security investment.

Figure 3 elaborates the causal loop diagram to include the dynamics of operator behavior. In this diagram the two kinds of platform user have been collapsed into one in order to simplify the picture. Operators receive revenue through platform use. They invest in both expanding platform capacity and in securing access and traffic from this revenue. The linkage between revenue and investment reflects the possibility of direct reinvestment as well as loans secured by future revenue.

The platform operator can encourage growth by making investments that improve platform performance and that increase security. Both kinds of investment tend to lower the effective cost to a user of transacting on the platform. Performance improvements are shown as coming from an increased capacity, although other platform changes that facilitate use (such as redesigning interfaces) can have a similar effect. Capacity investments may be reactive – driven by performance problems with the existing system – or proactive – driven by trends in the current usage which anticipate constraints on performance.

Investments in security are motivated by threats of attack. The kind of attack that might be staged and the costs imposed by successful attacks depend on the specific platform being considered. A denial of service attack for example might delay users' business operations and might degrade the reputation of the platform operator. Theft of credit-card data might lead to financial losses to issuing banks and inconvenience costs to cardholders. A successful attack will increase (to some degree) both the operator's and users' estimated costs, leading to increased investment in security by operators and possible changes in usage. An increased investment in security will reduce the probability of successful attack to some degree, lessening users' perceived costs and encouraging growth in platform use. There can be significant delays in this process, both in deploying security measures and in changing users' perceptions, so that the return on security investments may come long after expenses are incurred. Heightened security can impose burdens on the system and its users. These effects are shown in Figure 3 as a possible reduction in capacity and a possible increase in user costs driven by increases in the level of security.

The threats faced by a particular platform are also dynamic, and the model includes two important factors influencing the attractiveness of the platform as an attack target. The current level of security can deter attack or cause it to be directed elsewhere. The amount of traffic on the platform is assumed to make it more attractive as a target, whether the object is financial gain or spectacle. Increased attractiveness leads to more frequent attack attempts, and a greater incidence of successful attack.

4   Model Analysis and Development Status

The causal model defined above represents the processes that can determine the outcome of competition among platforms when the operators of those platforms adopt different strategies. Even without a precise formulation of the relationships represented by the causal links, the basic feedback structures can produce insights into possible behavior. For example Figure 3 suggests two mechanisms by which an operator might try to expand their platform: investing in capacity to improve nominal performance; and investment in security that decreases users' expected costs. While there are potential delays in realizing performance improvements through new capacity, the delays between an investment in security and an improvement in users' perception may be much greater, especially in system characterized by infrequent but costly attack. This suggests that a strategy emphasizing capacity expansion might out-compete a strategy emphasizing platform security, particularly in a market with rapid growth rates.

A mathematical specification of the causal links illustrated in Figures 2 and 3 is necessary to study particular systems. Such specification allows simulation of possible histories of subscription and use resulting from different user dispositions, costs, and operator policies. We are currently developing an application to retail payment systems, using an agent-based framework that allows for heterogeneous populations of Consumers and Producers, price differentiation by Operators, and other properties that characterize the real system but that make analytical approaches intractable.

References

1. Rochet, J.-C., Tirole, J.: Platform Competition in Two-Sided Markets. J. Eur. Econ. Assn. 1(4), 990–1029 (2003)
2. Evans, D.S.: Identification The Antitrust Economics of Two-sided Markets. AEI-Brookings Joint Center for Regulatory Studies (2002)
3. Evans, D.S.: Some Empirical Aspects of Multi-sided Platform Industries. Rev. Network Econ. 2(3), 191–209 (2003)
4. Alexandrova-Kabadjova, B., Krause, A., Tsang, E.: An Agent-Based Model of Interactions in the Payment Card Market. In: Yin, H., Tino, P., Corchado, E., Byrne, W., Yao, X. (eds.) IDEAL 2007. LNCS, vol. 4881, pp. 1063–1072. Springer, Heidelberg (2007)
5. Creti, A., Verdier, M.: Fraud, Investments, and Liability Regimes in Payment Platforms. European Central Bank Working Paper Series No, 1390 (2011)
6. Katz, M.L., Shapiro, C.: Some Network Externalities, Competition, and Compatibility. Amer. Econ. Rev. 75(3), 424–440 (1985)

The Impact of Attitude Resolve on Population Wide Attitude Change

Craig M. Vineyard, Kiran Lakkaraju, Joseph Collard, and Stephen J. Verzi

Sandia National Laboratories
cmviney@sandia.gov

Abstract. Attitudes play a critical role in informing resulting behavior. Extending previous work, we have developed a model of population wide attitude change that captures social factors through a social network, cognitive factors through a cognitive network and individual differences in influence. All three of these factors are supported by literature as playing a role in attitude and behavior change. In this paper we present a new computational model of attitude resolve which incorporates the affects of player interaction dynamics that uses game theory in an integrated model of socio-cognitive strategy-based individual interaction and provide preliminary experiments.

Keywords: Computational model, attitude change, cognitive modelling, social modelling, game theory.

1 Introduction

Attitudes are "general and relatively enduring evaluative responses to objects" where objects can be "a person, a group, an issue or a concept" [1, Page 1]. Attitudes are shown to have an impact on, and can sometimes predict, the behaviors of individuals (e.g., voting behavior [2], consumer purchases [3,4]).

Understanding population wide attitude change is thus an important step to understanding the behavior of societies. For instance, consider the change in attitudes towards global warming and the environment that has resulted in a significant change in public policy and national priorities [5].

While there are a number of factors that influence attitude change [6], we will focus on three in this paper. The first is social – individuals are exposed to various attitudes and information through interaction with others. Family, friends, acquaintances, and the media all influence the attitudes of individuals by providing new information/opinions.

The second factor is cognitive – individuals tend to hold a set of attitudes that are consistent with each other [7,8,9]. According to *cognitive consistency theories*, an individual holding a strong positive attitude towards environmentalism should also hold a strong positive attitude towards recycling; if they do not, the attitudes are inconsistent with each other and could cause an uncomfortable feeling (i.e. *cognitive dissonance*) which tends to result in either attitude or behavior change [6].

The third factor is individual differences in influence. Intuitively, some individuals seem more likely to change than others. This intuition has been supported

S.J. Yang, A.M. Greenberg, and M. Endsley (Eds.): SBP 2012, LNCS 7227, pp. 322–330, 2012.

by research from the marketing and social psychology fields (see Section 2 for more details).

Previous work [10,11] has described a socio-cognitive model that captures the social and cognitive aspects of attitude diffusion in individual interaction in a social setting. This work extends previous work by incorporating individual differences in influence. In our model we have two types of individuals, "susceptibles" and "advocates". "Susceptibles" represent individuals that are strongly influenced by others. Thus in any interaction they change significantly. "Advocates" represent individuals that do not change as much.

In this paper we present preliminary work on a new model that captures these three elements together to help understand how information, in the form of attitude, is spread in a social setting. In our model, individual differences will be modeled using *game theory*. We describe the basis of this model, then show some simple simulations that explore the dynamics of this model.

2   Theoretical Basis

Intuition and common experiences seems to indicate that people vary in how they are influenced by others. Some people stand firm and rarely change their attitude opinion, while others often vacillate. Research in social psychology and marketing has provided some evidence to this folk psychological idea.

Decades of research has occurred in the marketing domain to understand how consumer purchasing decision are influenced by others. Two types of influence are usually described, (1) informational – where individuals are influenced by obtaining new information from peers; and (2) normative – where individuals are influenced to conform to others decision in order to be liked [12]. The unit of analysis here is on individuals, and the demographic characteristics that can make them change, such as age or gender [13].

Another branch of research has focused more on attitudes themselves rather than individuals. *Strong* attitudes are: "resistant to change, stable over time, and have powerful impact on information processing and behavior" [14, page 279]. Several characteristics of an attitude, such as its importance to an individual or its accessibility, can influence it's strength.

3   Model Description

An attitude dissemination model consists of various levels of interactions. At the social level, some mechanism selects which agents are to interact with one another. Numerous possibilities exist such as teacher-student interactions, co-worker interactions, or a discussion amongst friends/peers to name a few. Each of these interaction types may have subtle effects with regards to how the selected interacting agents affect one another. Upon selecting the interacting agents, their affect upon one another must be assessed in some manner. And finally, at the individual level, any internal changes brought about by the social interaction may lead to a cascade of changing beliefs or a resilience of the agent's original attitudes.

Game theory is a branch of mathematics pertaining to strategic interactions. As such, it is applicable to the development of a framework which can address each level of interaction in an attitude dissemination model. In this work, we have investigated applying game theory to the agent interaction level. Before describing our game theoretic approach to agent interaction, we will first give an overview of the cognitive and social aspects of our model (for a more detailed description see [11]).

Within each agent in the model is a Parallel Constraint Satisfaction (PCS) model to represent their individual cognitive network. A PCS model is a type of connectionist, attractor neural network. Nodes within an agent's cognitive network represent concepts, hypotheses, or information. These values are continuous and range from -1 (min) to 1 (max). Weighted links between concepts (w_{ij}) indicate the strength of influence between the concepts. PCS models utilize a connectionist approach to find a consistent set of concepts: the value of each concept is updated according to a non-linear *activation function*. One activation function that is often used is [15],[16]:

$$a_j(t+1) = a_j(t)(1-d) + \begin{cases} net_j(\max - a_j(t)) & \text{if } net_j > 0 \\ net_j(a_j(t) - \min) & \text{if } net_j \leq 0 \end{cases} \qquad (1)$$

where:

$$net_j = \sum_i w_{ij} a_i(t) \qquad (2)$$

and d is a decay term that is set to 0.05. This update rule modifies the value of a concept based on the values of other concepts. An agent synchronously updates all concepts in their network and repeats this process a finite number of times defined as the cognitive effort.

Given this basic inter-agent cognitive framework, our overall model is then composed of three elements:

$$< A, G, C > . \qquad (3)$$

A represents a set of n agents that represent individuals, C is a *cognitive consistency network* that represents the internal cognition that drives change in an attitude, and G is the social network graph that defines the social structure of the population. Each agent has the same number of concepts (m), and the same weights between concepts (W), however each agent can differ in the value of those concepts.

We assume turn based dynamics for simulation of individual interaction – where at each timestep the following actions are taken:

1. A single agent is randomly chosen from the population, uniformly across all agents. We call this agent the *speaker*. One random neighbor of the speaker is designated the *hearer*, chosen uniformly from all neighbors.
2. A single concept is chosen as the *topic* of communication.
3. The speaker and hearer communicate with one another regarding the topic of communication. This is the *communication* step.

4. The interacting agents engage in an *information integration* step.
5. The interacting agents modify the values of their individual PCS model cognitive networks in order to find a more consistent set of beliefs. This is the *agent update* step.

In the communication step, our game theoretic approach to agent interactions incorporates agent personality types such that each agent is either an *advocate* or *susceptible*. Much like the canonical Hawk-Dove game, this approach yields differing payoffs depending upon what types of players interact. If an advocate speaks to a susceptible agent, then the advocate will have great success in influencing the suspectible agent towards their attitude with little concession on their own part. However, if two advocates interact with one another, neither is able to persuade the other to budge in their attitude. When two susceptibles interact they are both moderately successful in influencing the other. And so, while being an advocate and aggressively trying to influence the agent you are interacting with is quite successful when dealing with susceptible agents it is a poor strategy in dealing with another advocate. The level of influence one agent has on another is the information integration step. Consequently, new strategic dynamics emerge regarding societal interactions.

Formally, the following update equations govern this agent interaction. Let $x_i(t+1)$ be the activation of node x for agent i at time $t + 1$. Let $u_a \in [0,1]$ be the strength of influence of an advocate. Let $u_s \in [0, 1]$ be the strength of influence of a susceptible agent. Let $u_a + u_s = 1$, where the update equations are:

$$x_i(t+1) = \begin{cases} u_a x_i(t) + u_s x_j(t) & \text{for Advocate } i \text{ and Susceptible } j \\ u_a x_j(t) + u_s x_i(t) & \text{for Susceptible } i \text{ and Advocate } j \\ 0.5x_i(t) + 0.5x_j(t) & \text{for Susceptibles } i \text{ and } j \\ x_i(t) & \text{for Advocates } i \text{ and } j \end{cases} \quad (4)$$

4   Experimental Results

4.1   Parameterizations

In order to select functionally realistic values for the influence strengths of both the advocate and susceptible agent types we performed an empirical value sweep and found intuitively plausible behaviour to occur with an advocate agent influence strength (u_a) of 0.7 and susceptible agent influence strength (u_s) of 0.3. Weights between concepts in each agent's PCS network were set at 0.05, and we selected a small percentage (10% to 30%) of the population of agents to be of the advocate type while the rest became susceptible agents.

4.2   Experiment 1: Number of Advocates

To visualize and more easily understand the influential power of advocates who may significantly change susceptible agents and are not easily influenced themselves, we experimented with small, randomly connected networks and varied the percentage of advocates in the overall population (see results below). We used the same three node cognitive consistency network in each agent and initialized all of their networks to be cognitively consistent with an attitude value ranging from zero to the minimum defined by the negative attractor value (determined by the weight values in the PCS model). Then we selected a minority percentage of agents to adopt the opposing positive attitude value (ranging between 0 and the maximum postive attractor value, again determined by the weights in the PCS model). The negative attitude value agents are susceptible agents and the positive attitude value agents are advocates.

Iterating over this initialization, we investigated whether 10, 20, or 30 percent positive attitude advocate agents would be sufficient to persuade the opposing majority of susceptible agents to change their attitude belief or conversely if the majority of susceptible agents would be able to overwhelm the minority advocates.

4.3   Experiment 2: Position of Advocates

Beyond investigating the affect of the number of influential advocates in a network, we also examined the significance of the particular position of an advocate within the social network topology. To place advocates in a meaningful location we used small hierarchical and ring network topologies in which we could manually determine specific agents as advocates due to their connectivity or position within the topology.

5   Discussion

5.1   Experiment 1

We have used a small ten node, randomly connected network so that it is tractable to visualize and analyze the interactions and attitude values over each time step of our simulation. A single advocate (ten percent of the population) was never able to convert a single susceptible agent and eventually is overwhelmed to adopt the negative attitude itself. Using two advocates (twenty percent of the population) allows for possible advocate-advocate interactions as well as doubles the possible conversion influence. We began by randomly selecting two random agents as advocates and as before, despite their increased influential capabilities, the two minority advocates were not sufficient to persuade the majority of susceptible agents. Selecting two connected agents, each with low social connectivity, as the advocates was sometimes able to convert the opposing eight susceptible agents. Increasing the advocate population to three (thirty percent) typically was sufficient to convert the seven opposing attitude susceptible agents simply by using randomly selected advocates.

5.2   Experiment 2

To study the significance of the position of advocates within a network, we used the small ring and hierarchical network topologies shown on the left and right side of Figure 1 respectively.

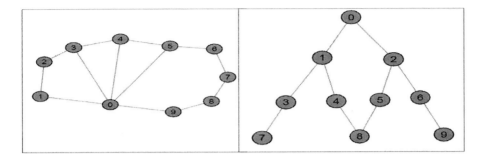

Fig. 1. Small ring and hierarchical network topologies

For the ring network we experimented with setting agent 0 as the sole advocate since it is the most connected agent. However, despite being the most connected agent it was unable to overcome the nine opposing attitude susceptibles. As one might expect, it was able to raise the attitude values of the agents it was connected to, but not sufficiently enough to convert any of them to the postive attitude. Additionally, we also experimented with setting agent 2 as the sole advocate. As a member of the perimeter of the ring which only interacts with its two adjacent neighbors agent 2 was also unable to convert the opposing 9 susceptibles. Finally, setting both agents 0 and 5 as advocates was typically enough to convert the other 8 agents despite the fact that neither is directly connected to many of the other agents in the ring. In this pairing, agents 0 and 5 were able to reinforce one another and eventually persuade the other agents despite being outnumbered.

Figure 2 depicts the affect of different advocate positions within a network. This image plots each agent's attitude value across time. On the left half of Figure 2, two arbitrary agents were selected as advocates, and were unable to persuade the other agents to adopt the positive attitude. In fact these advocates were themselves converted. This is illustrated by by the full convergence of all agent attitude values to the negative attractor at -0.33. Alternatively, the right half of Figure 2 plots the simulation results where agents 0 and 5 were selected as advocates. Although they were the only two agents initially in favor of the positive attitude value, over time they are successfully able to pull all the other agents attitude values up to the positive attractor at 0.33.

For the hierarchical network (seen in the right half of Figure 1), we likewise began by experimenting with single agents as the lone advocate and investigated whether the position at the topmost node in the network would be sufficient to propagate down and convert the rest of the nodes, or whether being a bottom node would be able to spread upwards and convert the others. In both cases, neither sole advocate was sufficient to overcome the other nine susceptibles. Alternatively, utilizing both a top down and bottom up influence by setting agents 0, 7, and 9 as advocates was largely sufficient to spread their attitude and convert all of the remaining agents despite the fact that their influence is

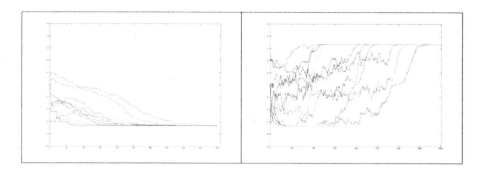

Fig. 2. Drastically different results for different advocate positions within a network

constrained such that they can only interact with one or two other agents whom they must convert and in affect those most continue the spread with their further neighbors.

6   Conclusion and Future Work

6.1   Conclusion

In this paper we have presented a new computation model of attitude resolve which incorporates the affects of player interaction dynamics using game theory in an integrated model of socio-cognitive strategy-based individual interaction. Running a variety of experiments on this model with different network topologies and population mixtures we have investigated attitude adoption and resolve within small networks. These experiments have indicated that selecting specific advocate locations within a population tends to result in a stronger spread of influence, even with only a few advocates in a small network, as opposed to random placement. And the enhanced spread of influence can be observed by the ability to convert agents with opposing attitude values that greatly outnumber the advocates as well as through lower diffusion time (iterations in the PCS network - also referred to as cognitive effort) for the adoption of the opposing attitude.

6.2   Future Work

Having investigated interaction dynamics on computationally tractable small networks, we would like to extend this work to larger real world networks to assess whether the same behavior occurs. Additionally, we can also extend the model to incorporate factors such as more sophisticated cognitive networks.

In addition to incorporating game theory at the agent level we are also interested in looking at the affects of applying game theory at both the social and individual, cognitive level. At the social level, agent interactions are typically dictated by given static social network topology. An alternative approach would be to investigate coalition formation games from coalitional game theory. This particular branch of game theory seeks to build network structure in a strategic manner. Furthermore, if a particular agent, selected as a speaker, is seeking to maximize the spread of their influence they might want to select the hearer they share their attitude view with strategically rather than randomly from everyone they know. Various aspects of game theory such as coalitional graph games or network routing could provide an alternative framework to model this interaction.

As an alternative mechanism to implement parallel constraint satisfaction (PCS) attitude updates within a single agent, a game may be played amongst two attitudes competing to influence one another. Rather than requiring multiple iterative updates the effect is achieved by a single instance of the game. For attitude nodes connected to (influenced by) several other nodes, the game is played simultaneously with all neighboring nodes and the net attitude change is a synchronous aggregated update of all games played in a pairwise interaction with neighbor nodes.

Acknowledgements. Sandia is a multiprogram laboratory operated by Sandia Corporation, a Lockheed Martin Company, for the United States Department of Energy under contract DE-AC04-94AL85000.

References

1. Visser, P.S., Clark, L.M.: Attitudes. In: Kuper, A., Kuper, J. (eds.) Social Science Encyclopedia. Routledge (2003)
2. Krosnick, J.A.: The role of attitude importance in social evaluation: A study of policy preferences, presidential candidate evaluation, and voting behavior. Journal of Personality and Social Psychology 55(2), 196–210 (1988)
3. Jones, C.R.M., Fazio, R.H.: Associative strength and consumer choice behavior. In: Haugtvedt, C.P., Herr, P.M., Kardes, F.R. (eds.) Handbook of Consumer Psychology, pp. 437–459. Psychology Press (2008)
4. Fazio, R.H., Olson, M.: Attitudes: Foundations, functions, and consequences. In: Hogg, M.A., Cooper, J. (eds.) The Sage Handbook of Social Psychology. Sage (2003)
5. Change, Y.P.O.C.: Climate change in the american mind: Americans' climate change beliefs, attitudes, policy preferences, and actions. Technical report, George Mason University (2009)
6. Visser, P.S., Cooper, J.: Attitude change. In: Hogg, M., Cooper, J. (eds.) Sage Handbook of Social Psychology. Sage Publications (2003)
7. Russo, J.E., Carlson, K.A., Meloy, M.G., Yong, K.: The goal of consistency as a cause of information distortion. Journal of Experimental Psychology: General (2008)
8. Simon, D., Holyoak, K.J.: Structural dynamics of cognition: From consistency theories to constraint satisfaction. Personality and Social Psychology Review 6(6), 283–294 (2002)
9. Simon, D., Snow, C.J., Read, S.J.: The redux of cognitive consistency theories: Evidence judgments by constraint satisfaction. Journal of Personality and Social Psychology 86(6), 814–837 (2004)

10. Lakkaraju, K., Speed, A.: Key parameters for modeling information diffusion in populations. In: Proceedings of the 2010 IEEE Homeland Security Technologies Conference. IEEE (2010)
11. Lakkaraju, K.,, S.: A cognitive-consistency based model of population wide attitude change. In: Proceedings of the 2010 AAAI Fall Symposium on Complex Adaptive Systems (2010)
12. Bearden, W.O., Netemeyer, R.G., Teel, J.E.: Measurement of consumer susceptibility to interpersonal influence. Journal of Consumer Research 15 (1989)
13. Girard, T.: The role of demographics on the susceptibility to social influence: A pretest study. Journal of Marketing Development and Competitiveness 5(1) (2010)
14. Krosnick, J.A., Smith, W.R.: Attitude strength. In: Ramachandran, V. (ed.) Encyclopedia of Human Behavior, vol. 1. Academic Press (1994)
15. Kunda, Z., Thagard, P.: Forming impressions from stereotypes, traits, and behaviors: A parallel-constraint-satisfaction theory. Psychological Review 103(2), 284–308 (1996)
16. Spellman, B.A., Ullman, J.B., Holyoak, K.J.: A coherence model of cognitive consistency: Dynamics of attitude change during the persian gulf war. Journal of Social Issues 49(4), 147–165 (1993)

The Impact of Network Structure on the Perturbation Dynamics of a Multi-agent Economic Model

Marshall A. Kuypers[1,2], Walter E. Beyeler[2], Robert J. Glass[2],
Matthew Antognoli[2,3], and Michael D. Mitchell[2]

[1] Center for Complex Systems Research, Univerity of Illinois at Urbana-Champaign,
Urbana-Champaign, IL, USA
kuypers1@illinois.edu, MarshallKuypers@gmail.com
[2] Complex Adaptive System of Systems (CASoS) Engineering
Sandia National Laboratories, Albuquerque, New Mexico, USA
{mkuyper,webeyel,rjglass,mantogn,micmitc}@sandia.gov
[3] School of Engineering, University of New Mexico
Albuquerque, New Mexico, USA
mantogno@unm.edu

Abstract. Complex adaptive systems (CAS) modeling has become a common tool to study the behavioral dynamics of agents in a broad range of disciplines from ecology to economics. Many modelers have studied structure's importance for a system in equilibrium, while others study the effects of perturbations on system dynamics. There is a notable absence of work on the effects of agent interaction pathways on perturbation dynamics. We present an agent-based CAS model of a competitive economic environment. We use this model to study the perturbation dynamics of simple structures by introducing a series of disruptive events and observing key system metrics. Then, we generate more complex networks by combining the simple component structures and analyze the resulting dynamics. We find the local network structure of a perturbed node to be a valuable indicator of the system response.

Keywords: Structure, Perturbation, Network, Complex Adaptive System.

1   Introduction

The increase in computing power seen in the past ten years has made agent-based modeling a viable option for studying complex systems. Recent work has begun to show the importance of network structure in the operation of complex adaptive systems (CAS), including how a system can be influenced using driver nodes [Liu et al. 2011]. Other research has focused on how structure can affect a diffusion algorithm as it propagates through a system [Ghoshal et al. 2011]. Researchers have also explored the implications of community structures in a network [Karrer et al. 2008]. While there is abundant research on CAS with a variety of structures, there has not been a systematic study of whether basic structural features could account for qualitative behavioral properties in large networks. This is an important question

S.J. Yang, A.M. Greenberg, and M. Endsley (Eds.): SBP 2012, LNCS 7227, pp. 331–338, 2012.

because structure has the potential to influence the robustness of a network. For example, supply networks and power grids losses due to perturbations could potentially be decreased by simply altering node connections. The inherent complexity of these systems makes analytical development of a perturbation theory difficult.

In this study we begin to explore dynamics within network structures resulting from perturbations. We compartmentalize complex networks into simple component structures whose dynamics are simply defined. We combine these compartmentalized networks and analyze their responses to perturbations. This experimental approach to system response provides valuable generalizations about complex networks.

1.1   Model Formulation

We study network structure and system dynamics using a configuration of the Interacting Specialist Model developed at Sandia National Laboratories [Beyeler et al. 2011]. In the interest of space, the model formulation is omitted but is rigorously defined in Beyeler et al. 2011. This model represents complex adaptive systems using coupled nonlinear first-order differential equations to describe the behavior of autonomous agents. The agents (or entities) must store, consume, and produce resources to maintain viability and competitiveness in their environment. The agents maintain their stability through a series of discrete interactions with markets, which create exchange pathways between agents. These interactions are facilitated through a money resource.

The model consists of a set of entities arranged in a hierarchy. Entities can be grouped into sectors, each of which is a collection of agents that produce and consume the same resources. Markets mediate transactions between sectors. A collection of sectors and markets makes up a Nation State.

Entities interact by joining a market and bidding to buy or sell resources. Consumers and producers are matched via a double auction. Entities make decisions about market transactions based on the entity health, resource reserves, and money levels. Health is defined as a scalar function that follows an agent's consumption with respect to a nominal consumption rate. Health abstractly represents a measure of an entity's success in a dynamic and competitive marketplace.

To study the dynamics of the model, we introduce perturbations and observe the system response. We simulate disruptive events by removing a certain percentage of an entity's produced resource in random events that occur with a defined frequency and duration. The resource is removed from the entity's production tank, preventing it from being sold to accrue a profit. This method can be used to represent a range of perturbation types, from an event analogous to a pipe bursting to smaller but more frequent perturbations, such as a 1% loss every time step, which simulates a leak in a pipe. This gives us considerable control over the perturbations we introduce into the model.

2   Structural Dynamics of Simple Networks

We would like to understand the dynamics of a complex network, such as an economy. Unfortunately, the complex feedback patterns created by a network of business relationships make it difficult to resolve causation. Thus, to make sense of the dynamics, we start by characterizing basic structures. These structures are idealized endpoints along axes of topological features commonly used to describe networks, such as path length and degree of connection. We use six sectors in each structure, and one entity in each sector. A connection from one sector to another means that a resource produced in the first sector is consumed in the second. We then consider a structure that superimposes several of these component structures and observe the resulting dynamics.

2.1   Fully Connected Networks

A fully connected network is defined as a network in which every node is connected in two directions to every other node in the system. This network is symmetrical and robust. Any perturbation will quickly reach every node, but because of the high connectivity, the impact is shared among several nodes and the system can cope with larger shocks.

Figure 1 illustrates the fully connected network and the response when node F is perturbed by a standard amount, removing 100% of its produced resource stores. By analyzing the responses at each node, we can characterize the nodal interactions very well.

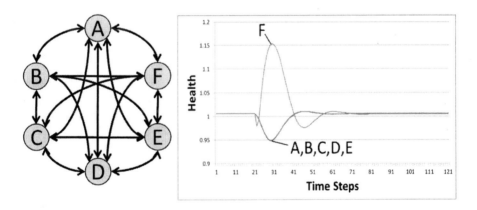

Fig. 1. A fully connected network and the perturbation response when node F is shocked

When node F is perturbed there is no product to sell and it immediately stops making money. As F lowers its consumption to preserve its money reserves, its health level is directly decreased. Because F's product is no longer available, all other sectors also begin to experience health deficits. Nodes A, B, C, D, and E see identical health trajectories because they are symmetric and identical.

Once F begins to produce again, it has a product that is very scarce with large demand. This causes the price to of F's product to spike, bringing in large profits, which spurs consumption leading to a large rise in health. Meanwhile, the other sectors are competing for a scarce resource at premium prices, causing them to consume less and continue to decrease in health. This behavior continues to a point at which F's product is overabundant, causing a trend reversal as the sectors readjust their consumption to changing prices. The health of the sectors then oscillates with a certain damping ratio until the system reaches a steady state again.

There are several notable features about this response. The oscillatory recovery response is a nontrivial trait of the fully connected network. Also, the perturbed node sees a maximum health deviation that is three times the maximum deviation in any other sector. Most importantly, the perturbed node sees a net health gain compared to the other sectors, which see a net health loss. This is similar to the competitive exclusion principle shown by Beyeler (2011). The perturbed sector is able to exploit the scarcity because our model has a fixed demand. In other systems, the market may have substitute goods and price would not spike.

2.2   Hub Networks

A hub network is asymmetrical, having a central node which consumes resources from every other node while also producing a resource that every other node requires. Although the vitality of this network depends upon the central node, the periphery nodes can also have a significant effect on the structure. Any perturbation quickly travels to the central node and then disperses through the rest of the network. It represents an extreme case of heterogeneity in connections.

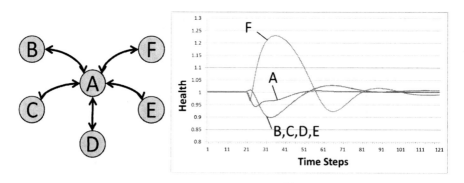

Fig. 2. A hub network and a perturbation response for a shock on node F

The response to a standard, 100% removal of a produced resource from a peripheral node is remarkably similar to the response of a fully connected network. The hub network exhibits three responses corresponding to the perturbed node (F), the center node (A) and the periphery nodes (B to E). The maximum health deviation is larger than the response observed in the fully connected structure.

The behavior becomes more interesting when we perturb the center node in the network.

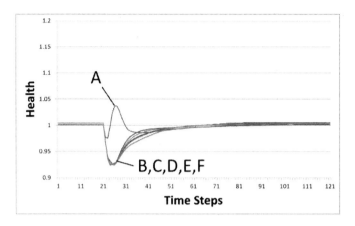

Fig. 3. A perturbation response for a hub network with the center node shocked

When a standard perturbation removes 100% of the center node's produced resource, the magnitude at the perturbed node and the recovery time are significantly reduced compared to the same perturbation introduced into the periphery nodes. This counterintuitive result shows that the local structure of the perturbed node in the hub network controls price spikes through negative feedbacks. The perturbed central node begins to see a health rise due to the price spike for its produced resource. The periphery nodes see a health decline that mirrors the perturbed node's increase. However, the hub structure causes the center node to be critically dependent on the periphery nodes. As the peripheral nodes decrease in health, they decrease production and the perturbed central node's consumption demand is not met by the other sectors production supply, due to their low health. This limit on consumption effectively caps the perturbation response.

2.3   Circular Networks

A circular network is made up of several nodes connected linearly in a circular pattern. This network is symmetrical and offers significant buffers to perturbations. Shocks must travel linearly through every node, taking more time to reach every node in the system. This structure has the longest path length and the minimum connection degree of any symmetrical network. The drawback to these features is that the magnitude of each perturbation is passed through each node, which can push fragile nodes to their death.

A circular network has a very distinct perturbation response. Figure 4 shows the response from removing 100% of node F's produced resource. Unlike the response seen in other network structures, there is no significant health gain in the perturbed node following this disruption. Instead, the node that consumes the perturbed node's resource sees a significant health loss. The two closest upstream nodes generally experience a health gain, with a phase lagging behind the perturbed node, but node C, sees a health loss. Node C is directly opposite from the perturbed node on this circular network: positive and negative responses ripple towards it from opposing directions. The negative ripple dominates and the node sees a net health loss.

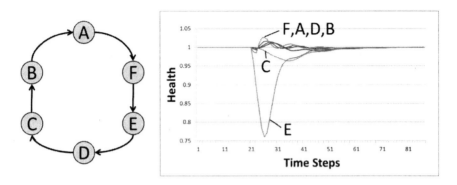

Fig. 4. A circular network and a perturbation response for a shock on node F

3   Combined Networks

In order to understand the dynamics of complex networks, we combine several simple component structures in a new structure which is shown in Figure 5.

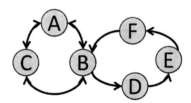

Fig. 5. A fully connected network combined with a circular network

This structure consists of two distinct parts. Nodes A, B, and C make up a fully connected network. Nodes B, D, E and F make up a circular network. Node B is a critical node that connects these two structures.

Perturbing each node will produce a unique system perturbation response since the structure is asymmetric. We expect the nodal response for A and C to be similar to nodal responses for a fully connected structure, and the nodal responses for B, D, E, and F should display circular features. This is an approximation, since the combination of structural components adds new feedbacks that cause the system dynamics to change. Generally, for this combination, we find that the component circular structure retains many qualitative and quantitative features of its dynamics, while the fully connected network sees a few distinct changes. In this regard and for this combination, we can say that the behavior of the circular structure is more robust to structural combination.

We can be more precise by observing that the system dynamics change depending on what node is perturbed. When nodes far upstream are shocked (node D), the entire structure behaves similarly to a circular network. Conversely, when we perturb nodes close to supplying the transition node B, the entire system begins to behave with fully connected network dynamics.

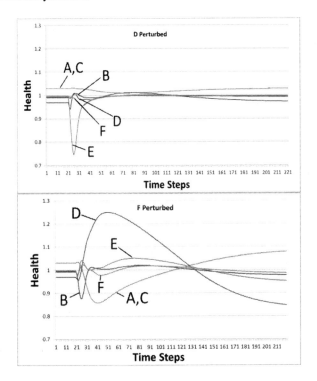

Fig. 6. Note that perturbations to node D have a response similar to a characteristic circular structure, while perturbations introduced at node F display features of a fully connected network

Since the system perturbation response varies as the perturbed node changes, the structural components immediately downstream of a perturbation are most critical to the system's response. This makes sense, since health is tied to consumption. Perturbations disrupt downstream consumption rates, which controls the perturbation response of the system.

To explore the differences between the component structures more rigorously, we can analyze individual node responses. The most interesting node is the transition node (B). It is unclear how the component dynamics of the full and circular structures will interfere at this point. When we perturb the transition node, the response models the circular structure. The shock hits the downstream node (D) the most severely, and the perturbed node does not see drastic net gain in health observed in a fully connected system. The health gain that ripples through the circular structure with some phase lag is clearly seen, and the characteristic fully connected network response of a significant health gain is only seen by node D, although it is muted.

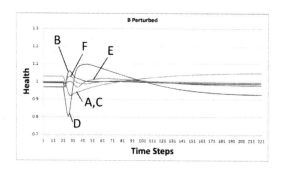

Fig. 7. A combined fully connected network and circular network, where the transition node is perturbed

This combined network provides interesting insight into structure's affect on perturbation responses. We see that downstream structure is a primary factor in determining a response. Also, we see circular structures retain their dynamics more effectively than a fully connected structure. Most importantly, we see that structural features are not additive and complex network structure dynamics cannot easily be analytically predicted.

4   Conclusion

Studying structure's affect on system dynamics presents many new challenges. By characterizing simple component structures and observing competing dynamics in combined structures, we can gain insight into to how these networks interact. The results presented here suggest that characterization and combination of sub-structural features will be inadequate and yield misleading indicators of the system's response to perturbations. We expect that system dynamics will become more complex as network structure becomes more complex. However, some general rules may be forthcoming. For example, perturbations to peripheral nodes may produce much larger responses than perturbations to central nodes as we found in the simple hub structure. It is the search for these general rules towards which we must focus future work to understand the dynamics within complex networks made of economic agents.

References

1. Liu, Y.-Y., Slotine, J.-J., Barabasi, A.-L.: Controllability of complex networks. Nature 43, 123–248 (2011)
2. Ghoshal, G., Barabasi, A.-L.: Ranking stability and super-stable nodes in complex networks. Nature Communications 2, 1–7 (2011)
3. Karrer, B., Levina, E., Newman, M.E.J.: Robustness of community structure in networks. Phsycial Review E 77, 46119 (2008)
4. Beyeler, W.E., Glass, R.J., Finley, P.D., Brown, T.J., Norton, M.D., Bauer, M., Mitchell, M., Hobbs, J.: Modeling systems of interacting specialists. In: 8th International Conference on Complex Systems (2011)

Robust Recommendations Using Regularized Link Analysis of Browsing Behavior Graphs

Shinya Naito and Koji Eguchi

Graduate School of System Informatics, Kobe University

Abstract. Recently, there has been a growing need for more sophisticated recommendation techniques with an increase in the amount of data available on the Web. In this study, we especially focus on recommending items with long text, and aim at achieving this using a method of link analysis of a user-item bipartite graph in a regularization framework based on Co-HITS algorithm. This method can integrate, via mutual reinforcement, the graph structure and the content of both user profiles and items. It has never been seen in the mainstream of conventional recommendation techniques. In our experiments, we used the data of Web browsing history, assuming Web news articles as target items. We evaluated the list of top-N items recommended based on the browsing history, using a test set that consists of a part of viewed items for each user. We demonstrate through the experiments that the proposed method outperformed several baseline methods in a situation where only a small amount of browsing behavior is observed.

1 Introduction

Recently, there has been a growing need for more sophisticated recommendation techniques with an increase in the amount of data available on the Web. Collaborative filtering, which makes use of similar users' behavior, is one of the most successful recommendation techniques [5,6]. One of the difficult problems for collaborative filtering is due to *data sparseness problem*, where the recommender system is not able to make good recommendations for the lack of a sufficient number of user-item records. *Cold-start problem* [7] is a more specific term referring to a situation where the system is required to give recommendations to any new users who have almost no stored preferences (*cold-start users*), or to recommend new items that almost no user has yet rated (*cold-start items*). This can be considered as a kind of data sparseness problem assuming the specific situations mentioned above.

In this paper, we especially focus on recommending items with long text, such as Web pages, news articles, academic papers, and any kinds of product items that are associated with text, and aim at addressing data sparseness problem using a method of link analysis of a user-item bipartite graph with regularization. For this purpose, we modified Co-HITS algorithm [1] that was proposed for query recommendations. By using this algorithm, consistency of text content between user profiles and target items are taken into accounts for recommendations.

S.J. Yang, A.M. Greenberg, and M. Endsley (Eds.): SBP 2012, LNCS 7227, pp. 339–347, 2012.

We compare the proposed method with content-based filtering and collaborative filtering using HITS algorithm [3]. We then demonstrate the proposed method works better than the others in a situation where only a small amount of browsing behavior is observed.

2   Co-HITS Algorithm

In this section, we briefly overview Co-HITS algorithm, which was proposed for query recommendations [1].

Suppose a bipartite graph $G = (U \cup V, E)$, where the nodes can be divided into two disjoint sets U and V such that every edge connects a node in U to one in V. When $U = \{u_1, \dots, u_\ell\}$ and $V = \{v_1, \dots, v_m\}$ and when a node u_i in U is linked to a node v_j in V, we can define a transition probability between u_i and v_j as w_{ij}^{uv} and w_{ji}^{vu}, taking into account all their connecting edges. From this transition probability, we can consider a hidden transition probability between two nodes in the same set, such as by $w_{ij}^{uu} = \sum_{k \in V} w_{ik}^{uv} w_{kj}^{vu}$. We indicate $W^{uu} \in \mathbf{R}^{m \times m}$ and $W^{vv} \in \mathbf{R}^{n \times n}$ as hidden transition matrices within U and V, respectively.

Co-HITS algorithm calculates relevance score of each node on the basis of both the text content and graph structure. Relevance scores x_i of u_i and y_k of v_k are defined as follows:

$$x_i = (1 - \lambda_u)x_i^0 + \lambda_u \sum_{k \in V} w_{ki}^{vu} y_k \tag{1}$$

$$y_k = (1 - \lambda_v)y_k^0 + \lambda_v \sum_{j \in U} w_{jk}^{uv} x_j \tag{2}$$

where $\lambda_u \in [0, 1]$ and $\lambda_v \in [0, 1]$ are weighting parameters and x_i^0 and y_k^0 are initial relevance scores for u_i and v_k, respectively. In the iteration above, Co-HITS algorithm regularizes the relevance scores with cost functions, such as:

$$R_1 = \frac{1}{2} \sum_{i,j \in U} w_{ij}^{uu} \left\| \frac{x_i}{\sqrt{d_{ii}}} - \frac{x_j}{\sqrt{d_{jj}}} \right\|^2 + \mu \sum_{i \in U} \| x_i - x_i^0 \|^2 \tag{3}$$

where $\mu > 0$ is a parameter, and $D = (d_{ij})$ is a diagonal matrix for normalization. In this cost function, the first term defines the global consistency of ranking scores, and the second term defines constraints from initial relevance scores. In the same manner, another cost function can be defines as:

$$R_2 = \frac{1}{2} \sum_{i,j \in V} w_{ij}^{vv} \left\| \frac{y_i}{\sqrt{d_{ii}}} - \frac{y_j}{\sqrt{d_{jj}}} \right\|^2 + \mu \sum_{i \in V} \| y_i - y_i^0 \|^2 . \tag{4}$$

The third cost function R_3 can also be defined in a way that penalizes a large difference between relevance scores in a node in U and one in V, as follows:

$$R_3 = \frac{1}{2} \sum_{i \in U, j \in V} w_{ij}^{uv} \left\| \frac{x_i}{\sqrt{d_{ii}}} - \frac{y_j}{\sqrt{d_{jj}}} \right\|^2 + \frac{1}{2} \sum_{j \in V, i \in U} w_{ji}^{vu} \left\| \frac{y_j}{\sqrt{d_{jj}}} - \frac{x_i}{\sqrt{d_{ii}}} \right\|^2 \tag{5}$$

An integrated cost function can be defined as below:

$$R = \lambda_\gamma (R_1 + R_2) + (1 - \lambda_\gamma) R_3 \tag{6}$$

where $\lambda_\gamma \in [0,1]$ is a parameter. Co-HITS minimizes this cost function R. To approximately solve this problem, the following equation can be obtained so that we can have a final relevance score vector F^* whose components represent relevance scores of respective nodes.

$$F^* = (1 - \mu_\alpha)(I - \mu_\alpha S)^{-1} F^0 \approx (1 - \mu_\alpha)(I - \mu_\alpha \hat{S})^{-1} \hat{F}^0 \tag{7}$$

where I represents an identity matrix. F^0 indicates an initial relevance score vector such that $F^0 = (x_1^0, \ldots, x_\ell^0, y_1^0, \ldots, y_m^0)^T$, and S indicates a (weighted) transition matrix. $\hat{F}^0$ and $\hat{S}$ indicate those which are based on a subgraph of the bipartite graph for approximation [1]. $\mu_\alpha \in [0,1]$ is another parameter.

3 Regularized Link Analysis of Browsing Behavior Graphs

As mentioned previously, we especially focus on recommending items with long text, such as Web pages, news articles, academic papers, and any kinds of product items that are associated with text. For convenience, we take an example of Web page recommendation. Suppose that a user $u \in U$ views Web page $v \in V$. The relationship between u and v is represented as an edge between the pair of nodes in a bipartite graph. What kind of Web pages should be recommended to a user, in this case? We suppose that:

1. The more similar a Web page's content is to a user profile's content, the more the user should prefer the Web page.
2. The larger number of or the more similar users prefer a Web page, the more the user should prefer the Web page.

Based on these assumptions, we apply Co-HITS algorithm to this task. When applying Co-HITS algorithm, we need to modify it at some points. We below describe how to apply and modify Co-HITS algorithm for recommending items with long text, such as Web pages.

A Method of Constructing a Bipartite Subgraph. In this part, we describe how to create a submatrix $\hat{S}$: the matrix based on a subgraph of the user-item bipartite graph, as in Eq.(7). We first describe how to construct a subgraph $\hat{G}$ of the user-item bipartite graph that is from browsing history, as below:

1. Based on the initial relevance scores according to each node's content (as we will describe later in this section), take top-10 nodes as seed sets: $\hat{U} = U_L$ for user nodes and $\hat{V} = V_L$ for item nodes.
2. Update the user set $\hat{U}$, adding the nodes that are linked to any nodes in $\hat{V}$.

3. Update the item set $\hat{V}$, adding the nodes that are linked from any nodes in $\hat{U}$.
4. Iterate through these two steps until $\hat{U}$ and $\hat{V}$ reach a desired size.

Once we get a subgraph of the user-item bipartite graph from the procedure above, we obtain a transition matrix for the subgraph. Deng et al. [1] computed the transition probability, between a query and a URL in their case, according to the number of times the item was clicked by any users when shown in response to the query. In our study, however, it does not often happen that a single user views the same Web page a number of times, except for special cases. Therefore, we compute the transition probability for all user-item pairs.

A Method of Computing Initial Relevance Scores. Each node's initial relevance score is computed by considering the content of the node. Deng et al. [1] computed the initial relevance scores of query nodes and URL nodes, in their case, using query likelihood [4,2]: the likelihood of generating a given query from the unigram language model of each node, assuming that the given query is short. In our study, however, either in the case of a user or an item is given as a query, their text information is usually long and thus the query likelihood is not suitable for the initial relevance scores. Instead, we use *Hellinger distance* between the given query and each node of Web pages or users, for computing the initial relevance scores. Given probability vectors $\mathbf{p} = (p_1, \ldots, p_n)$ and $\mathbf{q} = (q_1, \ldots, q_n)$ where $\sum_i^n p_i = 1$ and $\sum_j^n q_j = 1$, Hellinger distance can be defined as:

$$D_{HL}(\mathbf{q}, \mathbf{p}) = \sqrt{\frac{\sum_i^n (\sqrt{q_i} - \sqrt{p_i})^2}{2}}. \tag{8}$$

It satisfies $0 \leq D_{HL}(\mathbf{q}, \mathbf{p}) \leq 1$. We use the unigram language model for each Web-page node. We simply estimated the model via maximum likelihood estimation with the nouns that appear in each Web page, in this paper. We also use the unigram language model for each user's profile by computing the expectation of the unigram language models of the Web pages viewed by the user.

According to the discussions in this section, we modify Co-HITS algorithm for the use of information recommendation, as follows:

1. Input: user-item bipartite graph, and a user given as a query (*query user*)
2. Calculate initial relevance scores using (1-D_{HL}) where D_{HL} is given between the query user and each node, and take top-10 users U_L and top-10 items V_L as seed sets.
3. Construct a bipartite subgraph $\hat{G} = (\hat{U} \cup \hat{V}, \hat{E})$, using U_L and V_L in the manner mentioned previously.
4. Get a transition matrix $\hat{S}$ and initial relevance scores $\hat{F}^0$ from the subgraph.
5. Solve Eq.(7), and get final relevance scores F^*.
6. Output: Ranking of Web pages related to the query user.

4   Experiments

In our experiments, we use the data of Web browsing history, assuming Web news articles as target items. We evaluate the list of top-N items recommended based on the browsing history.

4.1   Data Sets

In this study, we experimented with the Japanese news article data gathered from Yahoo! News[1] and browsing history data provided from NetRatings Japan. This dataset contains 36218 users and 1731 news articles that were recorded in June 2010. For our experiments, we first removed the users who viewed less than 50 news articles. We then randomly selected 50 users from them. For evaluation assuming a situation where only a small amount of browsing behavior is observed, we randomly extracted a varying percentage of each user's history data. We set the percentage as 5% and 100%, for simplicity. We further randomly selected 20% of each user's history data in order to use the half (10%) for development data and the other half (10%) for test data. The remaining 80% of the data is used for running recommendation algorithms. We fixed the parameters of the proposed method using the development data, and then, using the test data, we evaluated the performance of the method with the fixed parameters.

4.2   Evaluation Metrics

We use the three evaluation metrics below that are commonly used in this field.

1. Precision at top-N ($P@N$): This is defined as:

$$P@N = \frac{\# \text{ of relevant items within top-}N}{N}$$

   We take the average of evaluated results over all users. In our experiments, we used P@5 and P@10, focusing on top-ranked recommended items.

2. Mean average precision (MAP): When the number of recommended items is n, average precision (AP) is defined as:

$$AP = \frac{1}{\# \text{ of all relevant items}} \sum_{i=1}^{n} r_i \times \frac{\# \text{ of relevant items within top-}i}{i}$$

   where r_i is a function that take a value 1 when i-th recommended item is relevant, and 0 otherwise. We take the average of AP over all users.

3. Mean reciprocal rank (MRR): Reciprocal rank (RR) is defined as:

$$RR = \frac{1}{\text{the rank of the highest ranked relevant item}}$$

   We take the average of RR over all users.

---

[1] We used *Igo* as a Japanese tokenizer that is available at
http://igo.sourceforge.jp/

4.3   Experimental Settings

We need to determine parameters λ_γ and μ_α, which appear in Eqs.(6) and (7), for the modified Co-HITS algorithm. We first optimized μ_α, setting λ_γ to 1, on the basis of Precision at top-5 (P@5). We then determined λ_γ with the optimized μ_α in the same manner. We briefly describe the baseline methods used, as below.

1. Content-based filtering: This is a method of ranking items with respect to Hellinger distance between a given user's profile and each item (the same as the initial relevance scores in the modified Co-HITS algorithm).
2. Modified HITS: This is a method that only considers global consistency in user-item bipartite graph structure in scoring, ignoring initial relevance scores by fixing $\lambda_\gamma = 0.5$ and $\mu_\alpha = 1$ in Eqs.(6) and (7), respectively. It corresponds to HITS algorithm [3] with some modifications. In the sense that this method is based on user-item bipartite graph, it can be considered as collaborative filtering.

4.4   Experimental Results

Experimental Results with 5% of History Data. We demonstrate P@5 and P@10 results, varying two parameters μ_α and λ_γ in Figures 1 and 2, respectively. We determined $\mu_\alpha = 0.1$ since the curves of both P@5 and P@10 have the peak at 0.1, as you can see in Figures 1. From Figure 2, you can see the peak at 0.2 in the curve of P@5 and almost similarly in the curve of P@10, and therefore we determined $\lambda_\gamma = 0.2$. Using these estimated parameters, we tested our Co-HITS-based method, compared with HITS and content-based filtering. The results are shown in Figures 3 and 4. From these figures, you can see that our methods outperformed the other methods in terms of all the evaluation metrics. Our method worked 150% better than HITS and 25% better than content-based filtering in terms of P@5. In terms of P@10, our method worked 140% better than HITS and 71% better than content-based filtering. In terms of MAP and MRR, our method outperformed the others with improvement ranging from 27.2% to 68.5%. The improvement achieved by the proposed method was statistically significant, compared with both HITS and content-based filtering, in terms of MAP using Wilcoxon signed rank test and pared t-test (significance level: 0.05). No significant difference was observed in terms of the other evaluation metrics; however, these metrics are known to require a larger number of samples for statistically stable evaluation. In our experimental results, the evaluated values were relatively small. This comes from our definition of relevant items. The relevant items were defined as those that were held out from users' actual browsing history data. Therefore, even if there was a web page that a user would prefer, it is deemed to be irrelevant unless the user actually viewed it in the past. From the discussions above, our method is shown to be effective in a situation where only a small amount of browsing behavior is observed.

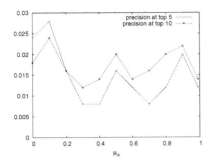

Fig. 1. The effect of varying μ_α using 5% of browsing history

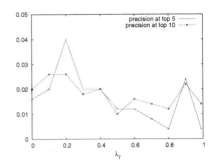

Fig. 2. The effect of varying λ_γ using 5% of browsing history

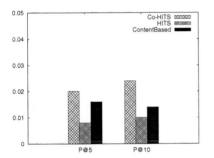

Fig. 3. Experimental results in terms of P@N using 5% of browsing history

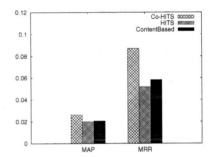

Fig. 4. Experimental results in terms of MAP/MRR using 5% of browsing history

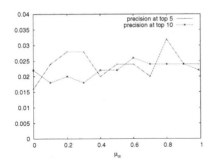

Fig. 5. The effect of varying μ_α using 100% of browsing history

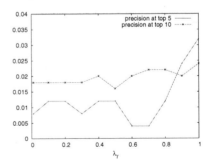

Fig. 6. The effect of varying λ_γ using 100% of browsing history

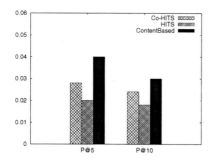
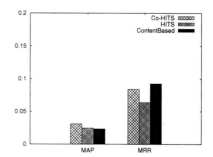

Fig. 7. Experimental results in terms of P@N using 100% of browsing history

Fig. 8. Experimental results in terms of MAP/MRR using 100% of browsing history

Experimental Results with All History Data. Similarly with the previous experiments, we estimated parameters using all browsing history data. We show the result of μ_α in Figure 5. As can be seen in this figure, the results of P@5 and P@10 are slightly different; however, we determined $\mu_\alpha = 0.8$ on the basis of P@5, emphasizing higher-ranked items. From the results shown in Figure 6, we determined $\lambda_\gamma = 1.0$ in the same manner. We demonstrate the test results in Figures 7 and 8. From these figures, you can see that content-based filtering worked best in this situation where data sparseness is explicitly eliminated.

5   Conclusions

In this paper, we proposed a method of link analysis of a user-item bipartite graph with regularization, considering consistency of text content among user profiles and target items, for recommending long text items. This method is based on Co-HITS algorithm that was originally proposed for recommending similar queries for a given query. We modified this algorithm for recommending items that are associated with relatively long text. We carried out the experiments of top-N recommendations of Web news articles based on browsing history, and demonstrated that the proposed method outperformed several baseline methods in a situation where only a small amount of browsing behavior is observed. Providing good recommendations in this situation is one of the most important issues in information recommendation research. Our proposed method is a robust approach to information recommendation in such a severe situation, by naturally integrating users' browsing behavior information and content information via regularized link analysis. Comparing with various conventional recommendation techniques is left for the future work.

Acknowledgments. We thank Dr. Hao Han, Dr. Hidekazu Nakawatase, and Prof. Keizo Oyama for providing us an experimental dataset. This work was supported in part by the Grants-in-Aid for Scientific Research (#23300039 and #22240007) from the Ministry of Education, Culture, Sports, Science and Technology, Japan.

References

1. Deng, H., Lyu, M.R., King, I.: A generalized Co-HITS algorithm and its application to bipartite graphs. In: Proceedings of the 15th ACM SIGKDD International Conference on Knowledge Discovery and Data Mining, pp. 239–248 (2009)
2. Hiemstra, D.: A Linguistically Motivated Probabilistic Model of Information Retrieval. In: Nikolaou, C., Stephanidis, C. (eds.) ECDL 1998. LNCS, vol. 1513, p. 569. Springer, Heidelberg (1998)
3. Kleinberg, J.M.: Authoritative sources in a hyperlinked environment. In: Proceedings of the 9th Annual ACM-SIAM Symposium on Discrete Algorithms, pp. 668–677 (1998)
4. Ponte, J.M., Croft, W.B.: A language modeling approach to information retrieval. In: Proceedings of the 21st Annual International ACM SIGIR Conference on Research and Development in Information Retrieval, pp. 275–281 (1998)
5. Resnick, P., Iacovou, N., Suchak, M., Bergstrom, P., Riedl, J.: GroupLens: An open architecture for collaborative filtering of NetNews. In: Proceedings of the ACM Conference on Computer Supported Cooperative Work, pp. 175–186 (1994)
6. Sarwar, B., Karypis, G., Konstan, J., Reidl, J.: Item-based collaborative filtering recommendation algorithms. In: Proceedings of the 10th International Conference on World Wide Web, pp. 285–295 (2001)
7. Schein, A.I., Popescul, A., Ungar, L.H., Pennock, D.M.: Methods and metrics for cold-start recommendations. In: Proceedings of the 25th Annual International ACM SIGIR Conference on Research and Development in Information Retrieval, pp. 253–260 (2002)

Whodunit? Collective Trust in Virtual Interactions

Shuyuan Mary Ho, Issam Ahmed, and Roberto Salome

Drexel University
3001 Market Street, Suite 100
Philadelphia, PA 19104
{smho,ina23,rcs48}@drexel.edu

Abstract. This paper presents a socio-technical study on how humans collectively perceive trustworthiness in a suspicious situation when evoked by computer-mediated communications (CMC). This research was designed through the use of an online game, entitled *"Whodunit"* to study how collective trust in a virtual scenario can be reflected in untrustworthy cloud environments. We propose that virtual dialogue can provide important clues to understanding an actor's inclination, using the cognitive process of observers' collective trust in a virtual collaborative group. The methodology proposed in this paper is built on research that demonstrates how human "sensors" can detect unusual or unexpected changes in a psychological construct: trustworthiness – based on observed virtual behavior. The research framework adopts the theory of trustworthiness attribution to model collective trust in virtual interactive environments. Humans "sensors," with limited access to information (e.g., online conversations) make assessments based on subtle observed changes to a person's disposition. This paper concludes that people that rely on "cognitive trust" can identify dispositional changes within untrustworthy scenarios more precisely than people that rely on "emotional trust." In addition, people who have a propensity to trust others more than themselves (lacking confidence in their own abilities) tend to do better identifying the correct murderer.

Keywords: Trustworthiness attribution, socio-technical system, online game simulation, empirical, psychological theory, human-computer interactions.

1   Introduction

This socio-technical study of collective trust is designed to understand how humans attribute trust collectively in a socially intelligent way. Based on the social theory of trustworthiness attribution, we posit that most humans have the capability to perceive when peers in the group are not trustworthy, even when a hidden agenda exists [1, 2]. Today, humans often interact with other humans through computer-mediated communication (CMC) supported environments. This relatively new communication channel affects our relationships, and how we interact. We extend our current research model to investigate collective trust by collecting a group of human "sensors" rating within social media, as part of a "whodunit" murder scenario in an online game setting.

S.J. Yang, A.M. Greenberg, and M. Endsley (Eds.): SBP 2012, LNCS 7227, pp. 348–356, 2012.
© Springer-Verlag Berlin Heidelberg 2012

Generally, humans interpret conversations based on either cognitive or emotional trust, which helps them make trust assertions in online conversations [3]. This socio-technical study addresses how humans perceive inclination change toward each other in a virtual computer-mediated group, and whether collective trust could provide more precise accuracy in assessing a target's dispositional change in a suspicious situation. This study advances our understanding as to how dispositional inclination in humans can change when observed by an intelligent computing network.

We propose a methodology; using the online *Whodunit* game - to simulate a mystery scenario that demonstrates how the cognitive process occurs with a human-like understanding based on chat conversations chronicled online. Human participants are challenged with a dilemma as part of the game scenario, and the study measures how participants react and judge a crime case though given limited information.

This paper contains five major sections describing the collective trust in dissecting dispositions during virtual interactions. The *Problem Statements* describes the need for such computational study aimed at understanding dispositional change in virtual collaborative environments. The *Theoretical Framework* encapsulates the logic behind how this model could be prototyped, and operates based on trustworthiness attribution theory, to illustrate with examples of the collective trust from the human "sensors." The *Online Gaming as Research Method* describes why online games can be used as an effective method to collect data, and how this data will be analyzed. The *Results and Findings* discuss the research results that correspond to the theoretical framework of this study. The *Conclusions, Contributions and Future Work* summarizes the study, and describes the contribution of this study to the larger research community, as well as the basic types of experimental online games, which we plan to conduct in future studies of virtual interactions.

2   Problem Statements

Difficulties with understanding and codifying an individual's social behavior can arise because individuals' intentions and methods of communication can vary over time. Face-to-face interactions have been the norm, but as more and more computer-mediated software and tools, e.g., Facebook, Twitter, Blogs, etc. are developed to foster virtual relationships, understanding virtual relationships have become increasingly important. One of the problems with virtual communications is the inability to accurately identify a person's disposition, which makes it hard to find out how trustworthy and trustable a virtual human being, especially when you are entering into a trust relationship via online commerce or virtual project collaboration. Detecting human dispositional change, and accurately classifying such change as untrustworthy has become a very significant problem. First, an individual may have a hidden agenda that is not explicitly known. Second, the resources available to uncover a hidden agenda may put the group at risk, and may be limited to only online communications - without access to background, shadow information, or contextual information on the individuals. Third, such issues are rarely studied explicitly in a virtual collaborative environment. The channel that sheds important clues lies in the manifestation of their conversations and the behavioral interactions that are observable online. The ability to characterize and understand virtual behavior and further detect dispositional changes in a scalable way becomes a critical research problem. The *Whodunit* game creates an interesting scenario with many hidden facts for players to address, just as many real-world virtual communications do. The purpose of this game is to unveil the psychological reasons behind decisions to trust - or distrust - a virtual entity. The murder situation allows human "sensors" to solve the mystery puzzle and identify possible suspects. We further study how human "sensors" attribute their trust regarding the suspect, and base collective trust on both rational and emotional grounds [3].

3   Theoretical Framework

The study adopts a framework of *trustworthiness attribution theory* to understand how collective trust plays a role in detecting suspicious behavior [1, 2]. We differentiate between collective trust and individual trust, while introducing the trustworthiness attribution framework in a group collaborative setting.

3.1   Individual and Collective Trust

Lewis [3] classified both individual and collective trust into two aspects; rational and emotional. These two aspects define what type of trust an individual uses to make decisions. Rational trust, or cognitive trust, is based on a high degree of rationality. Emotional trust is based on faith and intuition. For example, rational trust (e.g., Rational Prediction) must be involved when investing in a multi-million dollar business, while believing that your lover is faithful to you would be an example of emotional trust (e.g., Faith). Of course the two can also be mixed and matched. Lewis describes different types of trust using the following figure (Fig. 1).

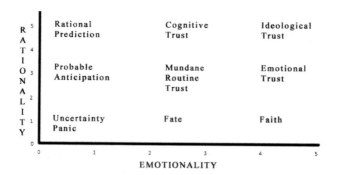

Fig. 1. Rationality & Emotionality Bases, Types of Trust, and Boundary States (from Lewis)

Collective trust poses a very important notion that the research community tries to define from different perspectives. One aspect of the collective trust was driven from the urge of economic changes. Organizational citizens interact with one another to maximize efficiency by reducing the expectation of opportunistic behavior and consequently lowering associated transaction costs [4]. McEvily [4] stressed the importance of collective trust as a "strategic resource" that potentially provides sustainable competitive advantages for organizations.

In a collaborative group, people trust one another depending on different rationalities and how they emotionally attach to one another [3]. Trust plays a major role in collective settings. For example, team members may blindly follow a leader by faith - based on the leaders past deeds, or trust the leader based on cognitive reasoning. The fact is that in either case, the leader may betray them. Under such circumstance, we focus our study on identifying the type of trust that helps to dissect betrayal of others. While rationality in trust can help supersede a person's emotional bonds, emotional bonds can provide a deeper understanding of another's disposition.

3.2   Trustworthiness Attribution

Trustworthiness is defined as a generalized expectancy concerning a target person's degree of correspondence between communicated intentions and behavioral activity that can be observed and evaluated, which remain reliable, ethical and consistent; as long as any fluctuation between target's intentions and actions does not exceed the observer's expectations over time [5-9]. Tyler [10] discovered that trustworthiness can serve as a strong predictor of fairness in authorities' decision-making. Tyler & Degoey [11] examined how a group attributes their authorities' trustworthiness, and how it shapes the willingness of group members to accept authoritative decisions. Mayer, Davis et al. [12] defined the three factors of perceived trustworthiness to be ability (competence), benevolence (kindness) and integrity (goodwill/ethics). Furthermore, Mayer and Davis [13] enhanced this trust model to validate that competence (ability), benevolence (kindness) and integrity (goodwill/ethics) are antecedent factors in trustworthiness.

Attribution theory is used to understand how humans assign meaning to either internal or external causes based on observed behavior [14, 15]. Ajzen [16, 17] stated that an individual's intention can be understood by his or her information behavior. In addition, the quality and trustworthiness of this behavior can be perceived and assigned with a meaning by a social network in a close relationship [18, 19].

Ho [1, 2, 6] develops the framework of *trustworthiness attribution*, which rigorously identifies how trustworthiness can be perceived - based on limited information behavior - using one's social network. In Ho's [6] work, an individual's social network is analogous to sensors in a network. The act of assigning meaning to an individual's behavior is characterized as attribution [20]. An actor's behavior is determined by an observers' judgment to be intentional or unintentional in a way that is attributable to either external (situational) causality or internal (dispositional) causality [15, 21]. Because all human beings are of the same species and born with similar types of features and functions, a man should "know" and be able to "sense" from his own perceptions and with his judgment how the world operates, despite the fact that sometimes those attributions may not be accurate or valid [14, 15, 21].

Moreover, an individual's observed behavior can be interpreted from a single observation point, or through multiple observations over time.

4   Online Gaming as Research Method

Online games have been constructed as data collection methods for socio-technical research [6], and can also be used to collectively solve large-scale computational problems [22-24]. In this study, we set up an online game to study how untrustworthy behavior is attributed in a virtual environment. Ho [6] simulated insider threat situation the "Leader's Dilemma" game which involved a series of group activities requiring a team leader to work closely with team players to achieve the team's goals. This game eventually presents an ethical dilemma to the leader, and observes how the leader responds and interacts with the team members during a process of behavioral change.

4.1   Research Design

The *Whodunit* game simulates the three difficulties (mentioned in the *Problem Statement*) to determine perceived trustworthiness of several suspects. Each of the possible "suspects" is given a back-story with possible motives. This creates the mystery and potential for a hidden agenda for each character. The second difficulty of not having enough information is trivial in this case. Since the player only knows what is being told about the suspects, one must infer "why" a suspect may have taken certain actions in the game. The conversations between characters may be the only information one has in determining trustworthiness. In the *Whodunit* game, it would be too tedious to have extended dialogue explaining the scenario. For this reason, a mix of dialogue and narrative facts are used to illustrate the scene of the crime as the detectives see it.

The *Whodunit* game was designed within a given scenario. Through a Facebook application, people are presented with a situation where a Detective must solve a murder. Based on the clues from the crime scene and information on the suspects, the players - as human "sensors" - must solve the mystery. In this game, we study how participants assign trust, and the categories of their trust towards their predicted suspects.

4.2   Data Collection

We collected data from 39 participants in Spring 2011[1]. Each participant was asked to decide which of six suspects committed the murder, and give reasons for their conclusions. Each participant was also asked to rate his or her decisions through a series of multiple questions. These questions were based on the suspect's ability to commit a crime, perceived benevolence, and dispositional integrity. Each participant

---

[1] The study was conducted during Spring 2011 under Drexel IRB protocol #19287.

was asked to assess the murder according to these three criteria factors, and their level of agreement to each question. Moreover, we collected participants' self-assessment based on two criteria; participants' propensity to trust and their trustfulness. Each criterion had several questions, and the responses ranged from strongly disagree to strongly agree.

In addition to the ratings provided by the participants, four data analysts were assigned to rate each decision - emotionality and rationality - toward participants' responses. The ratings were ranged from 1 (low) to 5 (high). This is then superimposed onto the scale which was used by Lewis [3] as illustrated in Fig. 1. Depending on the two ratings for each response, four data analysts labeled and categorized participants' responses according to Lewis chart and assigned with a trust type (Fig. 1).

4.3   Data Analysis

For the first part of the data collection, four analysts were used to obtain a rating for each of the participant's qualitative responses toward their assessment of "who the suspect is." The tabulated ratings for each murder choice and the average of the four analysts are presented in the following Table 1.

Table 1. Four Analyst Ratings based on Rationality (R) & Emotionality (E)

Murder Choice	Analyst								Average	
	1		2		3		4			
	R	E	R	E	R	E	R	E	R	E
Roger	3.9	2	3.4	3.2	4.2	2.7	4.6	1.3	4.025	2.3
George	3.3	2.3	0.7	5	4	2	3.4	3	2.85	3.075
Gerald	2.6	4.2	1	5	2.6	3.2	2	4	2.05	4.1
Janice	2.7	3	1.4	4.8	3.8	3	3.5	3.3	2.85	3.525
Jillian	3.6	2.6	0.8	5	3.8	3.6	2.5	4.3	2.675	3.875
Marissa	3.5	3	1	5	2.5	4	2	4.6	2.25	4.15

5   Results and Findings

The averages from Table 1 are plotted using Lewis' graph (Fig. 1). The Emotionality averages are taken as the x-value and Rationality averages are taken as the y-value. The result is the following Fig. 2.

In Fig. 2, the plots for each suspect choice are in range of a designated trust type, depending on their rationality or emotional response. Comparing responses along with the logic used to ascertain each response, the overall results demonstrated a higher rationality and mid-range emotionality ratings. These correct responses can be said to have been assessed using *cognitive trust* since it is bordering that area. The people that correctly identified the murderer used less emotional trust, and more rationality.

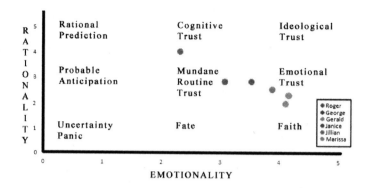

Fig. 2. Graph of Rationality Ratings vs. Emotionality Ratings

The rest of the responses (the group that did not successfully identify the murderer) can be seen to have used less rationalization and more emotional trust. To summarize, the people who correctly identified the murderer used more rationality in their trust analysis when compared to the people who chose the wrong suspect.

6   Conclusions, Contributions and Future Works

We seek to develop a methodology that can generalize the perceptions disposition from analysis of online interactions. Online games mimic real-world trust scenarios when designed in a systematic way to give us observations and inferences about each individual's disposition, and how it is reflected in an actors' information behavior. Trust scenarios are designed according to social-psychological principles within online situations where various threat cases were simulated and dissected [6].

While previous work establishes that humans can identify a number of trust factors pertaining to a single online situation, our methodology will further generalize the perceptions and identification of a human's virtual disposition based on individual and group online interactions. Using online gaming methodologies, one can engage people to contribute and chronicle their perceptions of trust. We model the cognitive process to infer a high level of human attribution based on observed behaviors. In future studies, a computational model could have the ability to assign a probable subjective rating, and assertions of target's disposition at both the individual and collective level.

The paper identifies the possibilities for a cognitive model to mimic how human "sensors" are able to detect downward shift of trustworthiness in threat situations over time. The conceptual framework of this socio-technical study seeks to capture intelligence in virtual interactions and scenarios, and identify factors of trustworthiness attribution according to our current research model [6]. Furthermore, this paper states the methodology and discusses the results of an online game scenario that captures virtual interactions among human beings via cyber infrastructure, and then processes those communications according to different psychological models.

The design of this study facilitates our understanding towards socially intelligent computing. Such a virtual system can be designed to be scalable and replicable among interactions of people in computer-mediated virtual groups in cyber infrastructure.

References

1. Ho, S.M.: Attribution-Based Anomaly Detection: Trustworthiness in an Online Community. In: Liu, H., Salerno, J.J., Young, M.J. (eds.) Social Computing, Behavioral Modeling, pp. 129–140. Springer, Tempe (2008)
2. Ho, S.M.: A Socio-Technical Approach to Understanding Perceptions of Trustworthiness in Virtual Organizations. In: Liu, H., Salerno, J.J., Young, M.J. (eds.) Social Computing and Behavioral Modeling, pp. 1–10. Springer, Tempe (2009)
3. Lewis, J.D., Wrigert, A.: Trust as a Social Reality. Social Forces 63(4), 967–985 (1985)
4. McEvily, B., et al.: Can groups be trusted? An experimental study of collective trust, Carnegie Mellon University Department of Social and Decision Sciences (2002)
5. Hardin, R.: Gaming Trust. In: Ostrom, E., Walker, J. (eds.) Trust and Reciprocity: Interdisciplinary Lessons From Experimental Research, pp. 80–101. Russell Sage Foundation, New York (2003)
6. Ho, S.M.: Behavioral Anomaly Detection: A Socio-Technical Study of Trustworthiness in Virtual Organizations. In: School of Information Studies 2009, p. 437. Syracuse University, Syracuse (2009)
7. Hosmer, L.T.: Trust: The Connecting Link Between Organizational Theory and Philosophical Ethics. Academy of Management Review 20(2), 379–403 (1995)
8. Rotter, J.B.: Interpersonal Trust, Trustworthiness and Gullibility. American Psychologist 35(1), 1–7 (1980)
9. Rotter, J.B., Stein, D.K.: Public Attitudes Toward the Trustworthiness, Competence, and Altruism of Twenty Selected Occupations. Journal of Applied Social Psychology 1(4), 334–343 (1971)
10. Tyler, T.R.: What is Procedural Justice? Law and Society Review 22, 301–355 (1988)
11. Tyler, T.R., Degoey, P.: Trust in Organizational Authorities: The Influence of Motive Attributions on Willingness to Accept Decisions. In: Kramer, R.M., Tyler, T.R. (eds.) Trust in Organizations: Frontiers of Theory and Research, pp. 331–356. Sage, Thousand Oaks (1996)
12. Mayer, R.C., Davis, J.H., Schoorman, F.D.: An Integrative Model of Organizational Trust. Academy of Management Review 20(3), 709–734 (1995)
13. Mayer, R.C., Davis, J.H.: The Effect of the Performance Appraisal System on Trust for Management: A Field Quasi-Experiment. Journal of Applied Psychology 84(1), 123–136 (1999)
14. Heider, F.: Social Perception and Phenomenal Causality. Psychological Review 51, 358–374 (1944)
15. Heider, F.: The Psychology of Interpersonal Relations. John Wiley & Sons, New York (1958)
16. Ajzen, I.: The Theory of Planned Behavior. Organizational Behavior and Human Decision Processes 50(2), 179–211 (1991)
17. Beck, L., Ajzen, I.: Predicting Dishonest Actions Using the Theory of Planned Behavior. Journal of Research in Personality 25(3), 285–301 (1991)
18. Holmes, J.G., Rempel, J.K.: Trust in Close Relationships. In: Hendrick, C. (ed.) Review of Personality and Social Psychology. Sage, Beverly Hills (1989a)

19. Rempel, J.K., Holmes, J.G., Zanba, M.D.: Trust in Close Relationship. Journal of Personality and Social Psychology 49, 95–112 (1985)
20. Bigley, G.A., Pearce, J.L.: Straining for shared meaning in organizational science: Problems of trust and distrust. Academy of Management Review 23(3), 405–421 (1998)
21. Kelley, H.H., et al.: The Process of Causal Attribution. American Psychology 28(2), 107–128 (1973)
22. von Ahn, L.: Games with a Purpose. IEEE Computer Magazine, 96–98 (2006)
23. von Ahn, L., Dabbish, L.: General Techniques for Designing Games with a Purpose. Communications of the ACM, 58–67 (2008)
24. von Ahn, L., et al.: reCAPTCHA: Human-based Character Recognition via Web Security Measures. Science, 1465–1468 (2008)

Towards Building and Analyzing a Social Network of Acknowledgments in Scientific and Academic Documents

Madian Khabsa[1], Sharon Koppman[3], and C. Lee Giles[1,2]

[1] Computer Science and Engineering
[2] Information Sciences and Technology
[1,2] The Pennsylvania State University, University Park, PA 16802
[3] Department of Sociology, University of Arizona, Tucson, AZ 85721
madian@psu.edu,
skoppman@email.arizona.edu,
giles@ist.psu.edu

Abstract. Acknowledgments in scientific and academic papers are a method of expressing appreciation and gratitude between scholars. Inside the acknowledgments section, authors usually thank different entities or individuals for their support, most commonly in terms of grants, suggestions and discussions. Social networks of authors and publications has been well studied, with an exhaustive study of nearly all network properties. However, to the best of our knowledge the social graph of acknowledgments have never been investigated. In this paper we propose building an acknowledgments graph, and study the relationships between the entities of this graph.

We first describe how to extract the acknowledgments section. We then extract the entities inside the acknowledgment section, classifying them into persons or organizations. After that we take a subset of the nodes and build a directed graph between the authors of these papers and the entities being acknowledged and study the characteristics of the resulting social network.

Keywords: Acknowledgments, Social Network Analysis.

1 Introduction

Research and academic papers have been considered as an example of social networks where the authors form a network with each others, and the type of relationship is limited to authoring a paper together. The literature is rich with studies about the co-authorship network and associated analysis [16,15,13,12,11,14,4]. Such analysis gives insights about the nature of collaboration between scientists along with findings on how authors tend to choose their collaborators.

Despite the success of studying the co-authorship network, we argue that such studies are missing important information available in the acknowledgments section. As an example authors tend to thank others whose contribution to the paper was significant but not significant enough to be considered as an author or

S.J. Yang, A.M. Greenberg, and M. Endsley (Eds.): SBP 2012, LNCS 7227, pp. 357–364, 2012.

entities that funded the work or provided resources. As such, a complete scientific social network should look into the acknowledgments along with the authorship relations to get a complete picture of the interactions between scientists. Edge [9] recognized that in 1979 calling acknowledgments a "super-citations". Later, Cronin advocated the collecting and indexing of acknowledgments in order to provide a complete " register of influence" [8]. He even argues that " a composite acknowledgment statement at the end of a multiauthor paper is no less an expression of functional interdependence between scholars than the list of coauthors" [6]. Therefore, the most helpful acknowledged person might have contributed to the paper more than the last author.

How common are acknowledgements? Our analysis showed that 65% of papers the top most cited 1000 papers in CiteSeerX have acknowledgments sections, hence the information included in that section is important enough to be studied (of all CiteSeerX papers, 33% have acknowledgements).

The problem of mining the acknowledgments network had the following interesting results. First, the graph resulting from this network is directed, unlike the authorship graph. Second, it's a heterogeneous graph containing nodes of persons, organizations and companies. Third, the edges in this graph may have different types - i.e. the edge might express discussions, help in writing a paper, or acknowledgement of funding. In this work, however, we collapse for simplicity all the edges into one class. Fourth, to the best of our knowledge, there has been no acknowledgments network dataset ever collected or made available.

The contributions of this paper can be categorized as the following: 1). we introduce a method to detect acknowledgment sections in research papers and books, 2) we then extract the entities in the acknowledgments section 3) we build a research paper acknowledgments graph, 4). we analyze this network and present key findings, 5). this data set will be made publicly available for other researchers to study.

The remainder of this paper is organized as follows. Section two surveys related work. In section three we discuss the data set and how we did the entity extraction. In section four we describe and analyze the resulting network. Finally, we conclude in section five.

2   Related Work

To the best of our knowledge, there hasn't been any study of an acknowledgments graph as a social network. Previously, Councill and Giles [5,10] studied how to identify acknowledgments passages in academic papers, how to extract the entities in the acknowledgments section and how to build a searchable acknowledgement index. Their work mainly uses a combinations of regular expressions and machine learning methods, support vector machines, to extract the passage and identify the entities based on a predefined dictionary. The goal of their work was to index the acknowledgments and make them searchable by the users.

The closest comparable work to ours is the study of co-authorship network in scientific papers. Newman [16] examined the authorship network of three fields, mathematics, physics, and biology in terms of the evolution on the network and how authors tend to form their authorship relations. He also studies the patterns of collaborations and the distance between the authors in the network. In other work [15] Newman studies the co-authorship network and suggest methods to find the most connected scientist in the network.

Although Cronin et. al [7] examined more than 100 volumes of two prominent journals in psychology and philosophy covering more than 100 years of publications, their work didn't explore the properties of acknowledgments social network. It was also a manual experiment which can't scale to study large sets of acknowledgments.

In [13] Liu et. al studies the co-authorship network of a digital library research community. Their work is similar to the work of Newman in [16,15] but they apply these measures on a different dataset. They introduce *AuthorRank* as a measure to capture the impact of an author in the network, which is a modified version of *PageRank*. Another study that examines the cross-country relations between authors was conducted by Glanzel and Schubert [12]. Glanzel also studied the implication of the co-authorship network on both searching and indexing of literature [11]. On a smaller scale, Nascimento et. al. [14] analyzed the co-authorship graph of SIGMOD conference. On a higher level than individual authors, Borner et. al [4] studied the interactions between research teams in the co-authorship network.

3   Dataset

We build our acknowledgment dataset using the papers in the CiteSeerX repository which contains more than 1.5 million papers in computer science and mathematics. We take a subset of the repository containing the top 1000 cited papers to build the graph. Constructing the dataset is divided into two tasks, acknowledgments extraction, and entities extraction.

3.1   Acknowledgments Extraction

After examining papers to get an intuition of how and where the authors tend to write the acknowledgments section, we manually tagged the top 200 cited papers in CiteSeerX for acknowledgments and entities. Among the 200 papers that we manually examined, 130 of them had an acknowledgment section which accounted for around 65% of the training dataset.

Based on the training dataset that we tagged, we devised regular expressions that would match most of the acknowledgments sections in the training set. The regular expressions extracted all the sections of the paper along with their headers; we then filter out sections that correspond to acknowledgments. Our regular expression extractor is a modified version of a previous work for detecting paper sections [17]. The regular expression extractor had precision and recall values at 92.3% and 91.6% with F1 measure of 91.9%. In calculating the precision and recall, we would consider the extracted passage to be correct if it's an exact match of the tagged acknowledgment, or if it has at most one extra line from the paper (i.e. journal name, received date ...). However, if there is more than one extra line beyond the original acknowledgment, we consider it to be incorrect. If we only consider verbatim match to be correct, the F1 measure drops into the 80s.

It's worth mentioning that our dataset comprises both books and papers, and the errors which our extractor makes are mostly on books, which is expected since books have more variation in their acknowledgement format than papers. If books are excluded, our precision and recall jumps to 98.3% and 97.6% respectively.

3.2   Named Entity Extraction

The next phase in building the acknowledgment database is to extract the entities inside the acknowledgment section. Named Entity Extraction (NER) has been studied extensively in the literature, and systems have been able to achieve 90% accuracy in competitions. Therefore we decided to leverage what's available in open source and in free web services. We have examined many open source tools and chose to use OpenCalais[1] and AlchemyAPI[2] because they provide a web service that can be called easily and provides very accurate extraction and disambiguation results.

Table 1. Top acknowledged entities in CiteseerX

Entity	Number of Acks
National Science foundation	67659
NSF	14718
National Aeronautics and Space Administration	12540
European Union	12199
NASA	10009
IBM	9644
DARPA	8976
National Institute of Health	8879

3.3   Resulting Dataset

We ran the section extraction on the entire CiteSeerX repository which contains more than 1.5 million papers. The section extractor identified acknowledgments in 526,930 of the papers in the repository. From the extracted acknowledgments we extracted 4,486,134 entities, 1,648,013 of them were unique. The distribution of entity types was as follows: 2,847,893 persons, 1,225,947 organizations, and 412294 companies. We ranked the top acknowledged entities, and our results are similar to [10,5] in showing that National Science Foundation is the top acknowledged entity. Table 1 shows the top acknowledged Organizations/Companies. The underlying NER provides disambiguation for the extracted entities, but as some entities are not merged together correctly (NSF and National Science Foundation) we realized that the NER disambiguation is not always correct. However, we leave this for future work.

As noted in table 1, the top acknowledged entities are all organizations and companies. We extracted the top acknowledged persons in our dataset, and noticed that there is a relation between the h-index[1] of the person and the number of time she/he has been acknowledged. We use the h-index values computed by CiteSeerX. We leave analyzing this observation statistically to future work.

Table 2. Top acknowledged persons in the top 1000 cited papers of CiteSeerx

Entity	Number of Acks	H-Index
Oded Goldreich	471	37
Olivier Danvy	350	24
Avi Wigderson	325	29
Vern Paxson	325	36
David Wagner	313	32
Jim Gray	309	21
Paul Taylor	292	14
Sally Floyd	288	40
Jon Crowcroft	270	25

We now take a subset of the repository containing the top 1000 cited papers, and build the social graph using their entities. This smaller graph is easier to analyze yet it's still representative of the bigger graph. Since CiteSeerX doesn't contain all the papers in computer science, the entire repository is not a complete graph, therefore the repository itself is a sub-graph, so for the time being we analyze a subset of the dataset. The graph is constructed using 7265 entities, 5963 of them unique. And the distribution of the type was as follows: 575 companies, 1041 organizations, and 5649 persons.

We build the directed graph in the following manner:

1. Create a node for each acknowledged entity
2. Create a node for each author of a paper that has acknowledgments section
3. Make an edge between nodes *(v,u)* if the author *v* writes a paper which acknowledged the entity *u*

As a result of the previous method, we built a graph containing 5983 nodes, and 17287 edges.

---

[1] An author would have h-index = X if he has X papers each of which has been cited at least X times

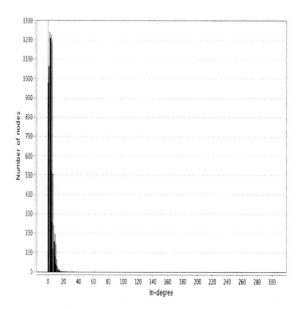

Fig. 1. Distribution of in-degree value against the number of nodes

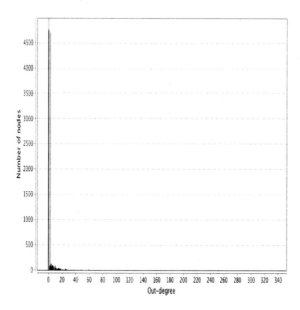

Fig. 2. Distribution of out-degree value against the number of nodes

4   Analyzing the Network

After building the acknowledgment graph, we use Cytoscape [3] which is a an open source tool for complex network analysis and visualization. The network had 67 connected components with largest components comprising more than 5000 nodes. The diameter of the network was 15, and the average number of neighbors is 5.75. In the following section we study different characteristics of the network including clustering coefficients and degree distributions.

4.1   Clustering Coefficients

Clustering Coefficient , CC, measures the tendency of the nodes to cluster together in a graph, and the higher CC value at one node the higher the probability of its neighbors to be connected together. It is defined as

$$ClusteringCoefficient = \frac{Number\ of\ closed\ triplets}{Number\ of\ connected\ triplets\ of\ vertices} \quad (1)$$

The average clustering coefficients for this network is 0.103, which is quite low and indicates that neighboring nodes don't tend to cluster.

4.2   Degree Distribution

We examine the in-degree and the out-degree distribution of the network using Cytoscape, and found out that it follows a power law distribution of the form $Y = aX^b$. In the case of in-degree distribution, we find out that the parameters of the power law distribution are $a = 1843.1$ and $b = -1.923$. While for the out degree distribution the parameters where: $a = 331.21$ and $b = -1.187$ Figure 1 shows the distribution of in-degree with respect to the number of nodes while figure 2 shows the out-degree distribution.

5   Conclusions and Future work

We have analyzed an acknowledgments network in scientific documents, extracted a sub-graph of the acknowledgments network and analyzed the resulting social network. We showed how to extract the acknowledgments sections in scientific papers with F1 measure ranging from 88% to 97%. We extracted the entities from the acknowledgment sections using web services of OpenCalais and AlchemyAPI. As a result we built a graph containing almost 6 thousand nodes and more than 17 thousand edges. Graph analysis showed that the degree distribution follows a power law for both in-degree and out-degree. The clustering coefficients of the network are quite low, which upon retrospection might be expected.

For future work we plan to analyze the complete acknowledgments graph of all the papers in CiteSeerX which contains more than 1 million nodes and 4 million edges. We also plan on improving the disambiguation techniques to identify the similar nodes. We intend on releasing an acknowledgement indexing search engine and making the data public for additional research.

Acknowledgments. We thank Pucktada Treeratpituk and Pradeep Terege-owda for providing the original section extraction code. We also thank Shaun Roach from AlchemyAPI for increasing our API usage limit.

References

1. http://www.opencalais.com/
2. http://www.alchemyapi.com/
3. http://www.cytoscape.org/
4. Borner, K., Dall'Asta, L., Ke, W., Vespignani, A.: Studying the emerging global brain: Analyzing and visualizing the impact of co-authorship teams. Complexity 10(4), 57–67 (2005)
5. Councill, I.G., Giles, C.L., Han, H., Manavoglu, E.: Automatic acknowledgement indexing: expanding the semantics of contribution in the citeseer digital library. In: Proceedings of the 3rd International Conference on Knowledge Capture (2005)
6. Cronin, B., McKenzie, G., Rubio, L., Weaver-Wozniak, S.: Accounting for influence: Acknowledgments in contemporary sociology. Journal of the American Society for Information Science 44(7), 406–412 (1993)
7. Cronin, B., Shaw, D., La Barre, K.: A cast of thousands: Coauthorship and sub-authorship collaboration in the 20th century as manifested in the scholarly journal literature of psychology and philosophy. Journal of the American Society for Information Science and Technology 54(9), 855–871 (2003)
8. Cronin, B., McKenzie, G., Rubio, L., Weaver-Wozniak, S.: Accounting for influence: acknowledgments in contemporary sociology. J. Am. Soc. Inf. Sci. 44, 406–412 (1993)
9. Edge, D.: Quantitative measures of communication in science: A critical review. History of Science 17, 102–134 (1979)
10. Giles, C., Councill, I.: Who gets acknowledged: Measuring scientific contributions through automatic acknowledgment indexing. Proceedings of the National Academy of Sciences of the United States of America 101(51), 17599 (2004)
11. Glanzel, W.: Coauthorship patterns and trends in the sciences (1980-1998): A bibliometric study with implications for database indexing and search strategies.. Library Trends 50(3), 461–473 (2002)
12. Glanzel, W., Schubert, A.: Analysing scientific networks through co-authorship. In: Handbook of Quantitative Science and Technology Research, pp. 257–276 (2005)
13. Liu, X., Bollen, J., Nelson, M.L., Van de Sompel, H.: Co-authorship networks in the digital library research community. Inf. Process. Manage. 41, 1462–1480 (2005)
14. Nascimento, M., Sander, J., Pound, J.: Analysis of SIGMOD's co-authorship graph. ACM SIGMOD Record 32(3), 8–10 (2003)
15. Newman, M.: Who is the best connected scientist? A study of scientific coauthorship networks. Complex Networks, 337–370 (2004)
16. Newman, M.: Coauthorship networks and patterns of scientific collaboration. Proceedings of the National Academy of Sciences of the United States of America 101(suppl.1), 5200 (2004)
17. Treeratpituk, P., Teregowda, P., Huang, J., Giles, C.: SEERLAB: A system for extracting key phrases from scholarly documents. In: Proceedings of the 5th International Workshop on Semantic Evaluation, pp. 182–185 (2010)

Author Index